QUANTUM FIELDS AND PROCESSES

Wick ordering of creation and annihilation operators is of fundamental importance for computing averages and correlations in quantum field theory and, by extension, in the Hudson–Parthasarathy theory of quantum stochastic processes, quantum mechanics, stochastic processes, and probability. This book develops the unified combinatorial framework behind these examples, starting with the simplest mathematically, and working up to the Fock space setting for quantum fields. Emphasizing ideas from combinatorics such as the role of the lattice of partitions for multiple stochastic integrals by Wallstrom–Rota and combinatorial species by Joyal, it presents insights coming from quantum probability. It also introduces a "field calculus" that acts as a succinct alternative to standard Feynman diagrams and formulates quantum field theory (cumulant moments, Dyson–Schwinger equation, tree expansions, 1-particle irreducibility) in this language. Featuring many worked examples, the book is aimed at mathematical physicists, quantum field theorists, and probabilists, including graduate and advanced undergraduate students.

John Gough is professor of mathematical and theoretical physics at Aberystwyth University, Wales. He works in the field of quantum probability and open systems, especially quantum Markovian models that can be described in terms of the Hudson–Parthasarathy quantum stochastic calculus. His more recent work has been on the general theory of networks of quantum Markovian input-output and their applications to quantum feedback control.

Joachim Kupsch is professor emeritus of theoretical physics at the University of Kaiserslautern, Germany. His research has focused on scattering theory, relativistic S-matrix theory, and infinite-dimensional analysis applied to quantum field theory. His publications have examined canonical transformations, fermionic integration, and superanalysis. His later work looks at open systems and decoherence and he coauthored a book on the subject in 2003.

CAMBRIDGE STUDIES IN ADVANCED MATHEMATICS

Editorial Board:
B. Bollobás, W. Fulton, F. Kirwan, P. Sarnak, B. Simon, B. Totaro

All the titles listed below can be obtained from good booksellers or from Cambridge University Press. For a complete series listing visit: www.cambridge.org/mathematics.

Quantum Fields and Processes

A Combinatorial Approach

JOHN GOUGH

Aberystwyth University

JOACHIM KUPSCH

University of Kaiserslautern

CAMBRIDGE
UNIVERSITY PRESS

University Printing House, Cambridge CB2 8BS, United Kingdom

One Liberty Plaza, 20th Floor, New York, NY 10006, USA

477 Williamstown Road, Port Melbourne, VIC 3207, Australia

314–321, 3rd Floor, Plot 3, Splendor Forum, Jasola District Centre,
New Delhi – 110025, India

79 Anson Road, #06–04/06, Singapore 079906

Cambridge University Press is part of the University of Cambridge.

It furthers the University's mission by disseminating knowledge in the pursuit of
education, learning, and research at the highest international levels of excellence.

www.cambridge.org
Information on this title: www.cambridge.org/9781108416764
DOI: 10.1017/9781108241885

First published 2018

Printed in the United States by Sheridan Books, Inc.

A catalogue record for this publication is available from the British Library

Library of Congress Cataloguing-in-Publication Data
Names: Gough, John, 1967– author. | Kupsch, Joachim, 1939– author.
Title: Quantum fields and processes : a combinatorial approach / John Gough,
Aberystwyth University, Joachim Kupsch, University of Kaiserslautern.
Description: Cambridge : Cambridge University Press, 2018. |
Series: Cambridge studies in advanced mathematics ; 171 |
Includes bibliographical references and index.
Identifiers: LCCN 2017036746 | ISBN 9781108416764 (Hardback : alk. paper)
Subjects: LCSH: Combinatorial analysis. | Quantum field theory. | Probabilities.
Classification: LCC QA165 .G68 2018 | DDC 530.14/3015116–dc23 LC record
available at https://lccn.loc.gov/2017036746

ISBN 978-1-108-41676-4 Hardback

To Margarita, Sigrid, and John Junior.

Contents

Preface

It is probably safe to say that very few people have developed an interest in physical sciences due to combinatorics. Yet combinatorics makes its presence felt in modern mathematics and physics in a fundamental and elegant manner, and goes far beyond standard school problems such as to determine how many ways to rearrange the letters of the word MISSISSIPPI.

It has been argued in many places that modern physics owes much to the work of Ludwig Boltzmann, who in many ways was the first scientist to think like a modern physicist. Certainly, he was the first to use probability in an essential way, and his ideas on the microscopic foundations of thermodynamics directly influenced Gibbs and primed both Planck and Einstein at the start of the twentieth century. The machinery needed for modern mathematical physics was assembled in the twentieth century, and concentrated on the areas of functional analysis and probability theory, moving inexorably toward the description of stochastic processes and quantized fields. Behind this, however, was the realization that combinatorics played an important role.

In writing this book, we have been influenced by several recurring ideas in mathematical physics that all have an underlying combinatorial core. Wallstrom and Rota were the first to notice that several disparate strands, such as multiple Itō integrals, Wick products, and normal ordering, could be conveniently expressed in terms of the combinatorics of the lattice of partitions. They, in turn, were influenced by Meyer's book, *Quantum Probability for Probabilists* (1995). In many respects, our focal point for this book has been (Bose and Fermi) Fock space: this is the framework for quantum field theory and the quantum stochastic calculus of Hudson and Parthasarathy, and is well known to probabilists. The creation and annihilation operators, satisfying canonical (anti-)commutation relations, are then the objects of much of our attention, and we give a good deal of attention to both their mathematical and physical aspects. The elements of combinatorics that we cover here are those arising from quantum fields and stochastic processes. However, this gives ample room to bring in modern approaches, especially the combinatorial species of Joyal, to give a more algebraic feel than the traditional Feynman diagram approach. We combine elements of species with the Guichardet notation for symmetric Fock spaces in order to construct a field calculus that leads to explicit combinatorial formulas in place of the diagrammatic expansions. We have tried to resist the temptation to show off the often surprising and striking combinatorial expressions that arise. In truth, combinatorics appears as a tool in many branches of mathematics – and very frequently in the proofs of mathematical physics propositions – but doing justice to these varied and multifaceted techniques would be beyond the scope of this book.

We are grateful for the input of many colleagues over the years. We would especially like to thank Robin Hudson, Luigi Accardi, Yun Gang Lu, Igor Volovich, Wilhelm von Waldenfels, Hans Maassen, Madalin Guta, Hendra Nurdin, Matthew James, Joe Pulé, and Aubrey Truman, and to acknowledge the enormous influence of the extended quantum probability community at large. The completion of the project is in no small part due to Professor Oleg Smolyanov, whose constant refrain, "you need to write a book," propelled us forward. We are also very grateful to Diana Gillooly and the staff at Cambridge University Press for their help and advice while writing. Finally, we would like to thank our families – Margarita and John Junior, and Sigrid – for their support and patience in waiting for a book dedicated to them.

Notation

General Mathematical

$\mathrm{ran}(f)$	range of a function f
$A^{\top}$	transpose
A^*	adjoint/Hermitian conjugate/complex conjugate
$\mathrm{spec}(A)$	spectrum of a matrix/operator A
$\Delta_n(t,s)$	the simplex $\{(t_n,\ldots,t_1) : t \geq t_n > t_{n-1} > \cdots t_1 \geq s\}$
$\Delta_n^\sigma(t,s)$	the simplex $\{(t_n,\ldots,t_1) : t \geq t_{\sigma(n)} > t_{\sigma(n-1)} > \cdots t_{\sigma(1)} \geq s\}$
$\Theta(x)$	Heaviside function

Spaces, Norms, Etc.

$\mathbb{C}$	complex numbers
$\mathbb{R}$	real numbers
$\mathbb{Z}$	integers
$\mathbb{N}$	the natural numbers $\{1,2,3,\ldots\}$
$\mathbb{N}_+$	nonnegative integers $\{0,1,2,3,\ldots\}$
$\mathbb{N}_-$	the set $\{0,1\}$
$\mathbb{M}$	Minkowski spacetime
$\hat{\mathbb{M}}$	dual Minkowski spacetime
$\mathfrak{h}, \mathfrak{H}$	Hilbert space (always assumed separable!)
$L^2(\mathbb{R},\mathbb{C})$	complex-valued square integrable functions on $\mathbb{R}$
$\ell^2(\mathbb{N},\mathbb{C})$	complex-valued square summable sequences
$\mathfrak{B}(\mathfrak{h})$	bounded operators on a Hilbert space $\mathfrak{h}$
$\mathfrak{T}(\mathfrak{h})$	bounded operators on a Hilbert space $\mathfrak{h}$
$\Gamma(\mathfrak{h})$	Fock space (bosonic) over a Hilbert space $\mathfrak{h}$
$\mathfrak{S}_n$	Permutations on $\{1,2,\ldots,n\}$
$\Gamma(\mathfrak{h})$	Fock space over a Hilbert space $\mathfrak{h}$

$\Gamma^+(\mathfrak{h})$	Boson Fock space over a Hilbert space $\mathfrak{h}$
$\Gamma^-(\mathfrak{h})$	Fermion Fock space over a Hilbert space $\mathfrak{h}$
$\otimes$	tensor product
$\vee$	symmetric tensor product
$\wedge$	antisymmetric tensor product

Combinatorics

$x^{\underline{n}}$	falling factorial powers, i.e., $x(x-1)(x-2)\cdots(x-n+1)$
$\begin{bmatrix} n \\ m \end{bmatrix}$	Stirling number of the first kind
$\begin{Bmatrix} n \\ m \end{Bmatrix}$	Stirling number of the second kind
$\mathrm{Perm}(X)$	permutations on a set X
$\mathrm{Part}(X)$	partitions of a set X
$\mathrm{Pair}(X)$	pair partitions of a set X
$\mathrm{Power}(X)$	power set (set of finite subsets) of a set X
$H_n(x)$	Hermite polynomial, $= \sum_{k=1}^{[n/2]} \frac{(-1)^k n!}{2^k k!(n-2k)!} x^{n-2k}$
$\diamond$	Wick product

Quantum Mechanics

$\psi(x)$	wavefunction
$\lvert\psi\rangle$	state vector/ket
ρ	density operator
$\langle\psi\lvert\phi\rangle$	Dirac notation for inner product
P	projection
$D(\beta)$	Weyl displacement operator
$\lvert\exp\beta\rangle$	Exponential vector (Bargmann state)
$\circledast$	group product for the Heisenberg group
$:X:$	Wick (normal) ordering of an operator X

Classical Probability

$\mathbf{P}$	probability measure
$\mathbf{E}$	expectation
$\mathcal{A}$	σ-algebra
Ω	sample space
χ_E	indicator function for event E
$[[X,Y]]_t$	the quadratic covariance, that is, $\int_0^t dX_s.dY_s$
$X_t^{[n]},$	the process $\int_0^t (dX_s)^n$

$\int_{[0,t]^n} dX_{t_n}^{(n)} \ldots dX_{t_1}^{(1)}$	the diagonal-free stochastic integral
$\mathcal{E}^{X_t}$	the diagonal-free stochastic exponential
$X \delta Y$	Stratonovich differential (i.e., $X\,dY + \frac{1}{2}dX\,dY$).
$\vec{T}_S e^{Z_t}$	Stratonovich time-ordered exponential

Quantum Field Theory

Φ	space of all field configurations
$\mathfrak{J}$	space of all source fields (= test functions in the dual of Φ)
$\phi_X = \prod_{x \in X} \phi(x)$	
G_X	Green's functions (moments = expectation of ϕ_X)
K_X	connected Green's functions (= cumulants)
$A(f)$	annihilation operator with test function f
$D(f)$	Weyl unitary with test function f
$W(f) \equiv D(if)$	
$\Gamma(U)$	second quantization of a unitary U
$d\Gamma(H)$	differential second quantization
$\hat{\Phi}(f)$	Segal's field operator, $= A(f) + A^*(f)$
$\mathcal{P}, \mathcal{P}^\uparrow$	Poincaré group, the orthochronous Poincaré group

1

Introduction to Combinatorics

How we count things turns out to have a powerful significance in physical problems! One of the oldest problems stems from undercounting and over-counting the number of possible configurations a particular system can have – mathematically, this is usually due to the fact that objects are mistakenly assumed to be indistinguishable when they are not, and vice versa. However, one of the great surprises of physics is that identical particles are funda-mentally indistinguishable. In this chapter, we will introduce some of the basic mathematical objects that occur in physical problems, and give their enumeration. Statistical mechanics is one of the key sources of ideas, so we spend some time on the basic concepts here, especially as partition functions are clear examples of generating functions that we will encounter later on. We will recall some of the basic mathematical concepts in enumeration, leading on to the role of generating functions. At the end, we make extensive use of generating functions, exploiting the methods for dealing with partition functions in statistical mechanics, but for specific combinatorial families such as permutations and partitions.

We start, however, with the touchstone for all combinatorial problems: how to distribute balls in urns.

1.1 Counting: Balls and Urns

Proposition 1.1.1 *There are K^N different ways to distribute the N distinguish-able balls among K distinguishable urns.*

The proof is based on the simple observation that there are K choices of urn for each of the N balls. Suppose next that we have more urns than balls.

1

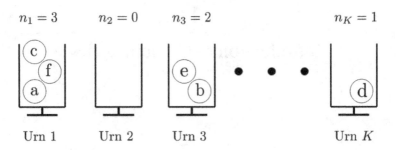

Figure 1.1 Occupation numbers of distinguishable balls in distinguishable urns.

Proposition 1.1.2 *The total number of ways to distribute the N distinguishable balls among K distinguishable urns so that no urn ends up with more than one ball is (later we will call this a falling factorial)*

$$K^{\underline{N}} \triangleq K(K-1)\cdots(K-N+1).$$

The argument here is simple enough: if we have already distributed M balls among the urns, with no urn having more than one ball inside, so if we now want to place in an extra ball we have these $K - M$ empty urns remaining to choose from; therefore, $K^{\underline{M+1}} = K^{\underline{M}}(K - M)$ with $K^{\underline{1}} = K$.

Let n_k denote the number of balls in urn k; we call this the **occupation number** of the urn. See Figure 1.1. We now give the number of possibilities leading to a given set of occupation numbers, subject to the constraint of a fixed total number of balls, $\sum_{k=1}^{K} n_k = N$.

Proposition 1.1.3 *The number of ways to distribute N distinguishable balls among N distinguishable urns so that we have a prescribed number n_k balls are in the kth urn, for each $k = 1, \ldots, K$, is the multinomial coefficient*

$$\binom{N}{n_1 \ldots n_K} = \frac{N!}{n_1! n_2! \ldots n_K!}.$$

The proof here is based on the observation that there are $\binom{N}{n_1}$ ways to choose the n_1 balls to go into the first urn, then $\binom{N-n_1}{n_2}$ ways to choose next n_2 balls to go into the second urn, and so on, leading to

$$\binom{N}{n_1}\binom{N-n_1}{n_2}\cdots\binom{N-n_1-\cdots-n_{K-1}}{n_K} \equiv \frac{N!}{n_1! n_2! \ldots n_K!}.$$

Suppose however that the balls are in fact indistinguishable as in Figure 1.2! Then we do not distinguish between distributions having the same occupation numbers for the urns.

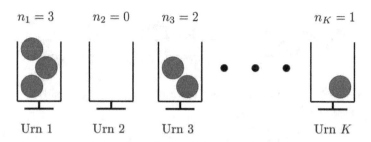

$n_1 = 3$ $n_2 = 0$ $n_3 = 2$ $n_K = 1$

Urn 1   Urn 2   Urn 3   Urn K

Figure 1.2 Occupation numbers of indistinguishable balls in distinguishable urns.

Proposition 1.1.4 *There are* $\binom{N+K-1}{N}$ *ways to distribute N indistinguishable balls among K distinguishable urns.*

Proof Take, for example, $K = 6$ urns and $N = 8$ balls and consider the distribution represented by occupation sequence $(1, 2, 0, 4, 1, 0)$, then encode this as follows:

$$\bullet \,|\, \bullet\, \bullet \,|\,|\, \bullet\, \bullet\, \bullet\, \bullet \,|\, \bullet \,|$$

which means one ball in urn 1, two balls in urn 2, no balls in urn 3, and so on. In this encoding, we have $N + K - 1$ symbols (balls and sticks), N of which are balls and $K - 1$ of which are sticks (separations between the urns). In any such distribution, we must choose which N of the $N + K - 1$ symbols are to be the balls, and there are $\binom{N+K-1}{N}$ different ways to do this. Each way of selecting these symbols corresponds to a unique distribution, and vice versa. $\qquad \square$

Proposition 1.1.5 *The number of ways to distribute N indistinguishable balls among K distinguishable urns, if we only allow at most one ball per urn, is* $\binom{K}{N}$.

That is, we must choose N out of the K urns to have a ball inside.

These enumerations turn out to be of immediate relevance to sampling theory in statistics. We note that if we have a set of K items and we draw a sample of size N, then if we make no replacement there will be $\binom{K}{N}$ such samples – imagine placing a ball into urn j if the jth element is selected! If replacement is allowed, then the number of samples is $\binom{N+K-1}{N}$.

1.2 Statistical Physics

Counting problems surfaced early on in the theory of statistical mechanics, and we recall the basic setting next.

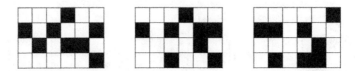

Figure 1.3 Three microstates, each corresponding to $N = 24$ particles, with 8 "on" (the black boxes).

1.2.1 The Microcanonical Ensemble

Two-State Model

Ludwig Boltzmann pioneered the microscopic derivation of laws of thermodynamics. To understand his ideas, we consider a very simple model of a solid material consisting of N particles, where each particle can be in either one of two states: an "off" state of energy 0 and an "on" state of energy ε. The total energy is therefore $U = \varepsilon M$, where M is the number of particles in the "on" state. In Figure 1.3, we have three typical examples where $N = 24$ and $U = 8\varepsilon$, that is, in each of these we have 8 "on" states of of a total of 24. Each of these configurations is referred to as a **microstate**, and we say that they are consistent with the **macrostate** ($U = 8\varepsilon, N = 24$).

Boltzmann's idea was that if the system was isolated, so that the energy U was fixed, the system's internal dynamics would make it jump from one microstate to another with only microstates consistent with the fixed macrostate (U, N) allowed. (In other words, the number of particles, N, and their energy U, are to be constants on the motion – whatever that happens to be.) Here the total number of microstates consistent with macrostate $(U = \varepsilon M, N)$ is then

$$W(U,M) = \binom{N}{M}.$$

He then made the **ergodic hypothesis**: *over a long enough period of time, each of these microstates was equally likely*: that is, the system may be found to be in a given microstate with frequency $1/W$. Therefore, long time averages would equate to averages over all the microstates consistent with the macrostate, with each microstate having equal probabilistic weight $1/W$. The latter probability system is known as the **microcanonical ensemble**. We note that the set of all microstates consistent with ($U = 8\varepsilon, N = 24$) also includes some less-than-random-looking configurations such as the ones shown in Figure 1.4. But, nevertheless, they each get equal weight: here $W = \binom{24}{8} = 735,471$.

At this resolution, we would expect to see the system run through the possible microstates, so that the picture over time would appear something like static on a TV screen. Configurations with a discernable pattern, as in Figure 1.4,

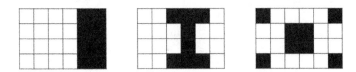

Figure 1.4 Another three microstates consistent with the $(U = 8\varepsilon, N = 24)$. Despite their apparent structure, each has the same 1/735,471 chance to occur as the more random ones in Figure 1.3.

may flash up briefly from time to time, but most of the time we are looking at fairly random-looking configurations such as in Figure 1.3. If N is large, and the particles are small, then we would expect to be looking most of the time at a uniform gray – the shade of gray determined by the ration M/N.

Boltzmann's remarkable proposal was that the entropy associated with a macrostate (U, N) was the logarithm of the number of consistent microstates

$$S(U, N) = k \ln W(U, N)$$

where k is a scale factor fixing our eventual definition of temperature scale. In the present case, we have $W(U = \varepsilon M, N) = \frac{N!}{(N-M)!M!}$. If we take N large with $U = Nu$ for some fixed ratio u, then using Stirling's approximation, $\ln N! = N \ln N - N + O(\ln N)$, we find that the entropy per particle in the bulk limit $(N \to \infty)$ is

$$s(u) = \lim_{N \to \infty} \frac{S(U = Nu, N)}{N}$$
$$= -kp_0 \ln p_0 - kp_1 \ln p_1$$

where $p_1 = \frac{M}{N} \equiv \frac{u}{\varepsilon}$ and $p_0 = 1 - p_1$. (Note p_1 is the proportion of particles that are "on" in each of these microstates, with p_0 the proportion "off".) Alternatively, we may write this as

$$s(u) = -k \left\{ \frac{u}{\varepsilon} \ln \frac{u}{\varepsilon} + \left(1 - \frac{u}{\varepsilon}\right) \ln \left(1 - \frac{u}{\varepsilon}\right) \right\}.$$

From thermodynamics, one should identify the temperature T via the relation $1/T = \frac{\partial s}{\partial u} \equiv -\frac{k}{\varepsilon} \ln \left(\frac{\varepsilon}{u} - 1\right)$, and so in this model we have

$$u = \frac{\varepsilon}{e^{\varepsilon/kT} + 1}.$$

Somewhat surprisingly, this artificial model actually shows very good qualitative behavior for small values of u – see, for instance, Callen (1985, chapter 15). (Note that for $0 \leq u < \varepsilon/2$, the temperature will be positive, but becomes negative for higher values $\varepsilon/2 < u \leq \varepsilon$. Negative temperatures do actually

0	0	0	0	1	0
0	0	2	0	0	0
0	3	0	0	0	1
0	0	0	1	0	0

Figure 1.5 A microstate consistent with the $(U = 8\varepsilon, N = 24)$ in the Einstein model: one of 7,888,725.

makes physical sense, however, and are encountered in the related model of a two-state ferromagnet.)

Einstein's Model

A related model is the Einstein model for a crystalline solid. The difference is that each particle can have allowed energies $0, \varepsilon, 2\varepsilon, 3\varepsilon, \ldots$. This time we may depict a microstate as in Figure 1.5, where the number n in each box tells us that the corresponding particle has energy $n\varepsilon$ (or, perhaps more physically, that there are n quanta in the box!). We see that the number of microstates consistent with a macrostate $(U = \varepsilon M, N)$ will be

$$W = \binom{N + M - 1}{M}.$$

In other words, the number of ways of distributing M is indistinguishable quanta among N distinguishable boxes. This time, again using Stirling's identity, one may show that the entropy per particle in the bulk limit is

$$s(u) = k \left\{ \ln \left(1 + \frac{u}{\varepsilon} \right) + \frac{u}{\varepsilon} \ln \left(1 + \frac{\varepsilon}{u} \right) \right\},$$

and that $u = \frac{\varepsilon}{e^{\varepsilon/kT} - 1}$.

1.2.2 The Canonical Ensemble

Boltzmann's ergodic hypothesis marks the introduction of probability theory into physics. So far we have not used probability explicitly; however, this changes if we consider the situation depicted in Figure 1.6. For simplicity, we work with the two-state model.

We fix the total energy U_{tot} of the $N_{tot} = N + N'$ particles, and then Boltzmann's principle tells us that each microstate of the total system is equally likely. However, the total number of "on" particles in total is constant, the number that are inside the system will vary with a probability distribution determined by the ergodic hypothesis. Let $M_{tot} = U_{tot}/\varepsilon$, and suppose that we have x particles in the "on" state in the system – then there are $\binom{N}{x}$ microstates

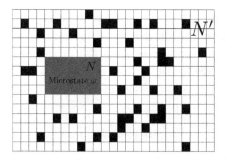

Figure 1.6 A system with N particles and microstate ω forms a subsystem of a larger system of $N + N'$ particles.

of the system consistent with this, and $\binom{N'}{M_{\text{tot}}-x}$ microstates possible for the complement. Allowing for all possible x leads to

$$\binom{N + N'}{M_{\text{tot}}} = \sum_{x=0}^{M_{\text{tot}}} \binom{N}{x}\binom{N'}{M_{\text{tot}} - x}.$$

If N, N' and M_{tot} are large of the same order, then it turns out that the largest term in the sum comes from $x \approx \rho N$, where $\rho = M_{\text{tot}}/N_{\text{tot}}$. In fact, this one term alone dominates to the extent that one may ignore all the other terms. This is related to the large deviation principle, which we discuss in Chapter 7.

The energy of the subsystem is now a random variable, and Boltzmann's hypothesis tells us its distribution. We now consider the situation where the number N of particles in our system is large but fixed, but we take $N' \to \infty$. We do this in such a way that the average number $\bar{u} = \varepsilon \, M_{\text{tot}}/N_{\text{tot}}$ is a constant. Suppose we have a fixed microstate ω of our subsystem with energy $E(\omega)$, i.e. the number of "on" particles in the subsystem. Then the probability of the particular microstate, ω, occurring is

$$p_{N'}(\omega) = W(U',N')/W(U_{\text{tot}},N_{\text{tot}})$$

$$= \left(\frac{N'}{\frac{\bar{u}}{\varepsilon}N' + \frac{\bar{u}-u}{\varepsilon}N}\right)\bigg/\left(\frac{N + N'}{\frac{\bar{u}}{\varepsilon}(N + N')}\right)$$

where $u = E(\omega)/N$ is the energy density of the subsystem. Now making the approximation that $W(uN,N) \approx e^{Ns(u)/k}$, we find

$$k \ln p_{N'}(\omega) = N' s\left(\bar{u} + (\bar{u} - u)\frac{N}{N'}\right) - (N + N') s(\bar{u})$$

$$= -N s(\bar{u}) + kN'\left[s\left(\bar{u} + (\bar{u} - u)\frac{N}{N'}\right) - s(\bar{u})\right]$$

$$\to -N s(\bar{u}) + kNs'(\bar{u})(\bar{u} - u),$$

since s is differentiable. Specifying the average energy per particle in the bulk limit $N' \to \infty$ to be $\bar{u}$ is equivalent to fixing the temperature T via the relation $1/T = s'(\bar{u})$, from which we see that $\ln p_{N'}(\omega) \to \frac{1}{kT}(F - uN)$, where $F = N(\bar{u} - Ts(\bar{u}))$. (Note that the relation between $\bar{u}$ and T is one-to-one. The variable $F = U - TS$ is the *Helmholtz free-energy* in thermodynamics.) That is, we obtain the probability

$$p_{can.}(\omega) = \frac{1}{Z}e^{-E(\omega)/kT},$$

where the normalization is given by the canonical partition function

$$Z_N = e^{-F/kT} = \sum_{\omega} e^{-E(\omega)/kT},$$

where the sum is over all microstates consistent with having a fixed number N of particles.

The probability distribution that we obtain in this way is called the **canonical ensemble** and is interpreted as saying that our subsystem is in thermal equilibrium with a heat bath at temperature T.

The derivation presented in the preceding is actually very general. We relied on the relation $W(uN, N) \approx e^{Ns(u)/k}$, but not the specifics of the entropy per particle, $s(u)$. So the same argument will go through so long as $s(u)$ exists and defines a monotone increasing, strictly concave function of u.

1.2.3 The Grand Canonical Ensemble

We now describe the situation in statistical mechanics where we have a gas consisting of a number of particles, N, each of which can have one of K distinguishable states with energy values $\varepsilon_1 \leq \varepsilon_2 \leq \cdots \leq \varepsilon_K$; see Callen (1985). We allow both the energy and the number of particles to vary: we allow microstates ω, which have $N(\omega)$ particles and energy $E(\omega)$, and introduce the **grand canonical ensemble**

$$p_{g.c.}(\omega) = \frac{1}{\Xi}e^{-\left(E(\omega) - \mu N(\omega)\right)/kT},$$

where the normalization is given by the **grand canonical partition function**

$$\Xi = \sum_{\omega} e^{-\left(E(\omega) - \mu N(\omega)\right)/kT},$$

where the sum is now over all microstates – unrestricted in both number and energy. We introduce the standard notation of the *inverse temperature* $\beta = 1/kT$, and the parameter μ is known as the chemical potential. The alternate parameter $z = e^{\beta\mu}$ is called the *fugacity*.

We will make extensive use of the following lemma.

Lemma 1.2.1 (The $\sum \prod \longleftrightarrow \prod \sum$ Lemma) *Let $\mathbb{M}$ and $\mathbb{K}$ be countable sets, and let $\mathbb{M}^{\mathbb{K}}$ denote the sequences $\mathbf{m} = (m_k)_{k \in \mathbb{K}}$ where $m_k \in \mathbb{M}$, and let $f \colon \mathbb{K} \times \mathbb{M} \to \mathbb{C}$. We have the formal series relation*

$$\sum_{\mathbf{m} \in \mathbb{M}^{\mathbb{K}}} \prod_{k \in \mathbb{K}} f(k, m_k) = \prod_{k \in \mathbb{K}} \left\{ \sum_{m \in \mathbb{M}} f(k, m) \right\}.$$

Proof If we expand out the right-hand side, we find that we get a sum over terms of the form $\prod_{k \in \mathbb{K}} f(k, m_k)$, where all possible values $m_k \in \mathbb{M}$ will occur. Written in terms of the $\mathbf{m} = (m_k)_{k \in \mathbb{K}}$ gives the left-hand side. $\qquad \square$

In many cases, the expression will be convergent.

1.2.4 Maxwell–Boltzmann Statistics

Here we assume that the particles are all distinguishable. Suppose that the jth particle has energy $\varepsilon_{k(j)}$, then the sequence of numbers $\mathbf{k} = (k(1), \ldots, k(N))$ determine the state of the gas. In particular, the set of all possible configurations is

$$\Omega_{N,K} = \{1, \ldots, K\}^N,$$

and we have $\#\Omega_{N,K} = K^N$. We give a Boltzmann weight to a state $\mathbf{k} \in \Omega_{N,K}$ of $e^{-\beta E(\mathbf{k})}$, where the total energy is $E(\mathbf{k}) = \sum_{j=1}^{N} \varepsilon_{k(j)}$. We shall be interested in the canonical partition function

$$Z_N(\beta) = \sum_{\mathbf{k} \in \Omega_{N,K}} e^{-\beta E(\mathbf{k})} = \sum_{\mathbf{k} \in \Omega_{N,K}} \prod_{j=1}^{N} e^{-\beta \varepsilon_{k(j)}} \equiv \left(\sum_{k=1}^{K} e^{-\beta \varepsilon_k} \right)^N,$$

where we used the $\sum \prod \longleftrightarrow \prod \sum$ Lemma for the last part. The associated grand canonical partition function is

$$\Xi(\beta, z) = \sum_{N=0}^{\infty} z^N Z_N(\beta) = \frac{1}{1 - z \sum_{k=1}^{K} e^{-\beta \varepsilon_k}}.$$

This was recognized as leading to an unphysical answer as some of the thermodynamic potentials (in particular, the entropy) ended up being nonextensive. This was known as the Gibbs paradox, and resolution was to apply a correction factor $1/N!$ to each $Z_N(\beta)$, nominally to account for indistinguishability of the gas particles and to crudely correct for overcounting of possibilities. This now leads to the physically acceptable form

$$\Xi\left(\beta,z\right) = \sum_{N=0}^{\infty} \frac{1}{N!} z^N Z_N\left(\beta\right) = \exp\left\{z\sum_{k=1}^{K} e^{-\beta\varepsilon_k}\right\}.$$

1.2.5 Bose–Einstein Statistics

Bosons are fundamentally indistinguishable particles. We are able to say, for instance, that n_k particles have the kth energy value ε_k, for each k, but physically there is no more detailed description to give – the particles have no identities of their own beyond that. The set of all possible configurations is therefore

$$\Omega_{N,K}^+ = \left\{(n_1,\ldots,n_K) \in (\mathbb{N}_+)^K : \sum_{k=1}^{K} n_k = N\right\}$$

where $\mathbb{N}_+ \triangleq \{0,1,2,\ldots\}$. In particular, we have $\#\Omega_{N,K}^+ = \binom{N+K-1}{N}$. The energy associated with a state $\mathbf{n} = (n_1,\ldots,n_K)$ is then $E(\mathbf{n}) = \sum_{k=1}^{K} \varepsilon_k n_k$, and we are led to the Boson canonical partition function

$$Z_N^+\left(\beta\right) = \sum_{\mathbf{n}\in\Omega_{N,K}^+} e^{-\beta E(\mathbf{n})} = \sum_{\mathbf{n}\in\Omega_{N,K}^+} \prod_{k=1}^{K} e^{-\beta\varepsilon_j n_k}.$$

This time the associated grand canonical partition function is

$$\Xi^+\left(\beta,z\right) = \sum_{N=0}^{\infty} z^N Z_N^+\left(\beta\right)$$

$$= \sum_{(n_1,\ldots,n_K)\in(\mathbb{N}_+)^K} \prod_{k=1}^{K} \left(ze^{-\beta\varepsilon_k}\right)^{n_k}$$

$$= \prod_{k=1}^{K} \sum_{n=0}^{\infty} \left(ze^{-\beta\varepsilon_k}\right)^n$$

$$= \prod_{k=1}^{K} \frac{1}{1-ze^{-\beta\varepsilon_k}}, \tag{1.1}$$

where again we use the $\sum\prod \longleftrightarrow \prod\sum$ Lemma at the last stage. The thermodynamic potentials have the correct scaling properties and we do not have to resort to any ad hoc corrections of the type needed for Maxwell–Boltzmann statistics.

1.2.6 Fermi–Dirac Statistics

Fermions are likewise indistinguishable particles; however, we can never have more that one in the same energy state ε_k. The set of all possible configurations is therefore

$$\Omega_{N,K}^- = \left\{ (n_1, \ldots, n_K) \in (\mathbb{N}_-)^K : \sum_{k=1}^K n_k = N \right\}$$

where $\mathbb{N}_- \triangleq \{0, 1\}$. In particular, we note that $\#\Omega_{N,K}^- = \binom{K}{N}$. The Fermion canonical partition function is

$$Z_N^- (\beta) = \sum_{\mathbf{n} \in \Omega_{N,K}^-} \prod_{k=1}^K e^{-\beta \varepsilon_j n_k},$$

and, similar to the Boson case, the grand canonical partition function is

$$\Xi^- (\beta, z) = \sum_{N=0}^\infty z^N Z_N^- (\beta) = \prod_{k=1}^K \sum_{n=0}^1 \left(z e^{-\beta \varepsilon_k} \right)^n \equiv \prod_{k=1}^K \left(1 + z e^{-\beta \varepsilon_k} \right).$$

1.2.7 Entropy of Statistical Ensembles

Let Ω be the set of all microstates. The set of all probability measures, Σ, is the set of vectors $\pi = (p(\omega))$ with $p(\omega) \geq$ with $\sum_{\omega \in \Omega} p(\omega) = 1$. Let $Y(\omega)$ be some real-valued function on Ω, then its average for a particular $\pi \in \Sigma$ is

$$\mathbf{E}[Y] \triangleq \sum_{\omega \in \Omega} Y(\omega) p(\omega). \tag{1.2}$$

Let us take $N(\omega)$ and $E(\omega)$ to be, respectively, the number of particles and the energy associated with a particular microstate ω. We introduce the following sets of microstates: $\Omega(N) = \{\omega \in \Omega : N(\omega) = N\}$ and $\Omega(U, N) = \{\omega \in \Omega : E(\omega) = U, N(\omega) = N\}$. We write Σ_N and $\Sigma_{U,N}$ for those measures supported on $\Omega(N)$ and $\Omega(U, N)$ respectively. In general, for $\pi \in \Sigma$ we define its entropy (Boltzmann's H-function) to be

$$\mathscr{H}(\pi) = -\sum_{\omega \in \Omega} p(\omega) \ln p(\omega). \tag{1.3}$$

The microcanonical ensemble is then distinguished as the measure in $\Sigma_{U,N}$ which maximizes $\mathscr{H}(\pi)$. Indeed, we have $W(U, N) = \#\Omega(U, N)$ and the maximizing π is the uniform measure $p(\omega) = \frac{1}{W(U,N)}$, with the maximum value then being

$$\max_{\pi \in \Sigma_{U,N}} \mathcal{H}(\pi) = \ln W(U,N). \tag{1.4}$$

The Boltzmann entropy is

$$S_{\text{micro.}}(U,N) = k \max_{\pi \in \Sigma_{U,N}} \mathcal{H}(\pi). \tag{1.5}$$

Similarly, we may define

$$S_{\text{can.}}(\bar{U},N) = k \max_{\pi \in \Sigma_N : \mathbb{E}[E] = \bar{U}} \mathcal{H}(\pi), \tag{1.6}$$

and

$$S_{\text{g.c.}}(\bar{U},\bar{N}) = k \max_{\pi \in \Sigma : \mathbb{E}[E] = \bar{U}, \mathbb{E}[N] = \bar{N}} \mathcal{H}(\pi), \tag{1.7}$$

and the maximizing probability measures are the canonical and grand canonical ensembles respectively. For instance, in the canonical ensemble case, we may employ Lagrange multipliers α, β and equivalently seek the maximizer over $\pi \in \Sigma_N$:

$$\Phi(\pi) = \mathcal{H}(\pi) - \alpha \sum_\omega p(\omega) - \beta \sum_\omega E(\omega) p(\omega).$$

Treating the $p(\omega)$ as independent, where $N(\omega) = N$, we have $\frac{\partial \Phi}{\partial p(\omega)} = 0$ implying $-1 - \ln p(\omega) - \alpha - \beta E(\omega) = 0$, or

$$p(\omega) \equiv \begin{cases} \frac{1}{Z_N} e^{-\beta E(\omega)}, & N(\omega) = N; \\ 0, & \text{otherwise.} \end{cases},$$

with $Z_N = \sum_{\omega \in \Omega_N} e^{-\beta E(\omega)}$. This is, of course, the canonical ensemble at inverse temperature $\beta = \frac{1}{kT}$. For this ensemble, we have

$$\mathbb{E}_{\text{can.}}^{\beta,N}[E] = \frac{1}{Z_N} \sum_\omega^{N(\omega)=N} E(\omega) e^{-\beta E(\omega)} = -\frac{\partial}{\partial \beta} \ln Z_N = \frac{\partial}{\partial \beta}(\beta F)$$

and we take the unique parameter value of β such that $\mathbb{E}_{\text{can.}}^{\beta,N}[E] = \bar{U}$. This then yields

$$S_{\text{can.}}(\bar{U},N) = -\frac{1}{T}(F - \bar{U}),$$

consistent with the definition $F = U - TS$ from thermodynamics.

We have made some tacit assumptions about the various thermodynamic variables appearing in the preceding. In practice, the limit values of the intensive variables per particle will be convex or concave functions, but not necessarily strictly so – that is, they may have a linear part. These are associated with phase transitions, and in such cases the various ensembles may be inequivalent. For more on this, see the review of Touchette (2009) and the references therein.

1.2.8 Integer Partitions

We digress slightly with a tale about a French aristocrat, Chevalier de Mere, who posed the following early question in probability theory: given a roll of three fair dice, is it more likely to get a total score of 11 or a total score of 12?

His answer was that both events were equally likely and here is the argument. We obtain 11 from three dice in $p\,(11,3) = 6$ ways

$$11 = 6 + 4 + 1 = 6 + 3 + 2 = 5 + 5 + 1$$
$$= 5 + 4 + 2 = 5 + 3 + 3 = 4 + 4 + 3,$$

and 12 from three dice in $p\,(12,3) = 6$ ways

$$12 = 6 + 5 + 1 = 6 + 4 + 2 = 6 + 3 + 3$$
$$= 5 + 5 + 2 = 5 + 4 + 3 = 4 + 4 + 4.$$

The problem, however, came to the attention of the famous mathematician Pascal, who gave a different answer. The situation where we roll a 6, a 5, and a 1 is not in fact a single outcome but, rather, is an event corresponding to six distinct outcomes – the three dice are distinguishable, so we can go to finer detail and say which dice take which value. Likewise rolling two 5's and a 1 is an event corresponding to three distinct outcomes, depending on which die is to be the 1. But the event corresponding to three 4's is a single outcome. Pascal argued that de Mere undercounted the events. The numbers of possibilities according to Pascal are $\tilde{p}\,(11,3) = 6 \; + \; 6 \; + \; 3 \; + \; 6 \; + \; 3 \; + \; 3 = 27$ and $\tilde{p}\,(12,3) = 6 \; + \; 6 \; + \; 3 \; + \; 3 \; + \; 6 \; + \; 1 = 25$, respectively. So a total score of 11 is more likely.

In retrospect, Pascal figured out the principle of distinguishability and applied Maxwell–Boltzmann statistics to the dice – as dice are macroscopic entities, this is the correct thing to do. De Mere, on the other hand, was using Bose–Einstein statistics for the dice.

Let us denote the number of de Mere events corresponding to rolling a total score of n from m dice by $p\,(n,m)$. Is there a nice formula for these numbers? Well, let us assume that each die has K faces with scores $1, 2, \ldots, K$ each with a probability $1/K$ to occur. Suppose that we roll n_1 1's, n_2 2's, n_3 3's, and so on, and obtain a total score of n, then

$$n = \underbrace{K + K + \cdots + K}_{n_K \text{ terms}} + \cdots + \underbrace{2 + 2 + \cdots + 2}_{n_2 \text{ terms}} + \underbrace{1 + 1 + \cdots + 1}_{n_1 \text{ terms}}, \qquad (1.8)$$

and in the process must have rolled $m = n_K + \cdots + n_2 + n_1$ dice. We refer to (1.8) as an **integer partition** – more specifically, an integer partition of n into m parts.

Lemma 1.2.2 *The number, $p(n, m)$, of integer partition of n into m parts have the following generating function:*

$$\sum_{n,m} p(n, m) z^m x^n = \prod_{k=1}^{K} \frac{1}{1 - zx^k}.$$

Proof Let $\mathbf{n} = (n_1, n_2, \ldots, n_K) \in (\mathbb{N}_+)^K$ give a de Mere event; then it corresponds to rolling $m = \sum_{k=1}^{K} n_k$ dice and getting a total score of $n = \sum_{k=1}^{K} kn_k$, from (1.8). Each de Mere event corresponds to an integer partition and we have

$$\sum_{n,m} p(n, m) z^m x^n = \sum_{\mathbf{n} \in (\mathbb{N}_+)^K} z^{\sum_{k=1}^{K} n_k} x^{\sum_{k=1}^{K} kn_k}.$$

However, this is just the Bose grand canonical partition function (1.1) with $x = e^{-\beta}$ and $\epsilon_k = k$. □

This result, in fact, goes back to Euler. Note that we may even take our dice to have $K = \infty$ faces if desired.

1.3 Combinatorial Coefficients

We will now introduce formal notations for some of the combinatorial objects appearing in the preceding.

1.3.1 Factorials

The nth **falling factorial power** is defined by[1]

$$x^{\underline{n}} \triangleq x(x-1)(x-2) \cdots (x-n+1) \qquad (1.9)$$

---

[1] We refer to these as the Pochhammer symbols, though the notation we use here is due to Graham, Knuth, and Patashnik (1988). The symbol $x^{\underline{n}}$ is pronounced as "x to the n falling."

as well as the *n*th **rising factorial power**

$$x^{\overline{n}} \triangleq x(x+1)(x+2)\cdots(x+n-1). \tag{1.10}$$

(We also set $x^{\underline{0}} = x^{\overline{0}} = 1$.) The **falling** and **rising binomial coefficients** may then defined to be[2]

$$\binom{n}{k}_- \triangleq \frac{n^{\underline{k}}}{k!} = \frac{n(n-1)\cdots(n-k+1)}{k!}$$

$$\binom{n}{k}_+ \triangleq \frac{n^{\overline{k}}}{k!} = \frac{n(n+1)\cdots(n+k-1)}{k!}$$

We remark that the falling binomial coefficients $\binom{n}{k}_-$ are just the familiar *n*-choose-*k* binomial coefficients $\binom{n}{k}$. It is easy to derive the following properties.

The falling and rising binomial coefficients satisfy the following properties:

1. $\binom{n}{k}_+ = \binom{n+k-1}{k}_-$
2. $\binom{n}{k}_+ = (-1)^k \binom{-n}{k}_-$
3. Recurrence relations,

$$\binom{n}{k}_- = \binom{n-1}{k-1}_- + \binom{n-1}{k}_-,$$

$$\binom{n}{k}_+ = \binom{n}{k-1}_+ + \binom{n-1}{k}_+,$$

with $\binom{n}{0}_\pm = 1$, and $\binom{0}{k}_\pm = 0$ if $k \neq 0$. The first of these can be used to generate the usual binomial coefficients through the Pascal triangle construction.

4. Generating relations,

$$\sum_{k\geq 0} \binom{n}{k}_\pm t^k = (1 \mp t)^{\mp n}.$$

The generating relations are both instances of the general binomial theorem $(1+t)^p = \sum_{k\geq 0} \binom{p}{k} t^k$, valid for $|t| < 1$.

1.3.2 Stirling Numbers

We now derive the connection between the ordinary and falling factorial powers. We begin by remarking that the right-hand sides of (1.9) and (1.10) may be expanded to obtain a polynomial in *x* of degree *n* with integer coefficients.

---

[2] There is no established notation for these, and we use + and − due to the eventual connection with Bose and Fermi counting statistics.

The **Stirling numbers of the first and second kind** are defined respectively as the coefficients $\left[{n \atop m}\right]$ and $\left\{{n \atop m}\right\}$ appearing in the relations

$$x^{\overline{n}} \equiv \sum_m \left[{n \atop m}\right] x^m; \qquad x^n \equiv \sum_m \left\{{n \atop m}\right\} x^{\underline{m}}. \tag{1.11}$$

The nth **Bell number** B_n, is then defined to be

$$B_n \triangleq \sum_{m=1}^n \left\{{n \atop m}\right\}. \tag{1.12}$$

The Stirling numbers satisfy the following properties:

1. $\left[{n \atop m}\right] = \left\{{n \atop n}\right\} = \left\{{n \atop 1}\right\} = 1$.
2. The Stirling numbers of the first kind are integers satisfying $\left[{n \atop m}\right] \geq 0$. It turns out that this is also true of the second kind numbers.
3. We have $\left[{n \atop m}\right] = 0 = \left\{{n \atop m}\right\}$ for integers $m \neq 1, \ldots, n$.
4. The set of rising power polynomials $\{x^{\overline{n}} : n = 0, 1, 2, \ldots\}$ is linearly independent, as is the set of falling power polynomials.
5. The Stirling numbers are dual in the sense that if we introduce the pair of infinite matrices s and S with entries $s_{nm} = \left[{n \atop m}\right]$ and $S_{nm} = \left\{{n \atop m}\right\}$ for $n, m \in \{0, 1, 2, \ldots\}$, then

$$sCS = SCs = I,$$

where C is the checkerboard signed matrix $C_{nm} = (-1)^{n+m}$. In particular, $s^{-1} = CS$ and $S^{-1} = Cs$. This follows from observing that

$$x^{\underline{n}} \equiv \sum_m (-1)^{n+m} \left[{n \atop m}\right] x^m,$$

$$x^n \equiv \sum_m (-1)^{n+m} \left\{{n \atop m}\right\} x^{\overline{m}},$$

and so, from the defining relations and linear independence, we see for instance that

$$\sum_k (-1)^{n+k} \left[{n \atop k}\right]\left\{{k \atop m}\right\} = \delta_{nm}. \tag{1.13}$$

6. The Stirling numbers satisfy the recurrence relations (Stirling's identities)

$$\left[{n+1 \atop m}\right] = \left[{n \atop m-1}\right] + n\left[{n \atop m}\right], \tag{1.14}$$

$$\left\{{n+1 \atop m}\right\} = \left\{{n \atop m-1}\right\} + m\left\{{n \atop m}\right\}, \tag{1.15}$$

with $\left[\begin{smallmatrix}1\\1\end{smallmatrix}\right] = 1 = \left\{\begin{smallmatrix}1\\1\end{smallmatrix}\right\}$. (This follows readily from the relations $x^{\overline{n+1}} = x^{\overline{n}} \times (x + n)$ and $x \times x^{\underline{m}} = (x - m + m) x^{\underline{m}} = x^{\underline{m+1}} + m x^{\underline{m}}$.)

7. $\sum_{m=1}^{\infty} \left[\begin{smallmatrix}n\\m\end{smallmatrix}\right] = (1)^{\overline{n}} = n!$. (Substitute $x = 1$ in the defining relation (1.11). Note that there is no such simple expression for the Bell numbers.)

From the Stirling identities, the Stirling numbers may then be generated recursively using a construction similar to Pascal's triangle, viz.,

$\left[\begin{smallmatrix}n\\m\end{smallmatrix}\right]$ $n\backslash m$	1	2	3	4	5	6
1	1					
2	1	1				
3	2	3	1			
4	6	11	6	1		
5	24	50	35	10	1	
6	120	274	225	85	15	1

$\left\{\begin{smallmatrix}n\\m\end{smallmatrix}\right\}$ $n\backslash m$	1	2	3	4	5	6
1	1					
2	1	1				
3	1	3	1			
4	1	7	6	1		
5	1	15	25	10	1	
6	1	31	90	65	15	1

The first few Bell numbers are

n	1	2	3	4	5	6	7	8
B_n	1	2	5	15	52	203	877	4140

1.4 Sets and Bags

In this section, we fix a set $\mathfrak{X}$. A sample of size n drawn from $\mathfrak{X}$ is a sequence $x_1, x_2, \ldots, x_n$ of elements taken one after the other from $\mathfrak{X}$. Ultimately we are only interested in the elements drawn and not the order in which they were drawn. We say that the sampling was done without replacement if, once an element was drawn, it was not available to be drawn again: in this case, we have that $\{x_1, \ldots, x_n\}$ will be a set with n distinct elements. If replacement is allowed, then some elements may be drawn more than once and we refer to such instances as **coincidences**.

A **bag**, or **multiset**, drawn from a set $\mathfrak{X}$ is an unordered (finite!) collection $x_1, \ldots, x_n \in \mathfrak{X}$ where we may have some coincidences, that is, several of the elements of the bag may be the same element of $\mathfrak{X}$. A **set** drawn from $\mathfrak{X}$ is a bag in which there are no coincidences, that is, no element of $\mathfrak{X}$ may appear more than once. We shall write Bag($\mathfrak{X}$) and Power($\mathfrak{X}$), respectively, for the collections of bags and sets[3] that we can draw from $\mathfrak{X}$

For a countable set $\mathfrak{X}$, the **occupation numbers for bags** drawn from $\mathfrak{X}$ are the functions

$$n_x \colon \mathrm{Bag}(\mathfrak{X}) \mapsto \{0, 1, 2, \ldots\} = \mathbb{N}_+,$$

with $n_x(X)$ counting the number of times an element $x \in \mathfrak{X}$ appears in the bag X.

Proposition 1.4.1 *There is a one-to-one correspondence between bags drawn from a set $\mathfrak{X}$ and sequences of occupation numbers* $\mathbf{n} = (n_x)_{x \in \mathfrak{X}}$, *that is,* $(\mathbb{N}_+)^{\mathfrak{X}}$.

Sets are then just the bags with no coincidences, and so the occupation numbers for sets are restricted to $(\mathbb{N}_-)^{\mathfrak{X}}$. In fact, if $\mathfrak{X}$ consists of N elements, we have the enumerations

- the number of bags of size m is $\binom{N}{m}_+ \equiv \binom{N}{N+m-1}$;
- the number of sets of size m is $\binom{N}{m}_- \equiv \binom{N}{m}$

Note that the total number of sets that can be drawn from a set of size N is $\sum_m \binom{N}{m} = 2^N$. The total number of bags will always be infinite! It is worthwhile comparing these observations to the original problem of having N indistinguishable urns into which we can distribute m balls!

Corollary 1.4.2 *Let us define the size function for a bag (or a set) as*

$$N(\cdot) = \sum_{x \in \mathfrak{X}} n_x(\cdot). \qquad (1.16)$$

Then

$$\sum_{A \in \mathrm{Bag}(\mathfrak{X})} t^{N(A)} = (1 - t)^{-N}, \qquad (1.17)$$

$$\sum_{A \in \mathrm{Power}(\mathfrak{X})} t^{N(A)} = (1 + t)^N. \qquad (1.18)$$

The proof in either case comes down to setting $f(x, n_x) = t^{n_x}$ in the $\sum \prod \longleftrightarrow \prod \sum$ Lemma and observing that the sum over bags/subsets

---

[3] The set of all subsets of a given set $\mathfrak{X}$ is traditionally called the power set of $\mathfrak{X}$.

equates with a sum over the corresponding occupation sequences. (In fact, it is just a special case of the calculations for the grand canonical Boson and Fermion partition functions!) Fixing $t \in (-1, 1)$, we have

$$\sum_{A \in \text{Bag}(\mathfrak{X})} t^{N(A)} = \left(1 + t + t^2 + \cdots\right)^N = (1 - t)^{-N}.$$

The argument for sets is similar.

We recognize the generating functions here for $\binom{N}{m}_{\pm}$, so these clearly give the number of bags/sets of size m that may be drawn from N elements.

1.5 Permutations and Partitions

A **permutation** of a set of elements is an arrangement of those elements into a particular order. The set of permutations over a set X will be denoted by Perm(X).

Suppose for concreteness that $X = \{x_1, \ldots, x_n\}$, then a permutation $\sigma \in$ Perm(X) is a bijective mapping from X to itself and this may be uniquely represented by the sequence $(\sigma(x_1), \ldots, \sigma(x_n))$, which of course must be a list of all elements of X with no repetition. Perm(X) forms a nonabelian group under composition. We shall use the notation $\sigma^0 = id, \sigma^1 = \sigma, \sigma^2 = \sigma \circ \sigma$, and so on.

Given a permutation $\sigma \in$ Perm(X), the **orbit** of an element $x \in X$ under σ is the sequence $\{x, \sigma(x), \sigma^2(x), \ldots\}$. As the orbit must lie within X, it is clear that we eventually must have $\sigma^k(x) = x$ for some $0 < k \leq n$: the smallest such value is called the **period** of the orbit and clearly the orbit repeats itself beyond this point ($\sigma^{n+k}(x) = \sigma^n(x)$). The ordered collection $\left[x, \sigma(x); \sigma^2(x); \ldots; \sigma^{k-1}(x)\right]$ is referred to as a **cycle** or more explicitly a k-**cycle**. Cycles will be considered to be equivalent under cyclic permutation in the sense that $[a_1; a_2; \ldots; a_k]$ is not distinguished from $[a_2; a_3; \ldots; a_k; a_1]$, etc. Thus each k-cycle is equivalent to k sequences depending on where on the orbit we choose to start. Clearly orbits arising from the same permutation σ either coincide or are completely disjoint; this simple observation leads to the cyclic factorization theorem for permutations: each permutation σ can be uniquely written as a collection of disjoint cycles.

An element $x \in X$ is a **fixed point** of the permutation σ if $x = \sigma(x)$ – that is, x forms a 1-cycle.

We would like to work out the number $s(n, m)$ of permutations on n elements that have exactly m cycles – these turn out to be the Stirling number of the first kind.

Lemma 1.5.1 (Counting Permutations) *Let* $\mathrm{Perm}_m(X)$ *be the set of permutations in* $\mathrm{Perm}(X)$ *having exactly m cycles. If X has size n, then the number of permutations in* $\mathrm{Perm}(X)$ *is given by the Stirling numbers of the first kind* $\begin{bmatrix} n \\ m \end{bmatrix}$.

Proof Let $X = \{x_1, \ldots, x_n\}$ and suppose $\mathrm{Perm}_m(X) = s(n, m)$. Suppose that we have a new element x_{n+1} not in X, then $\mathrm{Perm}(X \cup \{x_{n+1}\})$ has $s(n + 1, m)$ permutations. Some of these will have x_{n+1} as a fixed point – these permutations will then have $m - 1$ other cycles drawn from X, and so there will be $s(n, m - 1)$ of these permutations. The remaining permutations will have x_{n+1} included in a cycle of period two or more: now if we take any permutation in $\mathrm{Perm}_m(X)$, then we could insert the x_{n+1} before any one of the elements $x \in X$ in the cyclic decomposition – in this way, we would generate all permutations of this particular type, and so there are $ns(n, m)$ such possibilities. We therefore obtain $s(n+1, m) = s(n, m-1) + ns(n, m)$, which is the same recurrence relation (1.14) as the Stirling numbers of the first kind. Clearly $s(1, 1) = 1 = s(n, n)$ while $s(n, m) = 0$ if $m > n$. Therefore, $s(n, m) \equiv \begin{bmatrix} n \\ m \end{bmatrix}$. □

A **partition** of a set X is an unordered collection (set!) of nonempty subsets whose union is X. The subsets are called the blocks of the partition. We denote the set of all partitions of X by $\mathrm{Part}(X)$, and specifically the set of all partitions into exactly n blocks by $\mathrm{Part}_n(X)$. A partition will be called a **pair partition** if each of its blocks contains exactly two elements, and we shall denote the set of all pair partitions of X by $\mathrm{Pair}\,(X)$.

Lemma 1.5.2 (Counting Pair Partitions) *A finite set X will have no pair partitions if it has an odd number of elements. If X has 2k elements then*

$$|\mathrm{Pair}\,(X)| = \frac{(2k)\,!}{2^k k!}.$$

Proof Without loss of generality, take $X = \{1, 2, \ldots, 2k\}$. We have $2k \times (2k - 1)$ choices for the first pair, then $(2k - 2) \times (2k - 3)$ for the second, and so on. This gives a total of $(2k)\,!$. However, we have overcounted by a factor of $k!$, as we do not label the pairs, and by 2^k, as we do not label the elements within each pair either. On the other hand, if X has an odd number of elements, then there are no ways to partition into pairs. □

The number $S(n, m)$ of partitions of n elements into m parts turns out to be the Stirling numbers of the second kind. For instance, the set $\{1, 2, 3, 4\}$ can be partitioned into 2 blocks in $S(4, 2) = 7$ ways:

$$\{\{1,2\},\{3,4\}\},\qquad \{\{1\},\{2,3,4\}\},$$
$$\{\{1,3\},\{2,4\}\},\qquad \{\{2\},\{1,3,4\}\},$$
$$\{\{1,4\},\{2,3\}\},\qquad \{\{3\},\{1,2,4\}\},$$
$$\{\{4\},\{1,2,3\}\},$$

and into 3 blocks in $S(4,3) = 6$:

$$\{\{1\},\{2\},\{3,4\}\},\qquad \{\{1\},\{3\},\{2,4\}\},$$
$$\{\{1\},\{4\},\{2,3\}\},\qquad \{\{2\},\{3\},\{1,4\}\},$$
$$\{\{2\},\{4\},\{2,4\}\},\qquad \{\{3\},\{4\},\{1,2\}\}.$$

In combinatorial terms, $S(n,m)$ counts the number of ways to distribute n distinguishable balls among m urns, where we do not distinguish the urns and no urn is allowed to be empty.

Lemma 1.5.3 (Counting All Partitions) *The number of partitions of the set X of size n having exactly m blocks is the Stirling number of the second kind $\left\{{n \atop m}\right\}$. In particular, the Bell numbers B_n count the number of ways to partition a set X of size n.*

Proof Let $S(n,m)$ be the number of partitions of n items into exactly m blocks. We first of all show that we have the formula

$$S(n+1,m) = S(n,m-1) + mS(n,m).$$

This is relatively straightforward. We see that $S(n+1,m)$ counts the number of partitions of a set $X = \{x_1,\ldots,x_n,x_{n+1}\}$ having m blocks. Some of these will have the singleton $\{x_{n+1}\}$ as a block: there will be $S(n,m-1)$ of these as we have to partition the remaining elements $\{1,\ldots,n\}$ into $m-1$ blocks. The others will have x_{n+1} appearing with at least some other elements in a block: we have $S(n,m)$ partitions of $\{1,\ldots,n\}$ into m blocks, and we then may place x_{n+1} into any one of these m blocks yielding $mS(n,m)$ possibilities. Clearly $S(1,1) = 1$ and $S(n,n) = 1$ while $S(n,m) = 0$ if $m > n$. The numbers $S(n,m)$ therefore satisfy the same generating relations (1.15) as the $\left\{{n \atop m}\right\}$ and so are one and the same. $\qquad\square$

In combinatorial terms, $\left\{{n \atop m}\right\}$ counts the number of ways to distribute n distinguishable balls among m urns, where we do not distinguish the urns, and no urn is allowed to be empty.

The proofs of the enumerations followed from elementary arguments where we show that $s(n,m)$ and $S(n,m)$ satisfy the same recurrence relations as the Stirling numbers. We will reestablish this later using the powerful machinery of combinatorial species.

1.5.1 The Lattice of Partitions

A **partial ordering** of Part (X) is given by saying that $\pi \leq \varpi$ if every block of ϖ is a union of one or more blocks of π. In such situations. we say that π is finer than ϖ. We also write $\pi < \varpi$ if we have $\pi \leq \varpi$ and $\pi \neq \varpi$.

We then have a minimum (finest) element given by

$$\mathbf{0}(X) \triangleq \{\{x\} : x \in X\},$$

and a largest (coarsest) element

$$\mathbf{1}(X) \triangleq \{X\},$$

and for all $\pi \in$ Part (X) we have

$$\mathbf{0}(X) \leq \pi \leq \mathbf{1}(X).$$

We shall often encounter the collection of partitions that are strictly finer than $\mathbf{1}(X)$ (**proper partitions!**) and give this the following special notation:

$$\text{Part}_<(X) \triangleq \{\pi \in \text{Part}(X) : \pi < \mathbf{1}(X)\}.$$

Mathematically, the collection Part (X) forms a lattice, meaning that for every pair of partitions π and ϖ, there is a coarsest one $\pi \wedge \varpi$ (the **meet** of π and ϖ) that is finer than both, and a finest one $\pi \vee \varpi$ (the **join** of π and ϖ) that is coarser than both. Specifically, $\pi \vee \varpi$ is the partition whose blocks are all nonempty intersections of a block from π and a block from ϖ.

1.6 Occupation Numbers

1.6.1 Occupation Numbers for Partitions

Given $\pi \in$ Part (X), we let $n_k(\pi)$ denote the number of blocks in π having size k. We shall refer to the n_k as **occupation numbers** and we introduce the functions

$$N(\pi) = \sum_{k \geq 1} n_k(\pi), \qquad E(\pi) = \sum_{k \geq 1} k n_k(\pi). \qquad (1.19)$$

Note that a partition π partitions a set of $E(\pi) = n$ elements into $N(\pi) = m$ blocks.

It is sometimes convenient to replace sums over partitions with sums over occupation numbers. Recall that a partition π will have occupation numbers $\mathbf{n} = (n_1, n_2, n_3, \dots)$ and we have $N(\pi) = N(\mathbf{n}) = n_1 + n_2 + n_3 + \cdots$ and $E(\pi) = E(\mathbf{n}) = n_1 + 2n_2 + 3n_3 + \cdots$.

Lemma 1.6.1 *The number of partitions of a set of n elements leading to the same set of occupation numbers* **n** *is given by*

$$v_{\text{Part}}(\mathbf{n}) = \frac{1}{(1!)^{n_1}(2!)^{n_2}(3!)^{n_3}\cdots} \frac{n!}{n_1!\,n_2!\,n_3!\cdots}$$

$$= n! \prod_{k=1} \frac{1}{(k!)^{n_k}\,n_k!}. \tag{1.20}$$

The proof follows from the observation that there are $n!$ ways to distribute the n objects; however, we do not label the n_k blocks of size k, nor their contents.

Bell Polynomials

We remark that the multinomials defined by

$$\text{Bell}(n, m; \mathbf{z}) = \sum_{\mathbf{n}}^{E(\mathbf{n})=n, N(\mathbf{n})=m} v_{\text{Part}}(\mathbf{n})\, z_1^{n_1} z_2^{n_2}\cdots,$$

for complex sequences $\mathbf{z} = (z_1, z_2, \ldots)$, are known as the **Bell polynomials**.
In particular, the Stirling numbers of the second kind take the form

$$\left\{ \begin{matrix} n \\ m \end{matrix} \right\} = \text{Bell}(n, m; 1, 1, 1, \ldots) = \sum_{\mathbf{n}}^{E(\mathbf{n})=n, N(\mathbf{n})=m} v_{\text{Part}}(\mathbf{n}),$$

or

$$\left\{ \begin{matrix} n \\ m \end{matrix} \right\} = n! \sum_{\mathbf{n}}^{E(\mathbf{n})=n, N(\mathbf{n})=m} \prod_{k=1}^{\infty} \frac{1}{(k!)^{n_k}\,n_k!}.$$

Theorem 1.6.2 (The composition formula) *Let* $h(x) = \sum_{n=1}^{\infty} \frac{1}{n!} h_n x^n$ *and* $g(x) = \sum_{n=1}^{\infty} \frac{1}{n!} g_n x^n$ *be a pair of analytic functions about the origin. Then their composition* $h \circ g$ *has the expansion* $\sum_{n=1}^{\infty} \frac{1}{n!} (h \circ g)_n x^n$, *where*

$$(h \circ g)_n = \sum_m h_m \,\text{Bell}(n, m; g_1, g_2, g_3, \ldots). \tag{1.21}$$

We introduce some notation based on this result: Let $\mathbf{h} = (h_n)_n$ and $\mathbf{g} = (g_n)_n$ be sequences of scalars where $n = 1, 2, 3, \ldots$; then we define $\mathbf{h} \copyright \mathbf{g}$ to be the sequence whose nth term is $(h \circ g)_n$ as in (1.21).

Proof We have

$$
\begin{aligned}
(h \circ g)(x) &= \sum_{m=1}^{\infty} \frac{1}{m!} h_m \left(\sum_{k=1}^{\infty} \frac{1}{k!} g_k x^k \right)^m \\
&= \sum_{m=1}^{\infty} \frac{1}{m!} h_N \sum_{\mathbf{n}}^{\sum_k n_k = m} \binom{m}{n_1, n_2, n_3, \ldots} \prod_{k=1}^{\infty} \left(\frac{1}{k!} g_k x^k \right)^{n_k} \\
&= \sum_{m=1}^{\infty} h_m \sum_{\mathbf{n}}^{N(\mathbf{n})=m} \frac{1}{E(\mathbf{n})!} \prod_{k=1}^{\infty} \left(\frac{1}{n_k! (k!)^{n_k}} (g_k)^{n_k} \right) x^{E(\mathbf{n})} \\
&= \sum_{m=1}^{\infty} h_m \sum_{\mathbf{n}}^{N(\mathbf{n})=m} \frac{1}{E(\mathbf{n})!} v_{\text{Part}}(\mathbf{n}) \left(\prod_{k=1}^{\infty} g_k^{n_k} \right) x^{E(\mathbf{n})} \\
&= \sum_{n,m=1}^{\infty} \frac{1}{n!} h_m \text{Bell}(n, m; g_1, g_2, g_3, \ldots) x^n.
\end{aligned}
$$

$\square$

Along similar lines, we may prove **Faà di Bruno's Formula** Let h and g be smooth functions, then

$$
\frac{d^n}{dx^n} h \circ g(x) = \sum_m \left. \frac{d^m h}{dx^m} \right|_{g(x)} \text{Bell}\left(n, m; \frac{dg}{dx}, \frac{d^2 g}{dx^2}, \ldots \right).
$$

1.6.2 Occupation Numbers for Permutations

In the same vein, we may introduce occupation numbers for permutations. Given a permutation $\sigma \in \text{Perm}(X)$, we let $n_k(\sigma)$ be the number of k-cycles in the decomposition of σ. Similarly, setting $N(\sigma) = \sum_{k \geq 1} n_k(\sigma)$ and $E(\sigma) = \sum_{k \geq 1} k n_k(\sigma)$, we have that σ is a permutation of $E(\sigma)$ elements into $N(\sigma)$ cycles. In general, several distinct permutations on the same set of n objects can lead to the same set of occupation numbers, and the degeneracy is now given by

$$
v_{\text{Perm}}(\mathbf{n}) = \frac{1}{(1)^{n_1}(2)^{n_2}(3)^{n_3}\ldots} \frac{n!}{n_1! n_2! n_3! \ldots} = n! \prod_{k=1}^{\infty} \frac{1}{k^{n_k} n_k!}.
$$

(This time we must compensate for the k different ways to represent a k-cycle as a sequence.)

For Stirling numbers of the first kind, we have

$$\begin{bmatrix} n \\ m \end{bmatrix} = n! \sum_{\mathbf{n}}^{E(\mathbf{n})=n, N(\mathbf{n})=m} \prod_{k=1}^{\infty} \frac{1}{k^{n_k} n_k!}$$

$$= \text{Bell}\,(n, m; 0!, 1!, 2!, \ldots)$$

1.7 Hierarchies (= Phylogenetic Trees = Total Partitions)

Given a set of n items, say $\{1, 2, \ldots, n\}$, with $n \geq 2$. We partition the set into two or more parts – that is, make a proper partition. We may repeat this procedure by making a proper partition each nonsingleton block until we have eventually "atomized" down to the finest partition $\mathbf{0}_n = \{\{1\}, \{2\}, \ldots, \{n\}\}$. We refer to this as a **total partition**.

For example, let $X = \{1, 2, \ldots, 10\}$; then an example of a total partition is given by

$$\pi_0 = \mathbf{1}_{10} \equiv \{1, 2, 3, 4, 5, 6, 7, 8, 9, 10\},$$

$$\pi_1 = \{\{1, 2, 5, 7, 8\}, \{3\}, \{4, 6, 9, 10\}\},$$

$$\pi_2 = \{\{1, 2\}, \{5\}, \{7, 8\}, \{3\}, \{4\}, \{6, 9, 10\}\},$$

$$\pi_3 = \mathbf{0}_{10} \equiv \{\{1\}, \{2\}, \ldots, \{n\}\}.$$

This gives a sequence $\mathbf{0}_{10} = \pi_3 < \pi_2 < \pi_1 < \pi_0 = \mathbf{1}_{10}$. This increasingly finer sequence of partitions can be can be described by Figure 1.7.

Alternatively, we may sketch this as a (phylogenetic) tree; see Figure 1.8.

Each node of the tree corresponds to a block B, and at each stage we have a proper partition $\pi = \{A_1, \ldots, A_m\} \in \text{Part}_<(B)$ and the branches are then labeled

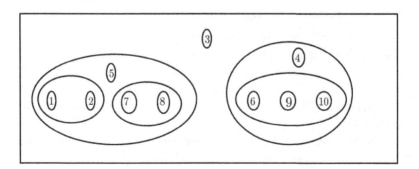

Figure 1.7 A hierarchy as an increasingly finer sequence of partitions of a set down to the finest partition $\mathbf{0}_n$.

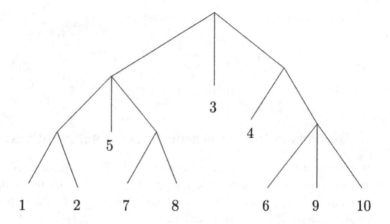

Figure 1.8 A hierarchy as a tree.

by the parts of π. We may draw this section of the tree as the following directed graph (a tree!) to sketch this:

The requirement that the partition be proper means that we have at least two branches! Eventually we come to a point where we are down to singletons and the process terminates. Note that the set of leaves of the tree is then just the original set X, and that for any part B appearing in the hierarchy (that is, any node) the actual set B is just the subset of leaves associated with each node. The object obtained in this way is termed a hierarchy.

Let X be a finite set. A **hierarchy** on X is a directed tree having subsets of X as nodes, with the property that the sets at nodes coming immediately from a parent node form a partition of the set at the parent node. X (or equivalently $\mathbf{1}(X)$) is the root of the tree, and the leaves (terminal nodes) are the singletons of X (equivalently $\mathbf{0}(X)$). We denote the collection of all hierarchies on a set X by Hier (X).

As we have seen, a hierarchy on X is a strictly order sequence of partitions

$$\mathbf{0}(X) = \pi_r < \cdots < \pi_2 < \pi_1 < \pi_0 = \mathbf{1}(X),$$

and, in the tree representation, the integer r gives the length of the longest branch.

We would like to know the values of h_n, the number of hierarchies on a set of n elements. We may work out the lowest enumerations:

When $n = 2$ we have the one tree . When $n = 3$, we have the possibilities

and, when we count the number of ways to attach the leaves, we have $h_3 = 1 + 3 = 4$ possibilities.

When $n = 4$, we have the topologically distinct trees

which implies that $h_4 = 1 + 4 + 6 + 12 + 3 = 26$. We find that

n	1	2	3	4	5	6	7	8
h_n	1	1	4	26	236	2752	39208	660032

We will say slightly more about the enumeration of hierarchies after we encounter the methods of combinatorial species later.

1.8 Partitions

Let Part(n) denote the class of partitions on a set of n items, which for definiteness we may take to be $\{1, \ldots, n\}$. The class of all partitions is then Part $= \cup_{n \geq 1}$Part(n). We have introduced the occupation numbers n_k as the functions on Part counting the number of blocks of a partition having exactly k elements. We let $\mathcal{N}$ denote the sigma-algebra on Part generated by the occupation number functions. We recall that we have defined already two $\mathcal{N}$-measurable functions

$$E(\pi) = \sum_k k n_k(\pi), \qquad N(\pi) = \sum_k n_k(\pi),$$

giving the number of elements being partitioned by π and the number of parts of π respectively. We say that a $\mathcal{N}$-measurable function ψ is **multiplicative** if it takes the product form

$$\psi(\pi) \equiv \prod_{k=1}^{\infty} \lambda_k^{n_k(\pi)} \tag{1.22}$$

for scalars $\lambda_1, \lambda_2, \lambda_3, \ldots$. The sequence λ of scalars uniquely determines the multiplicative function, and vice versa, so we may write $\psi = \psi_\lambda$ if we wish to emphasize this dependence.

1.8.1 Coarse Graining and Möbius Inversion

Let us fix a finite set X. We recall that the set of partitions of X is a partially ordered set, in fact a lattice.

A function $f \colon \text{Part}(X) \times \text{Part}(X) \mapsto \mathbb{C}$ is an **incidence kernel** if $f(\pi, \varpi)$ vanishes in all cases where we do not have $\pi \leq \varpi$. Incidence kernels can be added in the obvious way, and a product $*$ is given by the convolution

$$(f * g)(\pi, \varpi) = \sum_{\pi \leq \rho \leq \varpi} f(\pi, \rho) \, g(\rho, \varpi).$$

The algebra spanned by incidence functions over a general partially ordered set $\mathcal{P}$ is called the **incidence algebra**, $I(\mathcal{P})$, and its identity is the **Kronecker delta** $\delta(\pi, \varpi)$ defined by

$$\delta(\pi, \varpi) = \begin{cases} 1, & \pi = \varpi; \\ 0, & \text{otherwise.} \end{cases}$$

An important example of an incidence kernel is the **zeta kernel** $\zeta(\pi, \varpi)$ defined as

$$\zeta(\pi, \varpi) = \begin{cases} 1, & \pi \leq \varpi; \\ 0, & \text{otherwise.} \end{cases}$$

The zeta kernel will have an inverse in the incidence algebra that is called the **Möbius kernel** μ, and we shall construct μ in the following. We first remark that if $\pi \leq \varpi$, then we can introduce the occupation numbers $n_k(\pi, \varpi)$ counting the number of blocks of ϖ that occur as the union of exactly k blocks of π. We say that an incidence kernel Φ is **multiplicative** if we have

$$\Phi(\pi, \varpi) \equiv \prod_k \lambda_k^{n_k(\pi, \varpi)}$$

for scalars $\lambda_1, \lambda_2, \lambda_3, \ldots$. (We understand $n_k(\pi, \varpi) \equiv 0$ if $\pi \not\leq \varpi$.) We may write $\Phi(\pi, \varpi) \equiv \Phi_\lambda(\pi, \varpi)$ for emphasis.

We also have the identity

$$\Phi_\lambda(\mathbf{0}_n, \mathbf{1}_n) = \lambda_n \tag{1.23}$$

since $\mathbf{1}_n$ is a single block of size n, i.e. $n_k(\mathbf{0}_n, \mathbf{1}_n) = 1$ if $k = n$ and 0 otherwise.

The collection of partitions ρ such that $\pi \leq \rho \leq \varpi$ is called a **segment**, and we denote this as $[\pi, \varpi]$. We note that the segment is isomorphic to the set

$$[\pi, \varpi] = \times_{k=1}^{\infty} \text{Part}(k)^{n_k(\pi, \varpi)}$$

and we have $N(\pi) = \sum_k n_k(\pi, \varpi)$ and $N(\varpi) = \sum_k k n_k(\pi, \varpi)$. The number of partitions in a segment is $\#[\pi, \varpi] = \nu_{\text{Part}}(\mathbf{n}(\pi, \varpi))$.

Let $\mathcal{P}$ and $\mathcal{P}'$ be partially ordered sets and fix functions $\psi \in I(\mathcal{P})$, $\psi' \in I(\mathcal{P}')$. We define their product $\psi \times \psi'$ on $I(\mathcal{P} \times \mathcal{P}')$ by $\psi \times \psi' \colon (\pi, \pi'), (\varpi, \varpi') \mapsto \psi(\pi, \varpi) \psi'(\pi', \varpi')$, in which case $(\psi \times \psi') * (\phi \times \phi') = (\psi * \phi) \times (\psi' * \phi')$.

Lemma 1.8.1 *The convolution of two multiplicative kernels is again a multiplicative kernel.*

The proof follows directly from our remarks about segments of partitions and products of functions on the incident algebra. Now let us observe that if we are give a sequence $\mathbf{h} = (h_n)_{n=1}^{\infty}$, then we may define an associated

- Exponential generating function, $e_{\mathbf{h}}(x) = \sum_{k=1}^{\infty} \frac{1}{k!} h_k x^k$
- Multiplicative function, $\psi_{\mathbf{h}}(\pi) = \prod_{k=1}^{\infty} \lambda_k^{n_k(\pi)}$
- Multiplicative kernel, $\Psi_{\mathbf{h}}(\pi, \varpi) = \prod_{k=1}^{\infty} \lambda_k^{n_k(\pi, \varpi)}$

Theorem 1.8.2 *The map*: $\Psi_{\mathbf{h}} \mapsto e_{\mathbf{h}}$ *is an anti-isomorphism from the set of multiplicative kernels with convolution as product to the set of scalar functions having zero constant term and composition as products. In particular,* $e_{\mathbf{h}} \circ e_{\mathbf{g}} = e_{\mathbf{h} \odot \mathbf{g}}$ *and*

$$\Psi_{\mathbf{g}} * \Psi_{\mathbf{h}} = \Psi_{\mathbf{h} \odot \mathbf{g}},$$

where $\odot$ is the product of sequences introduced in (1.21).

Proof Let us remark first that the formula $e_{\mathbf{h}} \circ e_{\mathbf{g}} = e_{\mathbf{h} \odot \mathbf{g}}$ is just a restatement of Theorem 1.6.2. To prove the second part, we may compute the nth term of $\Psi_{\mathbf{g}} * \Psi_{\mathbf{h}}$ using equation (1.23) to be

$$\Psi_{\mathbf{g}} * \Psi_{\mathbf{h}}(\mathbf{0}_n, \mathbf{1}_n) \equiv \sum_{\pi \in \text{Part}(n)} \Psi_{\mathbf{g}}(\mathbf{0}_n, \pi) \Psi_{\mathbf{h}}(\pi, \mathbf{1}_n).$$

However, we have

$$\Psi_{\mathbf{g}}(\mathbf{0}_n, \pi) = \prod_{k=1}^{\infty} g_k^{n_k(\pi)}, \qquad \Psi_{\mathbf{h}}(\pi, \mathbf{1}_n) = h_{N(\pi)}.$$

(The first of these relations follows follows from the fact π is made up of $n_k\,(\pi)$ blocks of size k consisting of elements of $\mathbf{0}_n$. The second follows from the fact that $\mathbf{1}_n$ is made up of a single block consisting of the $N\,(\pi) = \sum_k n_k\,(\pi)$ parts of π.) We therefore get

$$\Psi_\mathbf{g} * \Psi_\mathbf{h}\,(\mathbf{0}_n, \mathbf{1}_n) \equiv \sum_{\substack{\pi \in \mathrm{Part}(n) \\ E(\pi)=n}} h_{N(\pi)} \prod_{k=1}^{\infty} g_k^{n_k(\pi)}$$

$$= \sum_{\mathbf{n}} \nu_{\mathrm{Part}}\,(\mathbf{n})\,h_{N(\mathbf{n})} \prod_{k=1}^{\infty} g_k^{n_k}$$

$$= \sum_{m} h_m \mathrm{Bell}\,(n, m; g_1, g_2, g_3, \ldots),$$

which by Theorem 1.6.2 equals the nth term of $\mathbf{h}\copyright\mathbf{g}$. $\square$

Corollary 1.8.3 *Suppose $e_\mathbf{g}$ has a compositional inverse, say $e_\mathbf{h} = e_\mathbf{g}^{-1}$, then $\Psi_\mathbf{g}$ has the convolutional inverse $\Psi_\mathbf{h}$, that is, $\Psi_\mathbf{g} * \Psi_\mathbf{h} = \delta$.*

Theorem 1.8.4 (Möbius inversion formula for partitions) *The Möbius kernel is given by*

$$\mu\,(\pi, \varpi) = \begin{cases} \prod_{k \geq 1} \left\{ (-1)^{k-1}\,(k-1)\,! \right\}^{n_k(\pi, \varpi)}, & \pi \leq \varpi; \\ 0, & \text{otherwise;} \end{cases}$$

where $n_k\,(\pi, \varpi)$ the number of blocks of ϖ that occur as the union of exactly k blocks of π.

Proof The zeta kernel is given by

$$\zeta = \Psi_\mathbf{g},$$

with $\mathbf{g} = (1, 1, 1, , \ldots)$, so $e_{\mathbf{g}(x)} = e^x - 1$.

Its inverse is therefore $\mu = \Psi_\mathbf{h}$, where $e_\mathbf{h}\,(e^x - 1) = x$, that is, $e_\mathbf{h}\,(x) = \ln\,(1 + x)$ and so $h_n = (-1)^{n-1}\,(n-1)\,!$. This gives the result. $\square$

The **zeta transform** of $\psi : \mathrm{Part}\,(X) \mapsto \mathbb{C}$, is $\mathcal{Z}\psi : \mathrm{Part}\,(X) \mapsto \mathbb{C}$, where

$$\mathcal{Z}\psi\,(\varpi) \triangleq \sum_{\pi \in \mathrm{Part}(X)} \zeta\,(\varpi, \pi)\,\psi\,(\pi) = \sum_{\pi \geq \varpi} \psi\,(\pi).$$

From the Möbius inversion formula for partitions, its inverse is then given by

$$\mathcal{Z}^{-1}\psi\,(\varpi) \triangleq \sum_{\pi \in \mathrm{Part}(X)} \mu\,(\varpi, \pi)\,\psi\,(\pi) = \sum_{\pi \geq \varpi} \mu\,(\varpi, \pi)\,\psi\,(\pi).$$

1.9 Partition Functions

Let $\lambda = (\lambda_1, \lambda_2, \lambda_3, \ldots)$ be a fixed sequence of complex numbers, and we take the multiplicative function ψ_λ. We consider the "canonical partition functions"

$$Z_n(\lambda) = \mathcal{Z}\psi_\lambda(\mathbf{0}_n) = \sum_{\pi \in \text{Part}(n)} \psi_\lambda(\pi),$$

$$\tilde{Z}_n(\lambda) = \mathcal{Z}^{-1}\psi_\lambda(\mathbf{0}_n) = \sum_{\pi \in \text{Part}(n)} \mu(\pi)\psi_\lambda(\pi),$$

along with the "grand canonical partition functions"

$$\Xi(z, \lambda) = \sum_n \frac{z^n}{n!}Z_n(\lambda), \qquad \tilde{\Xi}(z, \lambda) = \sum_n \frac{z^n}{n!}\tilde{Z}_n(\lambda).$$

Proposition 1.9.1 *The grand partition functions take the following form:*

$$\Xi(z, \lambda) = \exp\left\{\sum_{k \geq 1} \frac{1}{k!}\lambda_k z^k\right\}, \tag{1.24}$$

$$\tilde{\Xi}(z, \lambda) = \exp\left\{\sum_{k \geq 1} \frac{(-1)^{k-1}}{k}\lambda_k z^k\right\}. \tag{1.25}$$

Proof $\Xi(\lambda; z)$ can be written as $\sum_n \frac{z^n}{n!} \sum_{\pi \in \text{Part}(n)} \prod_{k=1}^{\infty} \lambda_k^{n_k(\pi)}$ which now involves an unrestricted sum over partitions of all sizes. We can alternatively describe this as a sum over all occupation sequences:

$$\Xi(\lambda; z) = \sum_{\mathbf{n}} \frac{\nu_{\text{Part}}(\mathbf{n})}{E(\mathbf{n})!} \prod_{k=1}^{\infty} \left(\lambda_k z^k\right)^{n_k}.$$

Using the expression (1.20) for $\nu_{\text{Part}}(\mathbf{n})$, and the $\sum \prod \leftrightarrow \prod \sum$ Lemma, we have

$$\Xi(\lambda; z) = \sum_{\mathbf{n}} \prod_{k=1}^{\infty} \frac{1}{(k!)^{n_k} n_k!} \left(\lambda_k z^k\right)^{n_k}$$

$$= \prod_{k=1}^{\infty} \sum_{n=0}^{\infty} \frac{1}{(k!)^n n!} \left(\lambda_k z^k\right)^n$$

$$= \prod_{k=1}^{\infty} \exp \frac{\lambda_k}{k!} z^k.$$

Likewise, we pick up the additional factors $\mu_k = (-1)^{k-1} (k-1)!$ when computing the Möbius transformed version:

$$\tilde{\Xi}(z,\lambda) = \sum_{\mathbf{n}} \frac{\nu_{\text{Part}}(\mathbf{n})}{E(\mathbf{n})!} \prod_{k=1}^{\infty} \left((-1)^{k-1} (k-1)! \lambda_k z^k \right)^{n_k}$$

$$\equiv \prod_{k \geq 1} \exp \frac{(-1)^{k-1}}{k} \lambda_k z^k. \qquad \square$$

The notion of a partition function is borrowed from statistical mechanics and relates to partitioning of energy among various energy states, as opposed to the term partition from combinatorics. The calculation in the preceding lemma is similar to the calculation of the grand canonical partition function for the free Bose gas at finite volume, where we have n_k particles in the kth energy, with $N = \sum_k n_k$ particles in total and an energy $E = \sum_k \varepsilon_k n_k$. In our case, we have $\varepsilon_k = k$.

1.9.1 The Hermite Polynomials

As a special case of the partition function for $\lambda = (\lambda_1, \lambda_2, 0, 0, \ldots)$, we consider the case where we truncate beyond pairs:

$$\psi_\lambda(\pi) = \begin{cases} \lambda_1^{n_1(\pi)} \lambda_2^{n_2(\pi)}, & n_k(\pi) = 0, \forall k \geq 3; \\ 0, & \text{otherwise.} \end{cases}$$

We write the partition function as

$$G(\lambda_1, \lambda_2, z) = \tilde{\Xi}(\lambda_1, \lambda_2, 0, 0, \ldots; z) = e^{\lambda_1 z - \frac{1}{2}\lambda_2 z^2},$$

and expand to get

$$G(\lambda_1, \lambda_2, z) \equiv \sum_{n=0}^{\infty} \frac{z^n}{n!} H_n(\lambda_1, \lambda_2).$$

The function $H_n(z, 1)$ is readily seen to be a polynomial of degree n in z called the nth **Hermite polynomial**.

Lemma 1.9.2 $H_n(\lambda_1, \lambda_2) = (-\lambda_2)^n e^{\lambda_1^2/2\lambda_2} \frac{\partial^n}{\partial \lambda_1^n} e^{-\lambda_1^2/2\lambda_2}.$

Proof From Taylor's theorem, one finds that the partition function is

$$G(\lambda_1, \lambda_2, z) = e^{\lambda_1/2\lambda_2} e^{-(\lambda_1 - \lambda_2 z)^2/2\lambda_2}$$

$$= e^{\lambda_1^2/2\lambda_2} \sum_{n=0}^{\infty} \frac{1}{n!} (-\lambda_2 z)^n \frac{\partial^n}{\partial \lambda_1^n} e^{-\lambda_1^2/2\lambda_2}$$

and the result follows by comparing coefficients of z^n. $\qquad\square$

Lemma 1.9.3 *For each n, the function $H_n(\lambda_1, \lambda_2)$ is analytic in λ_1 and λ_2 with*

$$H_n(\lambda_1, \lambda_2) = \sum_{k=1}^{[n/2]} \frac{(-1)^k n!}{2^k k! (n-2k)!} \lambda_1^{n-2k} \lambda_2^k$$

and conversely

$$\lambda_1^n = \sum_{k=1}^{[n/2]} \frac{n!}{2^k k! (n-2k)!} \lambda_2^{2k} H_{n-2k}(\lambda_1, \lambda_2).$$

Proof Observe that

$$G(\lambda_1, \lambda_2, z) = e^{\lambda_1 z} e^{-\frac{1}{2}\lambda_2 z^2}$$

$$= \left(\sum_{m=0}^{\infty} \frac{1}{m!} \lambda_1^m z^m \right) \left(\sum_{k=0}^{\infty} \frac{1}{k!} \left(-\frac{\lambda_2}{2} \right)^k z^{2k} \right)$$

and extracting the coefficients of z^n gives the result. The converse comes from likewise expanding $e^{\lambda_1 z} = G(\lambda_1, \lambda_2; z) e^{+\frac{1}{2}\lambda_2 z^2}$ and comparing coefficients of z^n. $\qquad\square$

Of course, the first of these relations just says that

$$H_n(\lambda_1, \lambda_2) = \sum_{\substack{n_1+2n_2=n \\ n_1, n_2 \geq 0}} (-1)^{n_2} \frac{n!}{1^{n_1} 2^{n_2} n_1! \, n_2!} \lambda_1^{n_1} \lambda_2^{n_2},$$

which we could have deduced directly.

Corollary 1.9.4 $H_n(\lambda_1, \lambda_2)$ *is a polynomial of degree n in λ_1 with leading coefficient unity. Moreover, $H_n(-\lambda_1, \lambda_2) = (-1)^n H_n(\lambda_1, \lambda_2)$.*

Note the rescaling $H_n(\lambda_1, \lambda_2) = \lambda_2^{n/2} H_n \left(\frac{\lambda_1}{\sqrt{\lambda_2}} \right)$. The first few instances are

$H_0(\lambda_1, \lambda_2) = 1;$

$H_1(\lambda_1, \lambda_2) = \lambda_1;$

$H_2(\lambda_1, \lambda_2) = \lambda_1^2 - \lambda_2;$

$H_3(\lambda_1, \lambda_2) = \lambda_1^3 - 3\lambda_1\lambda_2;$

$H_4(\lambda_1, \lambda_2) = \lambda_1^4 - 6\lambda_1^2\lambda_2 + 3\lambda_2^2;$

$H_5(\lambda_1, \lambda_2) = \lambda_1^5 - 10\lambda_1^3\lambda_2 + 15\lambda_1\lambda_2^2.$

Lemma 1.9.5 *The differential recurrence relation for the Hermite polynomials are:*

$$\frac{\partial}{\partial \lambda_1} H_0(\lambda_1, \lambda_2) = 0,$$

$$\frac{\partial}{\partial \lambda_1} H_n(\lambda_1, \lambda_2) = nH_{n-1}.$$

Proof Note that, on one hand,

$$\frac{\partial}{\partial \lambda_1} G(\lambda_1, \lambda_2, z) = zG(\lambda_1, \lambda_2, z) = \sum_{n=0}^{\infty} \frac{z^{n+1}}{n!} H_n(\lambda_1, \lambda_2)$$

while on the other

$$\frac{\partial}{\partial \lambda_1} G(\lambda_1, \lambda_2, z) = \sum_{n=0}^{\infty} \frac{z^n}{n!} \frac{\partial}{\partial \lambda_1} H_n(\lambda_1, \lambda_2)$$

and so comparing powers of z gives the result.   □

Lemma 1.9.6 *The algebraic recurrence relations:*

$$H_1(\lambda_1, \lambda_2) = \lambda_1 H_0(\lambda_1, \lambda_2),$$
$$H_n(\lambda_1, \lambda_2) = \lambda_1 H_{n-1}(\lambda_1, \lambda_2) - n\lambda_2 H_{n-2}(\lambda_1, \lambda_2), \qquad (n > 1)$$

Proof Note first of all that $\frac{\partial}{\partial z} G(\lambda_1, \lambda_2, z) = (\lambda_1 - \lambda_2 z) G(\lambda_1, \lambda_2, z)$, then use the Hermite power series expansion to obtain the result.   □

By combining the algebraic and differential recurrence relations, one can deduce readily that

$$\left(-\lambda_2^2 \frac{\partial^2}{\partial \lambda_1} + \frac{\partial}{\partial \lambda_1} \right) H_n = nH_n. \tag{1.26}$$

This gives a second characterization of the Hermite polynomials: $y = H_n(x)$ is uniquely defined as the polynomial solution with leading coefficient unity to the ordinary differential equation (ODE)

$$y'' - zy + ny = 0 \tag{1.27}$$

wherever n is a nonnegative integer.

Let γ be standard Gaussian measure, $\gamma(x) dx = (2\pi)^{-1/2} E^{-x^2/2} dx$. Then the Hermite polynomials are orthogonal as elements of $L^2(\mathbb{R}, \gamma)$ and in particular

$$\langle H_n, H_m \rangle_{L^2(\mathbb{R}, \gamma)} = n! \, \delta_{nm}; \tag{1.28}$$

they are moreover a complete basis for $L^2(\mathbb{R}, \gamma)$.

The **Hermite functions** $h_n(x)$ are defined to be

$$h_n(x) = \frac{1}{\sqrt{n!}} \frac{e^{-x^2/4}}{(2\pi)^{1/4}} H_n(x). \tag{1.29}$$

They consequently satisfy the ODEs

$$\left(-\frac{d^2}{dx^2} + \frac{1}{4}x^2\right) h_n = \left(n + \frac{1}{2}\right) h_n. \tag{1.30}$$

The mapping $L^2(\mathbb{R}, \gamma) \to L^2(\mathbb{R})$ given by $f(x) \to \sqrt{\rho(x)f(x)}$, where $\rho(x) = (2\pi)^{-1/2} \exp\left(-x^2/2\right)$ is the standard Gaussian density, is an isometry, viz.

$$\langle f | g \rangle_{L^2(\mathbb{R}, \gamma)} = \int \bar{f}(x) g(x) \rho(x) \, dx$$
$$= \langle \sqrt{\rho} f | \sqrt{\rho} g \rangle_{L^2(\mathbb{R})} \tag{1.31}$$

and extends to a unitary map. The set of functions $\{h_n : n = 0, 1, \ldots\}$ form a complete orthonormal basis for $L^2(\mathbb{R})$.

1.9.2 The Charlier Polynomials

Another situation of interest is when we take $\lambda_1 = x - \sqrt{\lambda}$ and $\lambda_k = \frac{x}{\sqrt{\lambda}^k}$ for all $k \geq 2$, then

$$\psi_\lambda(\pi) = \left(x - \sqrt{\lambda}\right)^{n_1(\pi) N(\pi) - n_1(\pi)}$$

and we obtain

$$\tilde{\Xi}(\lambda; z) = \exp\left\{-\sqrt{\lambda}z + x\sum_{k \geq 1} (-1)^{k-1} \frac{1}{k} \left(\frac{z}{\sqrt{\lambda}}\right)^k\right\}$$
$$= e^{-\sqrt{\lambda}z} \left(1 + \frac{z}{\sqrt{\lambda}}\right)^x \equiv \sum_{n=0}^{\infty} \frac{z^n}{n!} C_n(x, \sqrt{\lambda}).$$

The functions $C_n(x, \sqrt{\lambda})$ are polynomials in z of degree n called the **Charlier polynomials**.

More generally,

$$C_n(x, \sqrt{\lambda}) = \sum_m \binom{n}{m} \left(-\sqrt{\lambda}\right)^{n-2m} z^m$$
$$= \sum_{m,k} (-1)^{n-k} \sqrt{\lambda}^{n-2m} \binom{n}{m} \begin{bmatrix} m \\ k \end{bmatrix} z^k.$$

The Charlier polynomials have the property that they are orthogonal with respect to the Poisson measure of intensity λ, that is,

$$\sum_{x \geq 0} C_n(x, \sqrt{\lambda}) C_m(x, \sqrt{\lambda}) p(x, \lambda) = n! \, \delta_{n,m}$$

with $p(x, \lambda) = e^{-\lambda} \frac{\lambda^x}{x!}$. In particular, they form a complete basis for $L^2(\mathbb{N}, P_\lambda)$, the set of functions on $\mathbb{N} = \{0, 1, 2, \ldots\}$ with the Poisson measure P_λ.

2

Probabilistic Moments and Cumulants

In this chapter we will look at moments and cumulants of random variables and, more generally, stochastic processes. It turns out that the two most important families, the Gaussian and the Poissonian random variables, have moments that involve the combinatorial enumerations encountered in the introductory chapter. Moreover, the relationship between moments and cumulants is also deeply combinatorial and we study this is the first half of the chapter. In the second part, we extend this to stochastic processes and, in particular, moments of stochastic integrals.

We will assume the standard Kolmogorov setting of a probability space $(\Omega, \mathcal{F}, \mathbf{P})$ consisting of a measurable space $(\Omega, \mathcal{F})$, with Ω being the sample space and $\mathcal{F}$ the σ-algebra of events, and $\mathbf{P}$ being a probability measure on the events.

2.1 Random Variables

2.1.1 Moments

A random variable X on a probability space $(\Omega, \mathcal{F}, \mathbf{P})$ is a real-valued measurable function and we define expectations according to

$$\mathbf{E}\left[f(X)\right] = \int_{\Omega} f(X(\omega))\mathbf{P}\left[d\omega\right] = \int_{\mathbb{R}} f(x)\mathbf{K}_X[dx].$$

Here $\mathbf{K}_X$ is the probability distribution of X, and is defined as the pullback $\mathbf{K}_X = \mathbf{P} \circ X^{-1}$, where $X^{-1}[A] = \{\omega \in \Omega : X(\omega) \in A\}$ for any Borel subset A. That is, the event "X takes a value in A" has probability

$$\text{Prob}[X \in A] = \mathbf{K}_X[A] = \mathbf{P}[X^{-1}[A]].$$

The expectation of the nth power of X, when it exists, is called its nth moment:

$$\mu_n = \mathbf{E}\left[X^n\right] = \int_{\mathbb{R}} x^n \mathbf{K}_X[dx]. \tag{2.1}$$

The (exponential) moment generating functions is defined to be

$$M_X(t) = \mathbf{E}\left[e^{tX}\right] = \sum_{n \geq 0} \frac{1}{n!} \mu_n t^n, \tag{2.2}$$

whenever convergent. Note that M_X is the Laplace transform of $\mathbf{K}_X$.

2.1.2 Cumulants

Cumulants κ_n are defined through the relation

$$\sum_{n=1}^{\infty} \frac{1}{n!} \kappa_n t^n = \ln M_X(t) \tag{2.3}$$

or $\sum_{n=0}^{\infty} \frac{1}{n!} \mu_n t^n = \exp\left\{ \sum_{n=1}^{\infty} \frac{1}{n!} \kappa_n t^n \right\}$. One sees from Theorem 1.6.2 (by taking $h = \exp$ so that $h_m = 1$) that the relation between the moments and cumulants is

$$\mu_n = \sum_m \text{Bell}(n, m; \kappa_1, \kappa_2 \ldots). \tag{2.4}$$

The first few terms are

$$\mu_1 = \kappa_1,$$
$$\mu_2 = \kappa_2 + \kappa_1^2,$$
$$\mu_3 = \kappa_3 + 3\kappa_1\kappa_2 + \kappa_1^3,$$
$$\mu_4 = \kappa_4 + 4\kappa_1\kappa_3 + 3\kappa_2^2 + 6\kappa_1^2\kappa_2 + \kappa_1^4,$$
$$\text{etc.}$$

Conversely, from $\sum_{n=1}^{\infty} \frac{1}{n!} \kappa_n t^n = \ln\left(1 + \sum_{n=1}^{\infty} \frac{1}{n!} \mu_n t^n\right)$, we may again use Theorem 1.6.2 (taking $h(x) = \ln(1 + x)$ so that $h_m = (-1)^{m-1}(m-1)!$) to get

$$\kappa_n = \sum_m (-1)^{m-1}(m-1)! \, \text{Bell}(n, m; \mu_1, \mu_2 \ldots). \tag{2.5}$$

so that the inverse relationship is

$$\kappa_1 = \mu_1,$$
$$\kappa_2 = \mu_2 - \mu_1{}^2$$
$$\kappa_3 = \mu_3 - 3\mu_2\mu_1 + 2\mu_1{}^3$$
$$\kappa_4 = \mu_4 - 4\mu_3\mu_1 - 3\mu_2{}^3 + 12\mu_2\mu_1{}^2 - 6\mu_1{}^4,$$

etc.

Note that $\mu_1 = \kappa_1$ is the mean, and κ_2 is the variance of the random variable!

2.2 Key Probability Distributions

We now look at three of the most important examples.

2.2.1 Standard Gaussian Distribution

A continuous real-valued random variable has a **standard Gaussian distribution** if it has a probability density $\rho_X(x) = \gamma(x)$ given by

$$\gamma(x) = (2\pi)^{-1/2} e^{-x^2/2}, \tag{2.6}$$

leading to the moment generating function

$$M_X(t) = \frac{1}{\sqrt{2\pi}} \int_{-\infty}^{\infty} e^{tx} e^{-\frac{1}{2}x^2} dx = e^{t^2/2}.$$

We see that all cumulants vanish except $\kappa_2 = 1$. Expanding the moment generating function yields

$$\mu_n = \begin{cases} \dfrac{(2k)!}{2^k k!}, & n = 2k; \\ 0, & n = 2k+1. \end{cases}$$

The nth moment has the combinatorial interpretation as the number of pair partitions of n items.

2.2.2 Poisson Distribution

We take X to be discrete with

$$\text{Prob}[X = n] = \frac{1}{n!} \lambda^n e^{-\lambda}$$

for $n = 0, 1, 2, \ldots$, and where the parameter λ must be positive. This gives the **Poisson distribution of intensity** λ. The moment generating function is readily computed and we obtain

$$M_X(t) = \exp\{\lambda(e^t - 1)\}.$$

Taking the logarithm shows that all cumulants of the Poisson distribution are equal to λ:

$$\kappa_n = \lambda, \text{ for all } n = 1, 2, \ldots.$$

From (2.11), we see that they are polynomials of degree n in λ with the Stirling numbers of the second type as coefficients,

$$\mu_n = \sum_m \begin{Bmatrix} n \\ m \end{Bmatrix} \lambda^m, \tag{2.7}$$

and in particular λ is the mean (as well as the variance!). Note that we obtain the generating function for Stirling numbers of the second kind,

$$\sum_{n,m} \frac{1}{n!} \begin{Bmatrix} n \\ m \end{Bmatrix} t^n \lambda^m = e^{\lambda(e^t - 1)}, \tag{2.8}$$

and setting $\lambda = 1$ gives $\sum_n \frac{1}{n!} B_n t^n = e^{e^t - 1}$ the generating function for the Bell numbers. We may also expand the moment generating function $M_X(t)$ to obtain the explicit, though not particularly useful, identity

$$\begin{Bmatrix} n \\ m \end{Bmatrix} = \frac{1}{m!} \sum_{k=0}^m \binom{m}{k} (-1)^{m-k} k^n.$$

2.2.3 Gamma Distribution

Let X be a positive continuous variable with density

$$\rho_X(dx) = \begin{cases} \Gamma(\lambda)^{-1} x^{\lambda-1} e^{-x}, & x > 0, \\ 0, & x \le 0, \end{cases}$$

where $\lambda > 0$, and the Gamma function is $\Gamma(\lambda) = \int_0^\infty y^{\lambda-1} e^{-y} dy$ with $\Gamma(n+1) = (n)!$ for integer $n \ge 0$. We refer to this as a **Gamma distribution**, specifically the $\Gamma(\lambda)$ distribution. Its moment generating function is

$$M_X(t) = (1 - t)^{-\lambda}.$$

Now $(1-t)^{-\lambda} = \sum_{n=0}^{\infty} \binom{-\lambda}{n}(-t)^n \equiv \sum_{n=0}^{\infty} \frac{1}{n!}\lambda^{\bar{n}}t^n$ and so its moments are

$$\mu_n = \lambda^{\bar{n}} = \sum_m \begin{bmatrix} n \\ m \end{bmatrix} \lambda^m. \tag{2.9}$$

2.2.4 Factorial Moments

We refer to $\mathbf{E}[X^{\underline{n}}]$ as the n-th falling factorial moment. The moment generator can be written as $M_X(t) \equiv \mathbf{E}[(1+z)^X]$, where we set $z = e^t - 1$. Taylor expanding about $z = 0$ leads to $(1+z)^x = \sum_{n=0}^{\infty} \binom{x}{n}z^n = \sum_{n=0}^{\infty} \frac{1}{n!}x^{\underline{n}}z^n$ and so

$$\mathbf{E}[(1+z)^X] = \sum_{n=0}^{\infty} \frac{1}{n!}\mathbf{E}[X^{\underline{n}}]z^n. \tag{2.10}$$

Clearly $\mathbf{E}[(1+z)^X]$ acts as the falling factorial moment generating function.

From relation (1.11) for the Stirling numbers, we see that the ordinary and falling factorial moments are related by

$$\mathbf{E}[X^n] \equiv \sum_m \begin{Bmatrix} n \\ m \end{Bmatrix} \mathbf{E}[X^{\underline{m}}],$$

$$\mathbf{E}[X^{\underline{n}}] \equiv \sum_m (-1)^{n+m} \begin{bmatrix} n \\ m \end{bmatrix} \mathbf{E}[X^m]. \tag{2.11}$$

For example, in the case of the Poisson distribution with intensity λ we have $\mathbf{E}[(1+z)^X] = e^{\lambda z}$ and so we see that the falling factorial moments are just

$$\mathbf{E}[X^{\underline{n}}] = \lambda^n.$$

In general, we have

$$\sum_n \frac{1}{n!}\mathbf{E}[X^{\underline{n}}]z^n = \mathbf{E}[(1+z)^X] = M_X(\ln(1+z)),$$

$$\sum_n \frac{1}{n!}\mathbf{E}[X^{\overline{n}}]z^n = \mathbf{E}[(1-z)^{-X}] = M_X(-\ln(1-z)),$$

and from the di Bruno formula

$$\mathbf{E}[X^{\underline{n}}] = \sum_m \text{Bell}(n, m; 0!, -1!, 2!, -3!, \ldots)\mathbf{E}[X^m],$$

$$\mathbf{E}[X^{\overline{n}}] = \sum_m \text{Bell}(n, m; 0!, 1!, 2!, 3!, \ldots)\mathbf{E}[X^m].$$

2.3 Stochastic Processes

We now broaden our outlook by going from random variables to stochastic processes (families of random variables parameterized by a time variable). We start by recalling the basic definitions and properties of stochastic integrals. There are a number of equivalent formulations and we begin by recalling some basic concepts in this direction due to P. A. Meyer, especially from his appendix to Emery (1989). Our aim is to generalize some of the earlier generating function results to encompass the setting of stochastic integration. For definiteness, we shall work with semimartingale processes.

We fix a probability space $(\Omega, \mathcal{F}, \mathbf{P})$, then a real-valued stochastic process $X = X(t, \omega)$ is a jointly measurable function from $\mathbb{R}_+ \times \Omega$ to $\mathbb{R}$. We shall frequently write the process using the notation X or (X_t).

A process (X_t) is said to have **independent increments** if, for each pair of intervals $[s_1, t_1]$ and $[s_2, t_2]$ the increments $X_{t_1} - X_{s_1}$ and $X_{t_2} - X_{s_2}$ are independent random variables. It is furthermore said to have **stationary independent increments** if the distribution of $X_{t+h} - X_t$ depends only on $h > 0$. The two core examples are the standard **Wiener process**, (W_t), which has stationary independent increments $W_{t+h} - W_t$ with a mean zero Gaussian distribution of variance h, and the **Poisson process of rate** ν, (N_t), which has stationary independent increments $N_{t+h} - N_t$ with a Poisson distribution of intensity νh.

A celebrated theorem of Wiener gives an explicit construction of a probability space $(\Omega_W, \mathcal{F}_W, \mathbb{P}_W)$, called the **canonical Wiener space**, where $\Omega_W = C^0[0, \infty)$ is the space of all continuous paths $x = x(t)$ with $x(0) = 0$.

Our aim is to define stochastic integrals $\int_S^T Y_t dX_t$ for time intervals $[S, T]$ and suitable stochastic integrals. To this end, let $\mathcal{M}$ be a subdivision (mesh) of the time interval, say $t_0 = S < t_1 < t_2 < \cdots < t_N = T$ and let $|\mathcal{M}| = \max_k(t_{k+1} - t_k)$. We then define the finite sum

$$\mathcal{I}_{\mathcal{M}}(Y, X) = \sum_k Y_{t_k} (X_{t_{k+1}} - X_{t_k}).$$

The integrand Y is said to be **simple** if there exists a subdivision $\mathcal{M}$ such that $Y_t(\omega) = y_k(\omega)$ for $t_k < t \leq t_{k+1}$ for each k. In this case, $\mathcal{I}_{\mathcal{M}}(Y, X)$ defines the stochastic integral $\int_S^T Y_t dX_t$.

We denote the sigma-algebra on $\mathbb{R} \times \Omega$ generated by the simple processes as $\mathcal{F}(\mathcal{S})$. A process X is then said to be an **integrator** if there is map $\mathcal{I}(\cdot, X)$ extending the finite sum from the simple functions to all uniformly bounded $\mathcal{F}(\mathcal{S})$ functions, and satisfying the dominated convergence theorem in probability.

A **filtration** is a family, $(\mathcal{F}_t)_{t \geq 0}$, of sigma-algebras that are nested (that is, $\mathcal{F}_s \subset \mathcal{F}_t$ whenever $s < t$), more specifically future-continuous (that is,

$\mathcal{F}_s = \cap_{s<t}\mathcal{F}_t$) and complete (that is, every set of measure zero of $\mathcal{F}$ is in $\mathcal{F}_0$). A process X is then said to be **adapted** with respect to the filtration if $X(t, \cdot)$ is $\mathcal{F}_t$-measurable for each $t \geq 0$.

A integrator process that is adapted to a filtration is called a **semimartingale** for that filtration. A process (M_t) is said to be a martingale for a filtration $(\mathcal{F}_t)$ if it is adapted to the filtration, has $\mathbf{E}[|M_t|] < \infty$ for all tT, and has the conditional expectation property $\mathbf{E}[M_t|\mathcal{F}_s] = M_s$ for all $t > s > 0$.

We say that a process $(X)_t$ has future (=right) and past (=left) limits at t if the limits

$$X_{t^\pm}(\omega) = \lim_{\tau \to 0^\pm} X_{t+\tau}(\omega)$$

exist for each ω. In such cases, we often write $(X_{t^\pm})$ or even just $X_\pm$ for the corresponding processes. We also set

$$X_{t^*} \triangleq \frac{1}{2}[X_{t^+} + X_{t^-}], \qquad \Delta X_t \triangleq X_{t^+} - X_{t^-},$$

and these are the instantaneously averaged and the jump discontinuity processes respectively. These are sometimes shortened to X_* and ΔX when no confusion arises.

Semimartingales then have the property of possessing both limits and being continuous from the future, that is, $X_+ \equiv X$. These are known as cadlag processes, from the French designation *continue á droit, limite á gauch*. The version X_- will then be caglad. Moreover, for a semimartingale X the finite sums $\mathcal{I}_\mathcal{M}(Y, X)$ converge in probability for any uniformly bounded past-continuous adapted Y as the mesh size $|\mathcal{M}| \to 0$. We then denote the limit as $\int_S^T Y_t dX_t$.

Given a pair of semimartingales X and Y, it turns out that the following limit in probability exists:

$$\int_S^T dX_t dY_t \triangleq \lim_{|\mathcal{M}| \to 0} \sum_k (X_{t_{k+1}} - X_{t_k})(Y_{t_{k+1}} - Y_{t_k}),$$

and we define their **quadratic covariation** as[1]

$$[[X, Y]]_t \triangleq \int_0^t dX_s.dY_s.$$

In particular, $[[X, X]]_t$ is called the **quadratic variation** of X. For instance, the Wiener process satisfies $[[W, W]]_t = t$, and in fact a theorem of Lévy characterizes the Wiener process as the only local martingale with $W_0 = 0$ having this property. The Wiener process is a continuous semimartingale, so

---

[1] The standard probabilist notation is $[X, Y]_t$, but we avoid this due to potential confusion with commutators.

we have $W = W_- = W_+$. For the Poisson process N, we have the other remarkable identity $[[N,N]]_t = N_t$. Here we have jump process ΔN, which can equal 0 or 1. In any finite interval, the Poisson process will have a finite number of jumps only.

We may also define the processes

$$X_t^{[n]} \triangleq \lim_{|\mathcal{M}| \to 0} \sum_k (X_{t_{k+1}} - X_{t_k})^n \equiv \int_0^t (dX_s)^n,$$

where appropriate. In particular, $X^{[2]}$ is the quadratic variation $[[X,X]]$. For $n > 2$ we have $W^{[n]} = 0$ and $N^{[n]} = N$.

A key result concerning semimartingales is that we have the following integration by parts (IP) formula:

$$X_t Y_t = \int_0^t X_{s^-} dY_s + \int_0^t Y_{s^-} dX_s + [[X,Y]]_t.$$

It is useful to write this is in the differential notation

$$d(X_t Y_t) = X_{t^-} dY_t + (dX_t) Y_{t^-} + dX_t . dY_t. \tag{2.12}$$

The stochastic form of the Taylor series is

$$df(X_t) = f'(X_{t^-}) dX_t + \frac{1}{2} f''(X_{t^-}) (dX_t)^2 + \cdots$$

$$= \sum_{n=1}^{\infty} \frac{1}{n!} f^{(n)}(X_{t^-}) dX_t^{[n]},$$

The well-known Itō formula for the Wiener process is

$$df(W_t) = f'(W_t) dW_t + \frac{1}{2} f''(W_t) dt$$

while for the Poisson process we have

$$df(N_t) = \left[f(N_{t^-} + 1) - f(N_{t^-}) \right] dN_t.$$

From this point onward, we assume that all processes are semimartingales.

2.4 Multiple Stochastic Integrals

Let $X_t^{(j)}$ be stochastic processes for $j = 1, \ldots, n$, all taken to have $X_0^{(j)} = 0$ for convenience. We use the following self-explanatory conventions:

$$X_t^{(n)} \ldots X_t^{(1)} = \int_0^t dX_{t_n}^{(n)} \ldots \int_0^t dX_{t_1}^{(1)} \equiv \int_{[0,t]^n} dX_{t_n}^{(n)} \ldots dX_{t_1}^{(1)},$$

thereby representing the product as an integral over the hypercube $[0,t]^n$.

In the following, we denote by $\Delta_n^\sigma(t)$ the $n-$simplex in $[0, t]^n$ determined by a permutation $\sigma \in \mathfrak{S}_n$: that is,

$$\Delta_n^\sigma(t) = \left\{ (t_n, \ldots, t_1) \in (0, t)^n : t_{\sigma(n)} > t_{\sigma(n-1)} > \cdots > t_{\sigma(1)} \right\}.$$

The simplex determined by the identity permutation will be written simply as $\Delta_n(t)$, that is, $t > t_n > t_{n-1} > \cdots > t_1 > 0$. Clearly $\cup_{\sigma \in \mathfrak{S}_n} \Delta_n^\sigma(t)$ is $[0, t]^n$ with the absence of the hypersurfaces (diagonals) of dimension less than n corresponding to the t_j being equal. Moreover, the $\Delta_n^\sigma(t)$ are disjoint for different σ. Removing these diagonals makes no difference for standard integrals as we are ignoring a set of Lebesgue measure zero, however, this is not true for stochastic integrals!

We also define the **diagonal-free integral** "$\fint$" to be the expression with all the diagonal terms subtracted out. Explicitly

$$\fint_{[0,t]^n} dX_{t_n}^{(n)} \ldots dX_{t_1}^{(1)} \triangleq \sum_{\sigma \in \mathfrak{S}_n} \int_{\Delta_n^\sigma(t)} dX_{t_n}^{(n)} \ldots dX_{t_1}^{(1)}.$$

Take $s_1, s_2, \ldots$ to be real variables and let $\pi = \{A_1, \ldots, A_m\}$ be a partition of $\{1, \ldots, n\}$ then, for each $i \in \{1, \ldots, n\}$, define $s_\pi(i)$ to be the variable s_j where i lies in the part A_j.

Lemma 2.4.1 *The multiple stochastic integral can be decomposed as*

$$\int_{[0,t]^n} dX_{t_n}^{(n)} \ldots dX_{t_1}^{(1)} = \sum_{\pi \in \mathrm{Part}(n)} \fint_{[0,t]^{N(\pi)}} dX_{s_\pi(n)}^{(n)} \ldots dX_{s_\pi(1)}^{(1)}$$

Proof The $n = 1$ case is immediate as the $\int$ and $\fint$ integrals coincide. In the situation $n = 2$, we have by the IP formula (2.12)

$$X_t^{(2)} X_t^{(1)} = \int_0^t dX_{t_2}^{(2)} X_{t_2^-}^{(1)} + \int_0^t X_{t_1^-}^{(2)} dX_{t_1}^{(1)} + \int_0^t dX_s^{(2)} dX_s^{(1)}$$

$$= \int_{t > t_2 > t_1 > 0} dX_{t_2}^{(2)} dX_{t_1}^{(1)} + \int_{t > t_1 > t_2 > 0} dX_{t_2}^{(2)} dX_{t_1}^{(1)} + \int_0^t dX_s^{(2)} dX_s^{(1)}$$

$$\equiv \fint_{[0,t]^2} dX_{t_2}^{(2)} dX_{t_1}^{(1)} + \fint_{[0,t]} dX_s^{(2)} dX_s^{(1)}$$

and this is the required relation.

The higher-order terms are computed through repeated applications of the Itō formula. An inductive proof is arrived at along the following lines. Let $X_t^{(n+1)}$ be another process, then the preceding rule yields

$$X_t^{(n+1)} Y_t = \fint_{[0,t]^2} dX_{t_{n+1}}^{(n+1)} dY_{t_n} + \fint_{[0,t]} dX_s^{(n+1)} dY_s$$

and we take $Y_t = X_t^{(n)} \ldots X_t^{(1)}$. Assume the formula is true for n. The first term will be the sum over all partitions of $\{n+1, n, \ldots, 1\}$ in which $\{n+1\}$ appears as a singleton, and the second term will be the sum over all partitions of $\{n+1, n, \ldots, 1\}$ in which $n+1$ appears as an extra in some part of a partition of $\{n, \ldots, 1\}$. In this way, we arrive at the appropriate sum over Part (n). $\square$

Corollary 2.4.2 *The inversion formula for off-diagonal integrals is*

$$\fint_{[0,t]^n} dX_{t_n}^{(n)} \ldots dX_{t_1}^{(1)} \equiv \sum_{\pi \in \text{Part}(n)} \mu\,(\pi) \int_{[0,t]^{N(\pi)}} dX_{s_\pi(n)}^{(n)} \ldots dX_{s_\pi(1)}^{(1)}$$

where $\mu\,(\pi) = (-1)^{N(\pi)}$ is the Möbius function for partitions.

2.4.1 Expectation of Products of Wiener Integrals

Let us define random variable $W_t\,(f)$, for square-integrable test function f, by

$$W_t\,(f) \triangleq \int_0^t f\,(s)\,dW_s$$

and set $W\,(f) = W_\infty\,(f)$. We have that

$$W_t\,(f)\,W_t\,(g) = \fint_{[0,t]^2} f\,(s_1)\,g\,(s_2)\,dW_{s_1}\,dW_{s_2} + \fint_{[0,t]} f\,(s)\,g\,(s)\,(dW_s)^2 .$$

As the Wiener process has independent increments, we see that the diagonal-free integrals will vanish:

$$\mathbb{E}\left[\fint_{[0,t]^2} f\,(s_1)\,g\,(s_2)\,dW_{s_1}\,dW_{s_2}\right] = \fint_{[0,t]^2} f\,(s_1)\,g\,(s_2)\,\mathbb{E}\left[dW_{s_1}\right]\mathbb{E}\left[dW_{s_2}\right]$$
$$= 0.$$

Therefore, we obtain (taking $t \to \infty$) the following identity known as the Itō isometry:

$$\mathbb{E}\left[W\,(f)\,W\,(g)\right] = \int_0^\infty f\,(s)\,g\,(s)\,ds. \tag{2.13}$$

More generally, we see that

$$\mathbb{E}\left[W_t\,(f_n) \ldots W_t\,(f_1)\right] = \sum_{\pi \in \text{Part}(n)} \mathbb{E}\Big[\fint_{[0,t]^{N(\pi)}} f_n\,(s_\pi\,(n)) \ldots f_1\,(s_\pi\,(1))$$
$$\times dW_{s_\pi(n)} \ldots dW_{s_\pi(1)}\Big].$$

As $(dW_t)^2 = dt$ with higher powers vanishing, we only encounter partitions into singletons and pairs. Furthermore, the integrals over a single dW_s will average to zero, leaving a sum over pair partitions, which is known as the Isserlis formula[2]

$$\mathbb{E}\left[W\left(f_{2k}\right)\ldots W\left(f_1\right)\right] = \sum_{\text{Pairs}(2k)} \left(\int f_{p_k} f_{q_k}\right) \cdots \left(\int f_{p_1} f_{q_1}\right). \quad (2.14)$$

The average over an odd number of terms will vanish. In particular, we see that

$$\mathbb{E}\left[e^{W(f)}\right] = \sum_{k=0}^{\infty} \frac{1}{(2k)!} \frac{(2k)!}{2^k k!} \left(\int f\left(t\right)^2 dt\right)^k = \exp\left\{\frac{1}{2}\int f\left(t\right)^2 dt\right\}.$$

We remark that we may write the integrals in the form $W\left(f\right) = \int f\left(t\right) \dot{W}_t dt$. Formally then we have the mnemonic

$$\mathbb{E}\left[\dot{W}_{t_{2k}}\ldots \dot{W}_{t_1}\right] = \sum_{\text{Pairs}(2k)} \delta\left(t_{p_k} - t_{q_k}\right)\ldots \delta\left(t_{p_1} - t_{q_1}\right).$$

The formal derivative $\dot{W}$ is not a genuine stochastic process, but may be thought of as a singular process called **white noise**.

2.4.2 Expectation of Products of Poisson Integrals

Let us similarly define Poisson integrals of the form

$$N_t\left(f\right) = \int_0^t f\left(s\right) dN_s.$$

Again we see that

$$\mathbb{E}\left[N_t\left(f_n\right)\ldots N_t\left(f_1\right)\right] = \sum_{\pi \in \text{Part}(n)} \mathbb{E}\left[\oint_{[0,t]^{N(\pi)}} f_n\left(s_\pi\left(n\right)\right)\ldots f_1\left(s_\pi\left(1\right)\right)\right.$$
$$\left. \times dN_{s_\pi(n)}\ldots dN_{s_\pi(1)}\right].$$

This time each part of size m will have integrators $(dN_s)^m = dN_s$, which will be independent of the remaining integrators (as they correspond to nonoverlapping increments by the diagonal-free construction) and average to νds. The situation where the f_j are equal is tractable; here each part of size k makes a contribution $\nu \int f\left(s\right)^k ds$ and we have

$$\mathbb{E}\left[N_t\left(f\right)^n\right] = \sum_{\pi \in \text{Part}(n)} \prod_{k=1}^{\infty} \left(\nu \int f\left(s\right)^k ds\right)^{n_k(\pi)}.$$

---

[2] The right-hand side is sometimes referred to as the Haffnian of the matrix F, here with components $F_{ij} = \int f_i f_j dt$. This terminology is due to E. Caianiello (1973), who realized that it is essentially the permanent of F: a determinant would yield a Pfaffian!

In particular, using Proposition 1.9.1, we find

$$\mathbb{E}\left[e^{z N_t(f)}\right] = \Xi\left(z; \left(\nu \int f(s)^k \, ds\right)_k\right)$$

$$= \exp\left\{\sum_{k=1}^{\infty} \frac{z^k}{k!} \nu \int f(s)^k \, ds\right\}$$

$$= \exp\left\{\nu \int \left(e^{zf(t)} - 1\right) dt\right\}.$$

The nth cumulant of $N_t(f)$ is therefore $\nu \int_{[0,t]} f(s)^n \, ds$. For the shot noise process $I_t = \int h(s-t) \, dN_s$ we deduce that its nth cumulant[3] is $\nu \int_0^\infty h(s)^n \, ds$.

We remark that we can likewise introduce a singular process $\dot{N}_t$ as the formal derivative of the Poisson process. This time we have the mnemonic

$$\mathbb{E}\left[\dot{N}_{t_n} \ldots \dot{N}_{t_1}\right] = \sum_{\text{Part}(n)} \delta^{n_a}\left(t_{a(1)}, \ldots, t_{a(n_a)}\right) \delta^{n_b}\left(t_{b(1)}, \ldots, t_{b(n_b)}\right) \ldots,$$

where the sum is over partitions $\{A, B, \ldots\}$ of $\{1, \ldots, n\}$ with $A = \{a(1), \ldots, a(n_a)\}$, and so on, and where

$$\delta^m(s_1, \ldots, s_m) = \delta(s_1 - s_2)\delta(s_2 - s_3)\ldots\delta(s_{m-1} - s_m)$$

for $m > 2$, but with $\delta^1(s) \equiv 1$.

2.5 Iterated Itō Integrals

Let $\{X_t : t \geq 0\}$ be a classical process. Its **diagonal-free exponential** is defined to be

$$\phi^{X_t} \triangleq \sum_{n \geq 0} \frac{1}{n!} \oint_{[0,t]^n} dX_{t_n} \ldots dX_{t_1}. \tag{2.15}$$

By symmetry, we have

$$\phi^{X_t} = \sum_{n \geq 0} \frac{1}{n!} \sum_{\sigma \in \mathfrak{S}_n} \int_{\Delta_n^\sigma(t)} dX_{t_n} \ldots dX_{t_1}$$

$$= \sum_{n \geq 0} \int_{\Delta_n(t)} dX_{t_n} \ldots dX_{t_1}$$

---

[3] The statement that the mean is $\nu \int_0^\infty h(s) \, ds$ and the variance is $\nu \int_0^\infty h(s)^2 \, ds$ is the content of Campbell's theorem. However, this formula gives all cumulants.

It is clear that e^{X_t} is the solution to the integro-differential equation

$$e^{X_t} = 1 + \int_0^t dX_s \, e^{X_{s^-}}, \tag{2.16}$$

and we obtain the series expansion through iteration.

We may also use the notation of an **Itō time-ordered exponential**:

$$e^{X_t} \equiv \vec{T}_I e^{\int_0^t dX_s}. \tag{2.17}$$

Theorem 2.5.1 *Let $\{X_t : t \geq 0\}$ be a stochastic process, and recall the definition of the related processes $X_t^{[k]} = \int_{[0,t]} (dX_s)^k$, for $k = 1, 2, \ldots$. Then its diagonal-free exponential is given by*

$$e^{X_t} = e^{Y(t)}, \tag{2.18}$$

where

$$Y_t = \sum_{k \geq 1} \frac{(-1)^{k+1}}{k} X_t^{[k]}. \tag{2.19}$$

Inversely, we have

$$X_t = \sum_{k \geq 1} \frac{1}{k!} Y_t^{[k]}. \tag{2.20}$$

Proof We first observe that

$$\int_{[0,t]^{N(\pi)}} dX_{s_\pi(n)} \ldots dX_{s_\pi(1)} = \prod_{k=1}^{\infty} \left(X_t^{[k]} \right)^{n_k(\pi)}.$$

Therefore,

$$e^{zX_t} = \sum_{n \geq 0} \frac{1}{n!} z^n \sum_{\pi \in \text{Part}(n)} (-1)^{N(\pi)} \, \psi_\lambda(\pi),$$

where ψ_λ is the multiplicative function on the species of partitions, Part, with (stochastic) coefficients $\lambda_k = X_t^{[k]}$. We then recognize the "grand canonical partition function" from Proposition 1.9.1, that is,

$$e^{zX_t} = \tilde{\Xi}\left(z, X_t^{[1]}, X_t^{[2]}, \ldots \right) \equiv \exp\left\{ \sum_{k \geq 1} z^n \frac{(-1)^{k+1}}{k} X_t^{[k]} \right\},$$

which yields (2.19). To get the inverse relation, we note that $R_t = e^{Y(t)}$ satisfies the stochastic differential equation (SDE) equivalent to the equation (2.16):

$$dR_t = (e^{dY(t)} - 1) R_{t^-}, \qquad R_0 = 1. \tag{2.21}$$

As we have identified $R_t = e^{zX_t}$, it follows that R_t also solves the SDE $dR_t = (dX_t) R_{t-}$ with R_t. Therefore, we must have

$$dX_t = (e^{dY(t)} - 1) = \sum_{k \geq 1} \frac{1}{k!} (dY_t)^k \equiv \sum_{k \geq 0} \frac{1}{k!} dY_t^{[k]},$$

from which we deduce (2.20).   □

2.5.1 Iterated Wiener Integrals

Let us take $X^{(i)}$s to be the Wiener process W. Here we have $dW_s dW_s = ds$ with higher powers vanishing, therefore

$$W_t^{[1]} = W_t,$$
$$W_1^{[2]} = \int_0^t (dW)^2 = t,$$
$$W_t^{[k]} = 0, k \geq 0.$$

We therefore see that

$$e^{uW_t} \equiv \tilde{\Xi}(W_t, t, 0, \ldots; u) = \exp\left\{uW_t - \frac{1}{2}tu^2\right\}. \qquad (2.22)$$

This implies

$$\oint_{[0,t]^n} dW_{t_n} \ldots dW_{t_1} = t^{n/2} H_n\left(\frac{W_t}{\sqrt{t}}\right)$$

where H_n are the Hermite polynomials. The implication that

$$\int_{\Delta_n^\sigma(t)} dW_{t_n} \ldots dW_{t_1} = \frac{1}{n!} t^{n/2} H_n\left(\frac{W_t}{\sqrt{t}}\right), \qquad (2.23)$$

for any permutation σ, is due originally to Itō.

More generally, we could set $X_t = \int_0^t f(s) \, dW_s$ in which case $X_t^{[2]} = \int_0^t f^2(s) \, ds$ with $X_t^{[k]} = 0$ for $k \geq 3$, yielding

$$e^{\int_0^t f dW} \equiv \exp\left\{\int_0^t f dW - \frac{1}{2} \int_0^t f^2\right\}, \qquad (2.24)$$

for f locally square-integrable.

2.5.2 Iterated Poisson Integrals

Let N_t be the Poisson process. Then we have the differential rule $(dN_t)^k = dN_t$ for all positive integer powers k. Therefore,

$$N_t^{[k]} = N_t, \qquad k = 1, 2, \ldots.$$

We then obtain

$$\cancel{e}^{zN_t} = \tilde{\Xi}(z; mN_t, N_t, \ldots) = \prod_{k=1}^{\infty} \exp\left\{ (-1)^{k-1} \frac{z^k}{k} N_t \right\}$$

and using $\sum_{k=1}^{\infty} (-1)^{k-1} \frac{z^k}{k} = \ln(1+z)$, we end up with

$$\cancel{e}^{zN_t} = (1+z)^{N_t}. \tag{2.25}$$

Using the binomial theorem, we deduce that

$$\int_{\Delta_n^\sigma(t)} dN_{t_n} \ldots dN_{t_1} = \binom{N_t}{n}, \tag{2.26}$$

and so $\int_{[0,t]^n} dN_{t_n} \ldots dN_{t_1} = N_t^n$.

Let us remark that we could repeat the preceding exercise with N_t replaced by the compensated Poisson process $Y_t \triangleq N_t - t$. This has the modified rule $(dY_t)^k = dY_t + dt$ for all positive integer powers, except $k = 1$. This complication is easily dealt with, and the diagonal-free exponentiated random variable associated with the compensated Poisson process Y_t is found to be $\cancel{e}^{zY_t} = e^{-zt}(1+z)^{Y_t+t}$. We deduce that

$$\oint_{[0,t]^n} dY_{t_n} \ldots dY_{t_1} = C_n(N_t, t),$$

where $C_n(x, t)$ are the Charlier polynomials.

2.6 Stratonovich Integrals

An alternative to the Itō integral is to consider the limit of finite sum approximations of the form

$$\sum_k Y_{t_k^*} \left(X_{t_{k+1}} - X_{t_k} \right)$$

where we evaluate the integrand at the midpoint

$$t_k^* = \frac{1}{2} (t_{k+1} + t_k).$$

The limit is the Stratonovich, or symmetric, integral and we denote it by $\int Y_t \delta X_t$. For a pair of martingales X and Y, its relation to the Itô integral is

$$\int Y_- \delta X_+ \equiv \int Y_- dX_+ + \frac{1}{2}[[Y,X]]$$

or, in differential form,

$$Y_- \delta X_+ = Y_- \delta X_+ + \frac{1}{2} dY_+ dX_+.$$

The integration by parts formula now takes on the traditional Leibniz form:

$$d(XY)_+ = (\delta X_+) Y_- + X_- (\delta Y_+).$$

Let us look at a few examples.

First, we consider $Z = \int f(W) \delta W$. It follows that

$$dZ = f(W) \delta W = f(W) dW + \frac{1}{2} df(W) dW \equiv f(W) dW + \frac{1}{2} f'(W) dt$$

so that we obtain equivalent to an Itô integral of the form $\int f(W) \delta W = \int f(W) dW + \frac{1}{2} \int h'(W) dt$. More generally, for a diffusion determined by the SDE

$$dX_t = v(X_t) dt + \sigma(X_t) dW_t, \qquad X_0 = x_0,$$

we may try and write this as an equivalent Stratonovich SDE:

$$dX_t = w(X_t) dt + \varsigma(X_t) \delta W_t, \qquad X_0 = x_0.$$

We have exploited the fact diffusions are continuous semimartingales and so $X_+ = X_- = X$. We then have

$$dX_t = w(X_t) dt + \varsigma(X_t) dW_t + \frac{1}{2} d\varsigma(X_t) dW_t$$

$$= w(X_t) dt + \varsigma(X_t) dW_t + \frac{1}{2} \left[\varsigma'(X_t) dX_t + \frac{1}{2} \varsigma''(X_t) dX_t dX_t \right] dW_t$$

$$= \left[w(X_t) + \frac{1}{2} \varsigma''(X_t) \sigma(X_t) \right] dt + \varsigma(X_t) dW_t,$$

where we used the fact that $(dX_t)^2 = \sigma(X_t)^2 dt$. We see that $\varsigma \equiv \sigma$ and

$$v(x) \equiv w(x) + \frac{1}{2} \sigma(x) \sigma'(x).$$

We refer to v as the **Itô drift velocity** and w is the **Stratonovich drift velocity**.

Poisson processes are discontinuous, however, so we need to distinguish between $N = N_+$ and N_-. We have

$$f(N_-)\,\delta N = f(N_-)\,dN + df(N_-)\,dN_- = \frac{1}{2}\left[f(N_- + 1) + f(N_-)\right]dN.$$

2.6.1 Stratonovich Time-Ordered Exponentials

Let (Z_t) be a semimartingale. The **Stratonovich exponential** process of Z is defined as the solution $S_t =$ of the SDE

$$dS_t = S_{t-}\delta Z_t, \qquad S_0 = 1,$$

whenever it exists. The solution will be denoted as

$$S_t = \vec{T}_S e^{Z_t}.$$

We now relate this to earlier notions of exponentials of stochastic processes.

Proposition 2.6.1 *For a semimartingale Z, we have $\vec{T}_S e^{Z_t} \equiv \mathcal{E}^{X_t}$, where*

$$dX = dZ + \frac{1}{2}dZdX. \tag{2.27}$$

Proof Let $S_t = \vec{T}_S e^{Z_t}$ so that

$$dS = S_-\delta Z = S_-dZ + \frac{1}{2}dS_-dZ.$$

If we also require $S_t = \mathcal{E}^{X_t}$ then we also have $dS = S_-dX$. Substituting in the term $dS = S_-dX$, and dividing through by S_-, which will be positive when the SDE is well posed, we obtain the consistency condition (2.27). $\qquad\square$

Example: Wiener Stratonovich Exponentials Let us set $Z_t = zW_t$ then make the assumption that $dX_t \equiv \alpha W_t + \beta dt$. The consistency condition reads as

$$\alpha dW + \beta dt = kdW + \frac{1}{2}k\alpha dt$$

so that $\alpha = k$ and $\beta = \frac{1}{2}k$. Therefore, $\vec{T}_S e^{kW_t} = \mathcal{E}^{kW_t + \frac{1}{2}k^2 t} \equiv \mathcal{E}^{kW_t} e^{\frac{1}{2}k^2 t}$. However, using (2.22) we get $\vec{T}_S e^{kW_t} = e^{kW_t}$.

Example: Poisson Stratonovich Exponentials Now lets try $Z_t = kN_t$ and we make the assumption that $X_t = lN_t$. The consistency condition (2.27) then implies

$$l = k + \frac{1}{2}lk$$

and so $l = \frac{k}{1-\frac{1}{2}k}$. We therefore find, with the help of (2.25), that

$$\vec{T}_S e^{kN_t} = \phi^{\frac{k}{1-\frac{1}{2}k}N_t} \equiv \left(\frac{1+\frac{1}{2}k}{1-\frac{1}{2}k}\right)^{N_t}.$$

The problem is well posed so long as $-2 < k \neq 2$.

2.7 Rota–Wallstrom Theory

Finally, we discuss the relation of the preceding concepts of diagonal-free multiple stochastic integrals to the combinatorial methods introduced by Rota and Wallstrom (1997).

To begin with, let us fix a measurable space $(\mathfrak{X}, \mathcal{X})$. We consider vector-valued measures on $\mathcal{X}$, that is, a map $M \colon \mathcal{X} \mapsto V$, where V is a normed vector space with the property that $M[\emptyset] = 0$, and $M[\cup_n A_n] = \sum_n M[A_n]$, where the $\{A_n\}$ are pairwise disjoint elements of $\mathcal{X}$ and the sum converges in the norm topology. The problem of defining products of vector-valued measures is not as trivial as for real-valued measures.

In particular, we shall take $V = L^2(\Omega, \mathcal{F}, \mathbf{P})$, with the L^2-norm topology, where we fix a probability space $(\Omega, \mathcal{F}, \mathbf{P})$. In such circumstance, we shall refer to $L^2(\Omega, \mathcal{F}, \mathbf{P})$-valued measures as random measures.

Let π be a partition of $\{1, \ldots, n\}$, then we define a map $\mathcal{P}_\pi^{(n)}$ from the subsets of $\mathcal{X}^n$ to itself by

$$\mathcal{P}_\pi^{(n)} B \triangleq \left\{ (x_1, \ldots, x_n) \in B : i \overset{\pi}{\sim} j \Longleftrightarrow x_i = x_j \right\}$$

where $i \overset{\pi}{\sim} j$ means that the indices i and j lie in the same block. For instance, if the partition is $\pi = \{\{1, 2\}, \{3, 5, 7\}, \{4, 6\}\}$, then $\mathcal{P}_\pi^{(n)} B$ consist of the elements in B of the form (x, x, y, z, y, z, y) with x, y, z all distinct. We have that

$$\left(\mathcal{P}_\pi^{(n)} B\right) \cap \left(\mathcal{P}_\varpi^{(n)} B\right) = \emptyset$$

whenever $\pi \neq \varpi$. We also have the projective property

$$\mathcal{P}_\varpi^{(n)} \circ \mathcal{P}_\pi^{(n)} = \delta_{\varpi, \pi} \mathcal{P}_\pi.$$

It is convenient to introduce the map $\mathcal{P}_{\geq \pi}^{(n)}$ defined by (disjoint union!)

$$\mathcal{P}_{\geq \pi}^{(n)} B \triangleq \cup_{\varpi \geq \pi} \mathcal{P}_\varpi^{(n)} B, \tag{2.28}$$

so that $\mathcal{P}_{\geq \pi}^{(n)} B = \left\{ (x_1, \ldots, x_n) \in B : i \overset{\pi}{\sim} j \Longrightarrow x_i = x_j \right\}$. The map $\mathcal{P}_{\geq \mathbf{0}}^{(n)}$ is then just the identity map between subsets of $\mathcal{X}^n$. We note that

$$\mathcal{P}_0^{(n)} B \equiv \{(x_1, \ldots, x_n) \in B : \text{no coincidences!}\},$$

$$\mathcal{P}_1^{(n)} B \equiv \{(x, \ldots, x) \in B\}$$

We shall make the assumption that each $\mathcal{P}_\pi^{(n)}$ is measurable from $\mathcal{X}^n$ to itself: this is not guaranteed, but will be case if $\mathfrak{X}$ is a Polish space and $\mathcal{X}$ is the corresponding Borel σ-algebra. The following notion of "good" measures is due to Rota and Wallstrom.

Let $M_1, \ldots, M_n$ be random measures, then we define $M_1 \otimes \cdots \otimes M_n$ on the product sets $\mathcal{X}^n$ by

$$M_1 \otimes \cdots \otimes M_n [A_1 \times \ldots A_n] = M_1 [A_1] \ldots M_n [A_n]$$

and we say that the measures are **jointly good** if $M_1 \otimes \cdots \otimes M_n$ extends to a (unique) random measure on $\mathcal{X}^n$, which we also denote as $M_1 \otimes \cdots \otimes M_n$. A random measure M is said to be n-good if its n-fold product $M^{\otimes n}$ extends in this way.

The notion of *jointly good* is due to Farré, Jolis, and Utzet (2008). We are then interested in the restricted measures

$$(M_1 \otimes \cdots \otimes M_n)_\pi \triangleq (M_1 \otimes \cdots \otimes M_n) \circ \mathcal{P}_{\geq \pi}^{(n)},$$

$$\text{St}_\pi (M_1 \otimes \cdots \otimes M_n) \triangleq (M_1 \otimes \cdots \otimes M_n) \circ \mathcal{P}_\pi^{(n)}.$$

These are then related by the identities

$$(M_1 \otimes \cdots \otimes M_n) \circ \mathcal{P}_{\geq \pi}^{(n)} \equiv \sum_{\varpi \geq \pi} (M_1 \otimes \cdots \otimes M_n) \circ \mathcal{P}_\varpi^{(n)}, \tag{2.29}$$

$$(M_1 \otimes \cdots \otimes M_n) \circ \mathcal{P}_\pi^{(n)} \equiv \sum_{\varpi \geq \pi} \mu (\pi, \varpi) (M_1 \otimes \cdots \otimes M_n) \circ \mathcal{P}_{\geq \varpi}^{(n)}, \tag{2.30}$$

which follow immediately from (2.28) and its Möbius inversion.

Now let X be an L^2-semimartingale, then we may define a random measure M_X on the Borel subsets of $\mathbb{R}_+$ by extension of

$$M_X \colon (s, t] \mapsto X (t) - X (s).$$

In particular, given a collection $X^{(1)}, \ldots X^{(n)}$ of such processes, their diagonal-free multiple integrals are identified with

$$\oint_B dX_{t_n}^{(n)} \ldots dX_{t_1}^{(1)} \equiv \left(M_{X^{(1)}} \otimes \cdots \otimes M_{X^{(n)}} \right) \circ \mathcal{P}_0^{(n)} [B].$$

We also note that the cumulant $X^{[n]}$ of a process X occurs as

$$M_{X^{[n]}} \equiv (M_X)^{\otimes n} \circ \mathcal{P}_1^{(n)}.$$

Further discussion can be found in the monograph of Peccati and Taqqu (2011).

3

Quantum Probability

Our ultimate aim is to study quantum fields; however, in this chapter we look at quantum mechanical systems with the intention of working through some of the combinatorial aspects relating to commonly occurring operators for quantum states, as an analogy to random variables. The combinatorics associated with creation and annihilation operators for the Harmonic oscillator turns out to be particularly relevant as a bridge between what we have already seen and the fields to come later on.

In quantum theory, physical variables are understood as **observables**, which are modeled as self-adjoint operators acting on a fixed Hilbert space $\mathfrak{h}$. Observables do not necessarily commute, and this is the main point of departure from classical probability. Probabilities enter as a **state** (expectation) $\mathbb{E}$, which is a map, assigning a complex number $\mathbb{E}[A]$ to each observable A, satisfying the following properties:[1]

1. Linearity $\mathbb{E}[\alpha A + \beta B] = \alpha \mathbb{E}[A] + \beta \mathbb{E}[B]$;
2. Positivity $X \geq 0 \implies \mathbb{E}[X] \geq 0$;
3. Normalization $\mathbb{E}[I] = 1$;

for all observables X, Y and scalars α, β. Here I denotes the identity operator.

A central result of functional analysis – see Reed and Simon (1972) – is that every self-adjoint operator has an associated projection valued measure P_X such that

$$X \equiv \int_{\mathbb{R}} x P_X(dx), \tag{3.1}$$

which is the spectral decomposition of X. The distribution of the observable is then $\mathbf{K}_X = \mathbb{E} \circ P_X$. This is equivalent to the Born rule, which states that the

---

[1] We will reserve the symbol $\mathbf{E}[\cdot]$ for classical (Kolmogorov) averages and use $\mathbb{E}[\cdot]$ for quantum mechanical averages.

probability that the observable X will be measured to have a value in the Borel subset A is

$$\mathbf{K}_X[A] = \mathbb{E}[P_X[A]].$$

Note that, despite its quantum origins, $\mathbf{K}_X$ is then just a classical probability distribution!

3.1 The Canonical Anticommutation Relations

We start with anticommutation relations because they occur in the simplest form for two-level systems, and here the operators are matrices on $\mathfrak{h} = \mathbb{C}^2$. We introduce **annihilation** and **creation** operators

$$a = \begin{bmatrix} 0 & 0 \\ 1 & 0 \end{bmatrix}, \qquad a^* = \begin{bmatrix} 0 & 1 \\ 0 & 0 \end{bmatrix}.$$

They satisfy the **canonical anticommutation relations**

$$aa^* + a^*a = I, \qquad a^2 = a^{*2} = 0.$$

We recall the Pauli matrices

$$\sigma_x = \begin{bmatrix} 0 & 1 \\ 1 & 0 \end{bmatrix}, \qquad \sigma_y = \begin{bmatrix} 0 & -i \\ i & 0 \end{bmatrix}, \qquad \sigma_z = \begin{bmatrix} 1 & 0 \\ 0 & -1 \end{bmatrix}.$$

They satisfy the commutation relation $[\sigma_x, \sigma_y] = 2i\sigma_z$, et cyclia. We then have

$$a = \sigma_- = \sigma_x - i\sigma_y$$
$$a^* = \sigma_+ = \sigma_x + i\sigma_y$$

with $\sigma_\pm$ being standard notation as raising and lowering operators in quantum mechanics.

An orthonormal basis for the two-dimensional Hilbert space $\mathfrak{h}$ is then given by

$$|0\rangle = \begin{bmatrix} 0 \\ 1 \end{bmatrix}, \qquad |1\rangle = a^*|0\rangle = \begin{bmatrix} 1 \\ 0 \end{bmatrix}.$$

$|0\rangle$ is called the **ground state vector**, and we note that

$$a|0\rangle = 0.$$

The **ground state expectation** is then defined to be

$$\mathbb{E}[X] = \langle 0|X|0\rangle = [0, 1] \begin{bmatrix} x_{11} & x_{10} \\ x_{01} & x_{00} \end{bmatrix} \begin{bmatrix} 0 \\ 1 \end{bmatrix} = x_{00}.$$

In general,

$$\mathbb{E}\left[e^{tX}\right] = e^{x_+ t}p_+ + e^{x_- t}p_-,$$

where $x_\pm$ are the eigenvalues of X and (with $\phi_\pm$ the eigenvectors)

$$p_\pm = |\langle\phi_\pm|0\rangle|^2.$$

That is, $\mathbb{P}[dx] = p_+\delta_{x_+}[dx] + p_-\delta_{x_-}[dx]$ with

$$\text{Prob}\left[X = x_\pm\right] = p_\pm.$$

3.1.1 The Standard Fermionic Gaussian Distribution

The observable

$$Q = a + a^* = \begin{bmatrix} 0 & 1 \\ 1 & 0 \end{bmatrix}$$

has eigenvalues ± 1 with equal probability $p_\pm = \dfrac{1}{2}$ in the ground state. This is the distribution of a fair coin. For reasons that will become apparent later, we may refer to its as the standard Fermionic Gaussian distribution.

The observable

$$\Lambda \triangleq a^*a = \begin{bmatrix} 1 & 0 \\ 0 & 0 \end{bmatrix}$$

which has spectrum $\mathbb{N}_-$. In particular, it has 0,1 as eigenvalues, and $|0\rangle$, $|1\rangle$ as corresponding eigenvectors.

A related observable is

$$N = a^*a + a + a^* + I = (a + I)^*(a + I) = \begin{bmatrix} 2 & 1 \\ 1 & 1 \end{bmatrix}$$

and it has eigenvalues $\dfrac{3 \pm \sqrt{5}}{2}$, and these have associated probabilities $p_\pm = \dfrac{1}{2} \pm \dfrac{1}{2\sqrt{5}}$ for the ground state. It is not difficult to see that the moments of N in the ground state are the Fibonacci numbers $\mathbb{E}[N^n] = F_n$: $F_0 = F_1 = 1$, and $F_n = F_{n-1} + F_{n-2}$.

A little more work shows that (for $\lambda > 0$)

$$N_\lambda = a^*a + \sqrt{\lambda}a + \sqrt{\lambda}a^* + \lambda$$
$$= (a + \sqrt{\lambda}I)^*(a + \sqrt{\lambda}I) = \begin{bmatrix} 1+\lambda & \sqrt{\lambda} \\ \sqrt{\lambda} & \lambda \end{bmatrix} \qquad (3.2)$$

has ground state distribution $\mathbb{P}[dx] = p_+ \delta_{\mu_+}[dx] + p_- \delta_{\mu_-}[dx]$, where

$$\mu_\pm = \frac{2\lambda + 1 \pm \sqrt{4\lambda + 1}}{2},$$

$$p_\pm = \frac{1}{2}\left(1 \mp \frac{1}{\sqrt{4\lambda + 1}}\right). \tag{3.3}$$

We may similarly refer to this as the Fermionic Poisson distribution of intensity λ.

3.2 The Canonical Commutation Relations

We shall be interested in a pair of mutually adjoint operators b and b^* on a Hilbert space $\mathfrak{h}$ satisfying the canonical commutation relations

$$[b, b^*] = I, \tag{3.4}$$

where I is the identity operator. Alternatively, we introduce the quadrature operators Q, P defined to be

$$Q = b + b^*,$$

$$P = \frac{1}{i}(b - b^*),$$

and these satisfy the equivalent canonical commutation relations

$$[Q, P] = 2iI.$$

It is clear that we cannot realize these as operators on a finite dimensional Hilbert space, as taking the trace of equation (3.4) would lead to $0 = \dim(\mathfrak{h})$.

The operators may be realized concretely on $\mathfrak{h} = L^2(\mathbb{R}, dq)$ as

$$Q\psi(q) = q\psi(q),$$

$$P\psi(q) = -2i\psi'(q).$$

We shall call this the *q-representation*. From time to time, we will adopt the Dirac bra-ket notation $|\psi\rangle$ for the "ket" associated with a vector ψ, in which case the q-representation is $\langle q|\psi\rangle \equiv \psi(q)$.

It is convenient to adopt a complex phase space picture where we set $\beta = \frac{1}{2}(x + iy) \in \mathbb{C}$. The quadrature coordinates are then $x = 2\mathrm{Re}\beta$, $y = 2\mathrm{Im}\beta$; see Figure 3.1.

We may introduce a symplectic area on the complex phase space

$$\mathrm{Im}[\beta_1^* \beta_2] = \frac{1}{4}(x_1 y_2 - y_1 x_2), \tag{3.5}$$

for $\beta_j = \frac{1}{2}x_j + \frac{1}{2}y_j$.

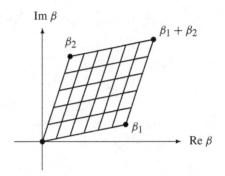

Figure 3.1 Signed phase (symplectic) area $\mathrm{Im}[\beta_1^* \beta_2]$.

We then view b, b^* and Q, P as the "quantized" versions of the variables β, β^* and x, y, respectively.

Proposition 3.2.1 *There is a unique up to normalization vector ψ_0 (with ket written as $|0\rangle$ and referred to as the vacuum vector) such that*

$$b|0\rangle = 0.$$

The observable Q in this state is a standard Gaussian.

Proof The relation becomes $(q + 2\frac{d}{dq})\psi_0(q) = 0$, which has the normalized solution

$$\psi_0(q) = \langle q|0\rangle = (2\pi)^{-1/4} e^{-q^2/4}.$$

The probability density of Q for this state is then

$$|\psi_0(q)|^2 = \gamma(q) \triangleq (2\pi)^{-1/2} e^{-q^2/2}.$$

□

3.2.1 Weyl Displacement Unitaries

The **Weyl displacement operator** with argument $\beta \in \mathbb{C}$ is the unitary operator

$$D(\beta) \triangleq e^{\beta b^* - \beta^* b}. \tag{3.6}$$

Note that we may alternatively write, with $\beta = \frac{1}{2}(x + iy)$,

$$D(\beta) = e^{2i\mathrm{Im}[b^* \beta]} = e^{\frac{i}{2}(Qy - Px)}.$$

We will derive the main properties of the displacement operators, and to this end the following (BCH) formula will be indispensable.

Lemma 3.2.2 (The Baker–Campbell–Hausdorff Formula) *Let A and B be operators such that* $[A, B]$ *commutes with both A and B, then*

$$e^{A+B} = e^A e^B e^{-\frac{1}{2}[A,B]}. \tag{3.7}$$

Proof Let $R_t = e^{t(A+B)}$ and $S_t = e^{-tA} R_t$, then $\frac{d}{dt} S_t = e^{-tA} B e^{tA} S_t = e^{-tA} B e^{tA} S_t \equiv B_t S_t$, where $B_t = e^{-tA} B e^{tA}$. Now we have that $\frac{d}{dt} B_t = e^{-tA} [B, A] e^{tA} = [B, A]$, since the commutator $[B, A]$ commutes with A. Therefore, $B_t = B + t [B, A]$. And so $\frac{d}{dt} S_t = (B + t [B, A]) S_t$. From the initial condition $S_0 = I$, one sees that

$$S_t = e^{tB} e^{-\frac{1}{2} t^2 [A,B]},$$

where the second part follows from the fact that $[A, B]$ commutes with B. Setting $t = 1$ yields the result. $\qquad\square$

As a corollary, if A and B satisfy the conditions of the preceding lemma, then

$$e^A e^B = e^B e^A e^{[A,B]}. \tag{3.8}$$

Proposition 3.2.3 (The Weyl Form of the Canonical Commutation Relations) *The Weyl displacement operators satisfy the relations*

$$D(\beta_2) D(\beta_1) = e^{-i\mathrm{Im}[\beta_2^* \beta_1]} D(\beta_1 + \beta_2). \tag{3.9}$$

The proof follows directly as an application of the BCH formula. As a corollary, the Weyl displacement operators are unitary and

$$D(\beta)^{-1} = D(\beta)^* = D(-\beta).$$

The collection of pairs (β, θ) with β complex and θ real forms a group with product $\circledast$

$$D(\beta_1, \theta_1) D(\beta_2, \theta_2) \equiv D((\beta_1, \theta_1) \circledast (\beta_2, \theta_2)).$$

This is the **Heisenberg group**. We may introduce related operators $D(\beta, \theta) = e^{-i\theta} D(\beta)$, where θ is a real phase. If we introduce the law

$$(\beta_2, \theta_1) \circledast (\beta_1, \theta_1) \equiv \left(\beta_1 + \beta_2, \theta_1 + \theta_2 + \mathrm{Im} \left\{ \beta_2^* \beta_1 \right\} \right)$$

Proposition 3.2.4 *Let* $\beta = \frac{1}{2}(x + iy) \in \mathbb{C}$, *then we may write*

$$\begin{aligned}
D(\beta) &\equiv e^{-|\beta|^2/2} e^{\beta b^*} e^{-\beta^* b} \\
&= e^{+|\beta|^2/2} e^{-\beta^* b} e^{\beta b^*} \\
&= e^{ixy/4} e^{-ixP/2} e^{iyQ/2} \\
&= e^{-ixy/4} e^{iyQ/2} e^{-ixP/2}.
\end{aligned}$$

They are all consequences of the BCH formula. The displacement vectors get their name from the following property:

$$D(\beta)^*bD(\beta) = b + \beta, \tag{3.10}$$

and likewise $D(\beta)^*b^*D(\beta) = b^* + \beta^*$. To see this, let us fix β and set $D_t = D(t\beta)$ for real t, and put $b_t = D_t^*bD_t$. Then

$$\frac{d}{dt}b_t = D_t^*[b, \beta b^* - \beta^* b]D_t = \beta$$

so $b_t = b + t\beta$. Setting $t = 1$ gives the result.

3.2.2 Number States

The **number operator** Λ is then defined as

$$\Lambda \triangleq b^*b, \tag{3.11}$$

and the trio b, b^* and Λ satisfy the algebraic relations

$$[b, b^*] = 1, \qquad [b, \Lambda] = b, \qquad [b^*, \Lambda] = -b^*. \tag{3.12}$$

The latter two equalities can be expressed as $\Lambda b = b(\Lambda - 1)$ and $\Lambda b^* = b^*(\Lambda + 1)$ respectively.

In the q-representation, we have

$$b \equiv \frac{1}{2}q + \frac{\partial}{\partial q}, \qquad b^* \equiv \frac{1}{2}q - \frac{\partial}{\partial q},$$

$$\Lambda \equiv -\frac{\partial^2}{\partial q^2} + \frac{1}{4}q^2 - \frac{1}{2}.$$

The eigenfunctions $|n\rangle$ of Λ in the q-representation are the square-integrable solutions h_n to the differential equation

$$\left(-\frac{\partial^2}{\partial q^2} + \frac{1}{4}q^2 - \frac{1}{2}\right)h_n(q) = nh_n(q)$$

and these are the Hermite functions

$$\langle q|n\rangle = h_n(q) = \frac{1}{\sqrt{n!}}\sqrt{\gamma(q)}H_n(q),$$

where we have $n = 0, 1, 2, \ldots$, and H_n are the Hermite polynomials. (Again γ is the standard Gaussian distribution function.)

The of vectors $|n\rangle$ are called the number vectors, and we exploit the fact that $\{h_n(x) : n = 0, 1, 2, \ldots\}$ form a complete orthonormal basis for the Hilbert space $L^2(\mathbb{R}, dx)$ to get the resolution of identity

$$\sum_{n \geq 0} |n\rangle\langle n| = 1. \tag{3.13}$$

The vectors $|n\rangle$, $n \in \mathbb{N}_+$ are then the complete set of eigenstates of Λ, specifically

$$\Lambda|n\rangle = n|n\rangle.$$

The case $n = 0$ is just the vacuum vector! It therefore follows that Λ has spectrum equal to $\mathbb{N}_+$.

We have that vector $b|0\rangle = 0$. From the identity $\Lambda b = b(\Lambda - 1)$, we also see that $b|n\rangle$ is an eigenvector of Λ with eigenvalue $n - 1$, for $n \geq 1$, since

$$\Lambda b|n\rangle = (b\Lambda - b)|n\rangle = (n - 1)b|n\rangle,$$

and therefore $b|n\rangle$ is proportional to $|n - 1\rangle$. In fact, since $\langle n|b^*b|n\rangle = \langle n|\Lambda|n\rangle = n$, we find that $b|n\rangle = \sqrt{n}|n - 1\rangle$. Likewise, $b^*|n\rangle$ is an eigenvector of Λ with eigenvalue $n + 1$ and a similar argument shows that $b^*|n\rangle = \sqrt{n + 1}|n + 1\rangle$. We summarize as follows

$$b^*|n\rangle = \sqrt{n + 1}|n + 1\rangle,$$
$$b|n\rangle = \begin{cases} \sqrt{n}|n - 1\rangle, & n \geq 1; \\ 0, & n = 0. \end{cases} \tag{3.14}$$

The operator b^* is therefore referred to as a **raising operator** and b as a **lower operator**. Indeed, we can obtain all number vectors by repeated applications of the raising operator

$$|n\rangle = \frac{(b^*)^n}{\sqrt{n!}}|0\rangle. \tag{3.15}$$

3.2.3 Bargmann Vectors

The **Bargmann vector** with argument $\beta \in \mathbb{C}$ is the vector defined by

$$|\exp(\beta)\rangle \triangleq e^{\beta b^*}|0\rangle. \tag{3.16}$$

Note that $|\exp(0)\rangle$ is the vacuum state $|0\rangle$.

Proposition 3.2.5 *The Bargmann vectors satisfy the properties*

$$|\exp(\beta)\rangle = e^{|\beta|^2/2}D(\beta)|0\rangle,$$
$$D(\alpha)|\exp(\beta)\rangle = e^{-\alpha^*\beta - |\alpha|^2/2}|\exp(\beta + \alpha)\rangle,$$
$$\langle\exp(\beta_1)|\exp(\beta_2)\rangle = e^{\beta_1^*\beta_2},$$
$$\langle q|\beta\rangle = (2\pi)^{-1/4}e^{-\frac{1}{4}q^2 + \beta q - \frac{1}{2}\beta^2}.$$

Proof The first relation follows from the fact that $e^{-\beta^*b}|0\rangle \equiv |0\rangle$. We therefore have $|\exp(\beta)\rangle \equiv e^{\beta b^*}e^{-\beta^*b}|0\rangle$, and using Proposition 3.2.4 we may rewrite in terms of $D(\beta)$. The second relation follows from

$$
\begin{aligned}
D(\alpha)|\exp(\beta)\rangle &= e^{|\beta|^2/2}D(\alpha)D(\beta)|0\rangle \\
&= e^{|\beta|^2/2}e^{-i\mathrm{Im}\alpha^*\beta}D(\alpha+\beta)|0\rangle \\
&= e^{|\beta|^2/2}e^{-i\mathrm{Im}\alpha^*\beta}e^{-|\alpha+\beta|^2}|\exp(\alpha+\beta)\rangle \\
&= e^{-\alpha^*\beta-|\alpha|^2/2}|\exp(\beta+\alpha)\rangle.
\end{aligned}
$$

To prove the third, we first note that

$$
\langle 0|D(\beta)|0\rangle = e^{-|\beta|^2/2}\langle 0|e^{\beta b^*}e^{-\beta^*b}|0\rangle = e^{-|\beta|^2/2}\langle 0|0\rangle = e^{-|\beta|^2/2},
$$

and then

$$
\begin{aligned}
\langle\exp(\beta_1)|\exp(\beta_2)\rangle &= e^{|\beta_1|^2/2}\langle 0|D(-\beta_1)D(\beta_2)|0\rangle e^{|\beta_2|^2/2} \\
&= e^{|\beta_1|^2/2}e^{i\mathrm{Im}[\beta_1^*\beta_2]}e^{|\beta_2|^2/2}\langle 0|D(\beta_2-\beta_1)|0\rangle \\
&= e^{\beta_1^*\beta_2}
\end{aligned}
$$

The Bargmann vectors may be computed in the q-representation. We use the factorization $D(\beta) = e^{-ixy/4}e^{iyQ/2}e^{-ixP/2}$, where $x = 2\mathrm{Re}\beta$ and $y = 2\mathrm{Im}\beta$. Now, in the q-representation

$$
e^{iyQ/2}\psi(q) = e^{iyq/2}\psi(q) \text{ and } e^{-ixP/2}\psi(q) = e^{-x\frac{\partial}{\partial q}}\psi(q) = \psi(q-x)
$$

so that

$$
\begin{aligned}
\langle q|\exp(\beta)\rangle &= e^{|\beta|^2/2}e^{-ixy/4}e^{iyq/2}(2\pi)^{-1/4}e^{-(q-x)^2/4} \\
&\equiv (2\pi)^{-1/4}e^{-\frac{1}{4}q^2+\beta q-\frac{1}{2}\beta^2}.
\end{aligned}
$$

$\square$

The set of Bargmann vectors $\{|\exp(\beta_j)\rangle : j = 1,\ldots,n\}$, for distinct complex numbers $\exp(\beta_j)$, is linearly independent. In general, the set of all exponential vectors is overcomplete, and we have the following resolution of identity.

Proposition 3.2.6 *The Bargmann vector $|\exp(\beta)\rangle$ may be expanded in the number basis as*

$$
|\exp(\beta)\rangle = \sum_{n\geq 0}\frac{\beta^n}{\sqrt{n!}}|n\rangle.
$$

We see again

$$\langle \exp(\alpha) | \exp(\beta) \rangle = \sum_{n \geq 0} \langle \exp(\alpha) | n \rangle \langle n | \exp(\beta) \rangle$$

$$= \sum_{n \geq 0} \frac{(\alpha^* \beta)^n}{n!} = e^{\alpha^* \beta}.$$

Lemma 3.2.7 *The exponential vectors give the following resolution of identity*

$$\int \frac{d\beta d\beta^*}{\pi} e^{-\beta\beta^*} |\beta\rangle\langle\beta| = 1. \tag{3.17}$$

Proof First of all, note that

$$| \exp(\beta) \rangle \langle \exp(\beta) | = \sum_{n,m \geq 0} \frac{\beta^n (\beta^*)^m}{\sqrt{n! \, m!}} |n\rangle\langle m|$$

$$= \sum_{n,m \geq 0} \frac{r^{n+m}}{\sqrt{n! \, m!}} e^{i\theta(n-m)} |n\rangle\langle m|$$

where $\beta = r e^{i\theta}$. The element of area in polar coordinates is $r dr d\theta$, and performing the θ-integration, one finds that the terms $n \neq m$ vanish. This leaves

$$\int_{\mathbb{C}} \frac{d\beta d\beta^*}{\pi} e^{-\beta\beta^*} | \exp(\beta) \rangle \langle \exp(\beta) | = \int_0^{\infty} 2r dr \, e^{-r^2} \sum_{n \geq 0} \frac{r^{2n}}{n!} |n\rangle\langle n|$$

$$= \sum_{n \geq 0} |n\rangle\langle n|,$$

using the substitution $t = r^2$ and the identity $n! = \int_0^{\infty} e^{-t} t^n dt$. The completeness relation for number states gives the result. $\qquad \square$

Lemma 3.2.8 *The number operator generates unitaries e^{itN} and these have the following action on Bargmann vectors:*

$$e^{it\Lambda} | \exp(\beta) \rangle = | \exp(e^{it}\beta) \rangle. \tag{3.18}$$

Proof Note that

$$e^{i\theta\Lambda} | \exp(\beta) \rangle = e^{i\theta\Lambda} \sum_{n \geq 0} \frac{\beta^n}{\sqrt{n!}} |n\rangle = \sum_{n \geq 0} \frac{(e^{i\theta}\beta)^n}{\sqrt{n!}} |n\rangle = | \exp(e^{i\theta}\beta) \rangle.$$

$\qquad \square$

For a one-dimensional harmonic oscillator model for a particle with mass m kg with harmonic frequency ω s^{-1}, the Hamiltonian operator is

$$\hat{H} = \frac{1}{2m}\hat{p}^2 + \frac{1}{2}m\omega^2\hat{q}^2, \tag{3.19}$$

where the canonical position and momentum operators satisfy the standard Heisenberg commutation relations $[\hat{q}, \hat{p}] = i\hbar$. In this problem, a natural length scale is set by

$$\sigma = \sqrt{\frac{\hbar}{2m\omega}}. \tag{3.20}$$

It is then convenient to introduce dimensionless operators

$$b = \frac{1}{2\sigma}\hat{q} + i\frac{\sigma}{\hbar}\hat{p},$$

$$b^* = \frac{1}{2\sigma}\hat{q} - i\frac{\sigma}{\hbar}\hat{p}, \tag{3.21}$$

and these satisfy the commutation relations $[b, b^*] = 1$. (Essentially the quadratures are $Q = \frac{1}{\sigma}q$ and $P = \frac{\sigma}{\hbar}p$.) The Hamiltonian may be written in terms of the corresponding b and b^* as

$$\hat{H} = \hbar\omega\left(b^*b + \frac{1}{2}\right) = \omega\left(\Lambda + \frac{1}{2}\right).$$

3.2.4 Expectations

The vacuum state expectation is defined by

$$\mathbb{E}_0[X] = \langle 0|X|0\rangle.$$

For instance, we average $D(z)$ to obtain

$$\mathbb{E}_0\left[e^{zb^* - z^*b}\right] = e^{-|z|^2/2}$$

and setting $z = it$ and $-s$, for t, s real, leads to

$$\mathbb{E}_0\left[e^{itQ}\right] = e^{-t^2/2}, \qquad \mathbb{E}_0\left[e^{isP}\right] = e^{-s^2/2}.$$

So both quadratures have standard Gaussian distribution.

A normalized Bargmann vector is called a coherent state, and we write

$$|\text{coh}(\beta)\rangle = e^{-|\beta|^2/2}|\exp(\beta)\rangle.$$

The coherent state with amplitude $\exp(\beta)$ likewise defines an expectation

$$\mathbb{E}_\beta[X] = e^{-|\beta|^2}\langle\exp(\beta)|X|\exp(\beta)\rangle.$$

The vacuum is of course the special case $\exp(\beta) = 0$, but we may relate the general coherent state to the vacuum via

$$|\exp(\beta)\rangle = e^{|\beta|^2/2}D(\beta)|0\rangle,$$

so that

$$\mathbb{E}_\beta[X] = \langle 0|D(\beta)^*XD(\beta)|0\rangle = \mathbb{E}_0\left[D(\beta)^*XD(\beta)\right].$$

In particular,

$$\begin{aligned}
\mathbb{E}_\beta\left[e^{zb^*-z^*b}\right] &= \mathbb{E}_0\left[D(\beta)^*e^{zb^*-z^*b}D(\beta)\right] \\
&= \mathbb{E}_0\left[e^{z(b+\beta)^*-z^*(b+\beta)}\right] \\
&\equiv e^{-|z|^2/2+z\beta^*-z^*\beta}.
\end{aligned}$$

From this we deduce that the quadratures Q and P are Gaussian with unit variance, but with displaced means $2\mathrm{Re}\beta$ and $2\mathrm{Im}\beta$, respectively.

The observable Λ has a Poisson distribution in the coherent state:

$$\begin{aligned}
\mathbb{E}_\beta\left[e^{it\Lambda}\right] &= e^{-|\beta|^2}\langle\exp(\beta)|e^{it\Lambda}|\exp(\beta)\rangle \\
&= e^{-|\beta|^2}\langle\exp(\beta)|\exp(e^{it}\beta)\rangle \\
&= e^{|\beta|^2(e^{it}-1)}. \tag{3.22}
\end{aligned}$$

Alternatively, we may say that the observable

$$D(\beta)^*\Lambda D(\beta) = (b+\beta)^*(b+\beta) = \Lambda + \beta^*b + \beta^*b + |\beta|^2 \tag{3.23}$$

has a Poisson distribution in the vacuum state.

3.2.5 Squeezing

Let us introduce the operators

$$\Sigma \triangleq b^2, \qquad \Sigma^* \triangleq (b^*)^2. \tag{3.24}$$

We find that

$$[b, \Sigma^*] = 2b^*, \qquad [\Sigma, b^*] = 2b \tag{3.25}$$

and

$$\begin{aligned}
[\Sigma, \Sigma^*] &= 4\Lambda + 2, \\
[\Lambda, \Sigma^*] &= 2\Sigma^*, \\
[\Sigma, \Lambda] &= 2\Sigma. \tag{3.26}
\end{aligned}$$

In particular, the set consisting of linear combinations of $I, \Lambda, \hat{\Sigma}$ and Σ^* is a Lie algebra with commutator as bracket.

For complex ε, we define the **squeezing operator** by

$$S(\varepsilon) \triangleq \exp\left\{\frac{1}{2}\varepsilon\Sigma^* - \frac{1}{2}\varepsilon^*\Sigma\right\}. \tag{3.27}$$

This is a unitary family and we note that

$$S(\varepsilon)^{-1} = S(\varepsilon)^* = S(-\varepsilon).$$

Lemma 3.2.9 *Let ε have the polar form $re^{i\theta}$, then*

$$S(\varepsilon)^* \, b \, S(\varepsilon) = \cosh(r) \, b + \sinh(r) \, e^{i\theta} b^* \tag{3.28}$$

Proof Let $b(u) = S(u\varepsilon)^* \, b \, S(u\varepsilon)$ for real u, then

$$\frac{d}{du}b(u) = \varepsilon b(u)^*,$$

and so $\frac{d^2}{du^2}b(u) = r^2 b(u)$. This is a simple second-order ODE with operator-valued initial conditions $b(0) = b$ and $\frac{d}{du}b(u)\,|_{u=0} = \varepsilon b^*$ yielding the solution (3.28). $\qquad\square$

The transformation $b \to \cosh(r) \, b + \sinh(r) \, e^{i\theta} b^*$ preserves the canonical commutation relations and is referred to as a **Bogoliubov transformation**.

Lemma 3.2.10 *Let ε have the polar form $re^{i\theta}$, then the squeezing operator may be placed in the following Wick ordered form*

$$S(\varepsilon) = \zeta^{\Sigma^*} (\cosh r)^{-\Lambda+\frac{1}{2}} (\zeta^*)^{-\Sigma}, \tag{3.29}$$

where $\zeta = \exp\left\{\frac{1}{2}e^{i\theta}\tanh r\right\}$.

Proof We shall map the Lie algebra generated by I, Λ, Σ and Σ^* to the Lie algebra of Pauli matrices as follows: $j(I) = I$ and

$$j(\Lambda) = \sigma_z - \frac{1}{2} = \frac{1}{2}\begin{bmatrix} 1 & 0 \\ 0 & -3 \end{bmatrix},$$

$$j(\Sigma) = 2i\sigma_- = \begin{bmatrix} 0 & 0 \\ 2i & 0 \end{bmatrix},$$

$$j(\Sigma^*) = 2i\sigma_+ = \begin{bmatrix} 0 & 2i \\ 0 & 0 \end{bmatrix}.$$

The map j preserves the Lie brackets since

$$[j(\Sigma), j(\Sigma^*)] = 4j(\Lambda) + 2,$$
$$[j(\Lambda), j(\Sigma^*)] = 2j(\Sigma^*),$$
$$[j(\Sigma), j(\Lambda)] = 2j(\Sigma).$$

In other words, j is a Lie algebra homomorphism. We note that $j(\Sigma)^* \neq j(\Sigma^*)$ so that j is not a *-map, but this does not affect our computation.

We see that

$$
j(S(\varepsilon)) = \exp\left\{\frac{1}{2}\varepsilon j(\Sigma^*) - \frac{1}{2}\varepsilon^* j(\hat{\Sigma})\right\}
$$

$$
= \exp\begin{bmatrix} 0 & ire^{i\theta} \\ -ire^{-i\theta} & 0 \end{bmatrix}
$$

$$
= \begin{bmatrix} \cosh r & \frac{1}{2}\sinh r\, e^{i\theta} \\ -\frac{1}{2}\sinh r\, e^{-i\theta} & \cosh r \end{bmatrix}.
$$

Alternatively, let us try and write $S(\varepsilon) = e^{s}e^{t\Sigma^*}e^{u\Lambda}e^{v\Sigma}$, then

$$
j(S(\varepsilon)) = e^{s}\exp\begin{bmatrix} 0 & 0 \\ 2it & 0 \end{bmatrix}\exp\begin{bmatrix} \frac{1}{2}u & 0 \\ 0 & -\frac{3}{2}u \end{bmatrix}\exp\begin{bmatrix} 0 & 2iv \\ 0 & 0 \end{bmatrix}
$$

$$
= e^{s}\begin{bmatrix} 1 & 2it \\ 0 & 1 \end{bmatrix}\begin{bmatrix} e^{\frac{1}{2}u} & 0 \\ 0 & e^{-\frac{3}{2}u} \end{bmatrix}\begin{bmatrix} 1 & 0 \\ 2iv & 1 \end{bmatrix}
$$

$$
= e^{s-\frac{3}{2}u}\begin{bmatrix} e^{2u} - 4tv & 2it \\ 2iv & 1 \end{bmatrix}.
$$

Comparing entries gives

$$
e^{s-\frac{3}{2}u} = \cosh r, \qquad e^{2u} - 4tv = 1,
$$

$$
te^{s-\frac{3}{2}u} = \frac{1}{2}\sinh r\, e^{i\theta}, \qquad ve^{s-\frac{3}{2}u} = -\frac{1}{2}\sinh r\, e^{-i\theta}
$$

with unique solution $e^{s} = \frac{1}{\sqrt{\cosh r}}$, $e^{u} = \frac{1}{\cosh r}$, $t = \frac{1}{2}\tanh r\, e^{i\theta}$ and $v = -\frac{1}{2}\tanh r\, e^{-i\theta}$. This yields the stated form. $\qquad\square$

3.3 Wick Ordering

Let $f(\beta, \beta^*)$ be a polynomial function in complex variables β and β^*. The **Wick (or normal) ordered operator** corresponding to P with β replaced by b and β^* by b^* is the operator $:f(b, b^*):$ where we make this replacement in each term putting all b^* to the left and b's to the right.

For instance, $f(\beta, \beta^*) = \beta^2\beta^* + 4\beta^{*2}\beta$ leads to $:f(b, b^*): = b^*b^2 + 4b^{*2}b$, while

$$
:(b + b^*)^2: = b^2 + 2b^*b + b^{*2}.
$$

We also have

$$:Q^n: = \sum_m \binom{n}{m} b^{*m} b^{n-m},$$
$$:\Lambda^n: = :(b^*b)^n: = b^{*n} b^n,$$
$$:D(\beta): = :e^{\beta b^* - \beta^* b}: = e^{\beta b^*} e^{-\beta^* b}.$$

The Wick ordering process is linear insofar as

$$:[c_1 f_1(b, b^*) + c_2 f_2(b, b^*)]: \equiv c_1 :f_1(b, b^*): + c_2 :f_2(b, b^*):. \qquad (3.30)$$

Proposition 3.3.1 *We have*

$$:Q^n: = H_n(Q), \qquad Q^n = (-1)^n H_n(-Q):,$$

and in particular $:(b + b^*)^n: = H_n(b + b^*)$.

Proof For t real, we have using Proposition 3.2.4,

$$:e^{tQ}: = :e^{t(b+b^*)}: = e^{tb^*} e^{tb} = e^{-t^2/2} e^{tb+tb^*} = e^{tQ - t^2/2}. \qquad (3.31)$$

Expanding in powers of t and using the generating function relation for Hermite polynomials gives the first result. Rearranging the preceding identity as $:e^{tQ + t^2/2}: = e^{tQ}$ gives the second result. $\qquad\qquad\square$

Proposition 3.3.2 *For $z \in \mathbb{C}$, set $N = (b+z)^*(b+z) = \Lambda + zb^* + z^*b + |z|^2$, then*

$$N^n = \sum_m \begin{Bmatrix} n \\ m \end{Bmatrix} :N^m: \text{ and } :N^n: = N^{\underline{n}}. \qquad (3.32)$$

Proof On the one hand, we have

$$\langle \exp(\alpha) | e^{itN} | \exp(\beta) \rangle = \langle \exp(\alpha) | D(z)^* e^{itN} D(z) | \exp(\beta) \rangle$$
$$= e^{-z\alpha^* - z^*\beta - |z|^2} \langle \exp(\alpha + z) | e^{itN} | \exp(\beta + z) \rangle$$
$$= e^{-z\alpha^* - z^*\beta - |z|^2} \langle \exp(\alpha + z) | \exp(e^{it}\beta + e^{it}z) \rangle$$
$$\equiv \exp\left\{ (\alpha + z)^*(\beta + z)(e^{it} - 1) + \alpha^*\beta \right\}.$$

While on the other

$$\langle \exp(\alpha) | :e^{isN}: | \exp(\beta) \rangle = e^{is(\alpha + z)^*(\beta + z) + \alpha^*\beta}.$$

The two expressions agree if $is + 1 = e^{it}$. Since the Bargmann vectors were arbitrary, we have

$$e^{itN} = :e^{(e^{it} - 1)N}:, \qquad (3.33)$$

which should be compared with the generating functions for the Poisson moments/Stirling numbers of the second kind (2.8). Expanding in t yields the identity $N^n = \sum_m \left\{ {n \atop m} \right\} :N^m:$. We may rearrange the equation (3.33) as $:e^{uN}: \equiv (1+u)^N$ from which we deduce that $:N^n: = n! \binom{N}{n} = N^{\underline{n}}$. $\qquad\square$

These identities hold true for $z = 0$ so we have, for instance,

$$\Lambda^n = \sum_m \left\{ {n \atop m} \right\} b^{*m} b^m. \tag{3.34}$$

3.3.1 Quantization

Let F be a function of the complex phase plane variable $\beta = \frac{1}{2}(x+iy)$, which we can write either as $F(\beta, \beta^*)$ or $f(x, y)$. The phase space Fourier transform is defined by

$$\tilde{F}(\beta_1, \beta_1^*) = \int e^{2i\mathrm{Im}[\beta_1^* \beta_2]} F(\beta_2, \beta_2^*) \frac{d\beta_2 d\beta_2^*}{\pi}$$

or equivalently

$$\tilde{f}(x_1, y_1) = \int e^{\frac{i}{2}(x_1 y_2 - y_1 x_2)} f(x_2, y_2) \frac{dx_2 dy_2}{4\pi}.$$

The inverse transform remarkably is given by

$$F(\beta_1, \beta_1^*) = \int e^{2i\mathrm{Im}[\beta_1^* \beta_2]} \tilde{F}(\beta_2, \beta_2^*) \frac{d\beta_2 d\beta_2^*}{\pi}$$

where the usual sign change is accounted for by the symplectic area. To see this, we note that

$$\int e^{\frac{i}{2}(x_1 y_2 - y_1 x_2)} \tilde{f}(x_2, y_2) \frac{dx_2 dy_2}{4\pi}$$

$$= \int \int e^{\frac{i}{2}[(x_1 - x_3) y_2 - (y_1 - y_3) x_2]} f(x_3, y_3) \frac{dx_2 dy_2}{4\pi} \frac{dx_3 dy_3}{4\pi}$$

$$= \int \delta(x_1 - x_3) \delta(y_1 - y_3) f(x_3, y_3)\, dx_3 dy_3$$

$$= f(x_1, y_1).$$

The **Weyl quantization** $\widehat{f(Q,P)}$ of a function $f = f(x, y)$ is the operator defined by

$$\hat{f} = \int e^{\frac{i}{2}(Qy - Px)} \tilde{f}(x, y) \frac{dx dy}{4\pi} \equiv \int D(\beta) \tilde{F}(\beta, \beta^*) \frac{d\beta d\beta^*}{\pi}. \tag{3.35}$$

The Weyl quantization has the property that

$$(aQ \widehat{+ bP})^n = \sum_m a^m b^{n-m} \widehat{Q^m P^{n-m}},$$

which implies that the Weyl quantization $\widehat{Q^n P^{n-m}}$ consists of all $\binom{n}{m}$ ways to symmetrically order the operators, e.g.

$$\widehat{Q^2 P^3} = \frac{1}{10}[Q^2 P^3 + QPQP^2 + QP^2 QP + QP^3 Q + PQPQP$$

$$+ PQ^2 P^2 + P^2 Q^2 P + PQP^2 Q + P^2 QPQ + P^3 Q^2].$$

It is not difficult to show that in the q-representation

$$\langle q_1 | \widehat{f(Q,P)} | q_2 \rangle = \int e^{\frac{i}{2}(q_1 - q_2)y} f\left(\frac{q_1 + q_2}{2}, y\right) dy$$

and inversely

$$f(x,y) = \int e^{\frac{i}{2}qy} \langle x - \frac{1}{2}q | \widehat{f(Q,P)} | x + \frac{1}{2}q \rangle \, dq.$$

We note that

$$\mathrm{tr}\left[\widehat{f(Q,P)}^* \, \widehat{g(Q,P)}\right] = \int f(x,y)^* \, g(x,y) \frac{dxdy}{4\pi}.$$

We shall now show that a similar principle can be applied to Wick ordering.

Proposition 3.3.3 $F(\beta, \beta^*) = \mathbb{E}_\beta \left[:F(b, b^*):\right].$

Proof This follows immediately from the fact that

$$\langle \exp(\beta) | :F(b, b^*): | \exp(\beta) \rangle = F(\beta, \beta^*) \langle \exp(\beta) | \exp(\beta) \rangle = e^{|\beta|^2} F(\beta, \beta^*).$$

$\square$

Lemma 3.3.4 *The Wick order form is given by*

$$:F(b, b^*): = \int e^{\beta b^*} e^{-\beta^* b} \tilde{F}(\beta, \beta^*) \frac{d\beta d\beta^*}{\pi}. \tag{3.36}$$

This is almost the same as (3.35) except that $D(\beta) = e^{\beta b^* - \beta^* b}$ is replaced by $e^{\beta b^*} e^{-\beta^* b}$ in the integrand!

Proof Taking coherent state expectations $\mathbb{E}_\alpha$ of the right-hand side leads to

$$\int e^{\beta \alpha^*} e^{-\beta^* \alpha} \tilde{F}(\beta, \beta^*) \frac{d\beta d\beta^*}{\pi} \equiv F(\alpha, \alpha^*),$$

which we identify with $\mathbb{E}_\alpha \left[:F(b, b^*):\right]$ through Proposition 3.3.3. $\square$

We also see that

$$\text{tr}\left\{:F(b,b^*):\right\} = \text{tr}\left\{:F(b,b^*): \int \frac{d\beta d\beta^*}{\pi}\, e^{-\beta\beta^*}\, |\exp(\beta)\rangle\,\langle\exp(\beta)|\right\}$$

$$= \int \frac{d\beta d\beta^*}{\pi}\, e^{-\beta\beta^*}\, \langle\exp(\beta)|:F(b,b^*):|\exp(\beta)\rangle$$

$$= \int \frac{d\beta d\beta^*}{\pi} F\left(\beta,\beta^*\right).$$

Proposition 3.3.5 *The projection onto a coherent state* $|\text{coh}(\alpha)\rangle$ *is* $P_\alpha = e^{-|\alpha|^2}|\exp(\alpha)\rangle\langle\exp(\alpha)|$. *We have* $P_\alpha =: F_\alpha(b,b^*)$, *where*

$$F_\alpha(\beta,\beta^*) = e^{-|\beta-\alpha|^2}.$$

Proof Again from Proposition 3.3.3, we will have

$$F_\alpha(\beta,\beta^*) = \mathbb{E}_\beta[P_\alpha]$$
$$= e^{-|\beta|^2-|\alpha|^2}|\langle\alpha|\beta\rangle|^2 = e^{-(\beta^*-\alpha^*)(\alpha-\beta)}.$$

$\square$

The special case $\alpha = 0$ yields the identification $:e^{-\Lambda}: = |0\rangle\langle0|$, but we may expand

$$:e^{-\Lambda}: = \sum_n \frac{(-1)^n}{n!}\, :\Lambda^n: = \sum_n \frac{(-1)^n}{n!}\Lambda(\Lambda-1)\cdots(\Lambda-n+1)$$

$$= \sum_n (-1)^n\binom{\Lambda}{m} = 0^\Lambda.$$

Interpreting $0^0 = 1$ shows that this surprising expression is indeed the correct one for projection onto the vacuum.

Corollary 3.3.6 *For* $l,m = 0,1,2,\ldots$, *we obtain the following representation of rank-one operators:*

$$|l\rangle\langle m| = \sum_{n\geq 0} \frac{(-1)^n}{n!}(b^*)^{n+l}b^{n+m}.$$

4

Quantum Fields

The aim of this chapter is to introduce standard combinatorial objects in a logical way to field theory. At the outset, we employ the shorthand notation of Guichardet for symmetric functions. Once this is done, we exploit the notion of combinatorial species (next chapter) as a mechanism for isolating the features, and to develop the powerful approach for characterizing and calculating moment generating functions.

4.1 Green's Functions

Let $\mathcal{X}$ be a Lusin space with a nonatomic measure, which we denote by dx. Formally a random field ϕ over the space $\mathcal{X}$ is determined by its moments[1]

$$G(x_1, \ldots, x_n) = \mathbf{E}[\phi(x_1) \ldots \phi(x_n)],$$

which we refer to as n-point **Green's functions**. Without loss of generality, we shall assume that the fields are real scalar. They must be completely symmetric in all arguments:

$$G(x_{\sigma(1)}, \ldots, x_{\sigma(n)}) = G(x_1, \ldots, x_n)$$

for any permutation σ of the indices. We may use the diagram in Figure 4.1 to denote a particular Green's function.

It is useful to assume that we have random variables

$$Q[J] = \int_{\mathcal{X}} \phi(x)J(x)dx$$

---

[1] To avoid unnecessary complications, we just take it for granted that $\mathbf{E}$ is an expectation (=state) on a sample space Φ consisting of all realizations of the fields. The exact mathematical realization of Φ as a measure space, as well as the construction of a probability measure equivalent to the expectation, is of course another matter; however, we wish to focus on algebraic properties in this chapter.

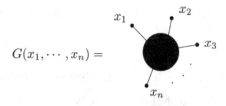

Figure 4.1 The diagram representing the Green's function $G(x_1, \ldots, x_n)$.

Figure 4.2 The series expansion appearing in (4.1) is drawn in terms of diagrams. The terms $J(x)$ are represented by a star at position x, while the general term $\int G(x_1, \ldots, x_n) J(x_1) \ldots J(x_n) \, dx_1 \ldots dx_n$ is denoted by the n-point Green's function with each leg terminated by a star.

for J in some suitable class of test function $\mathfrak{J}$. In this case, we introduce the moment generating functions $Z_G[J] \triangleq \mathbf{E}[e^{Q[J]}]$ and have formally

$$Z_G[J] \equiv \sum_{n \geq 0} \frac{1}{n!} \int G(x_1, \ldots, x_n) J(x_1) \ldots J(x_n) \, dx_1 \ldots dx_n \qquad (4.1)$$

This may be described diagrammatically as in Figure 4.2.

We now make the observation that, because the order of the variables in a completely symmetric function does not matter, we may write

$$G(x_1, \ldots, x_n) \equiv G(X) \qquad (4.2)$$

whenever these arguments are distinct, and X is the set $\{x_1, \ldots, x_n\}$.

Further, as $\mathfrak{X}$ is assumed to be a measurable Lusin space with nonatomic measure dx and we are assuming that G is measurable, we do not actually need to know the values of the Green's function for situations where we have coincidences, that is, two or more points equal. We may formulate this as follows: **(The No Coincidence Principle)** If $\mathfrak{X}$ is a Lusin space with nonatomic measure dx, then any sequence of functions $(G_n)_{n=0}^{\infty}$, with G_n measurable on $\times^n \mathfrak{X}$, is equivalent to a measurable function

$$G: \mathrm{Power}(\mathfrak{X}) \mapsto \mathbb{C}.$$

Note that the measurability on Power($\mathfrak{X}$) is the one induced from $\mathfrak{X}$. Note also that we include a value $G_0 = G(\emptyset)$.

4.1.1 Guichardet Shorthand Notation

We can also introduce a measure dX on Power($\mathfrak{X}$) as follows Guichardet (1970). A measure dX on Power($\mathfrak{X}$) is given by

$$\int F(X) dX = \sum_{n \geq 0} \frac{1}{n!} \int F(x_1, \ldots, x_n) \, dx_1 \ldots dx_n.$$

A further shorthand is the following: let f be measurable on $\mathfrak{X}$, then for X, a finite subset of $\mathfrak{X}$, we set

$$f^X \triangleq \prod_{x \in X} f(x). \tag{4.3}$$

The moment generating function (4.1) for a sequence G of Green's functions can be written in shorthand notation as

$$Z_G[J] = \int G(X) J^X \, dX, \tag{4.4}$$

for suitable test functions $J \in \mathfrak{J}$. We shall interpret (4.4) as saying a family G of Green's functions has generating functional Z_G. Conversely, we say that a functional $F = F[J]$ is analytic if it possesses a representation $F = \Phi \equiv Z_G$ for some measurable G.

Lemma 4.1.1 *Define the function* $\exp(f)$ *on the power set of* $\mathfrak{X}$ *by* $\exp(f) :$ $X \mapsto f^X$, *then*

$$Z_{\exp(f)}[J] = \exp \left\{ \int f(x) J(x) \, dx \right\}.$$

Proof In this case, we have

$$Z_G[J] = \int f^X J^X \, dX$$

$$= \sum_{n \geq 0} \frac{1}{n!} \int f(x_1) \ldots f(x_n) J(x_1) \ldots J(x_n) \, dx_1 \ldots dx_n$$

$$= \sum_{n \geq 0} \frac{1}{n!} \left(\int f(x) J(x) \, dx \right)^n$$

and summing the exponential series gives the result. $\square$

4.1.2 The Wick Product

Suppose that we have a pair F and G of Green's function sequences. Is there a sequence of Green's functions, say $F \diamond G$, such that

$$Z_F[J]Z_G[J] = Z_{F \diamond G}[J]?$$

The **Wick product** of functions on Power($\mathfrak{X}$) is defined to be

$$F \diamond G(X) \triangleq \sum_{X_1 + X_2 = X} F(X_1)G(X_2). \tag{4.5}$$

The sum is over all decompositions of the set X into ordered pairs (X_1, X_2) whose union is X.

Proposition 4.1.2 *We have*

$$Z_{F \diamond G} = Z_F Z_G, \tag{4.6}$$

where the $F \diamond G$ is the Wick product.

Proof Writing this out, we see that

$$\int FY)G(Z)J^{Y+Z}dYdZ = \int F \diamond G(X)J^X dX$$

or

$$\sum_{n_1, n_2 \geq 0} \frac{1}{n_1! n_2!} \int F(y_1, \ldots, y_{n_1})G(z_1, \ldots, z_{n_2})$$

$$\times J(y_1) \ldots J(y_{n_1})J(z_1) \ldots J(z_{n_2}) \, dy_1 \ldots dy_{n_1} dz_1 \ldots dz_{n_2}$$

$$= \sum_{n \geq 0} \frac{1}{n!} \int F \diamond G(x_1, \ldots, x_n) J(x_1) \ldots J(x_n) \, dx_1 \ldots dx_n$$

and comparing powers of J we get $F \diamond G(X) = \sum_{Y+Z=X} F(Y)G(Z)$. Here the sum is over all possible ways to decompose the set X into two disjoint subsets. $\qquad\square$

As an exercise, try to show that the Wick product of exponential vectors is

$$\exp(f) \diamond \exp(g) \equiv \exp(f + g). \tag{4.7}$$

Also try to show that the Wick product is a symmetric associative product with the general form

$$G_1 \diamond \cdots \diamond G_n(X) = \sum_{X_1 + \cdots X_n = X} G_1(X_1) \ldots G_n(X_n),$$

where the sum is now over all decomposition of X into n ordered disjoint subsets.

Some care is needed when the Green's functions are the same. Denoting the nth Wick power of F by $F^{\diamond n} = \underbrace{F \diamond \cdots \diamond F}_{n \text{ times}}$, we have

$$F^{\diamond n}(X) = \sum_{\substack{X_1 + \cdots + X_n = X \\ \text{unordered}}} F(X_1) \ldots F(X_n)$$

$$= n! \sum_{X_1 + \cdots + X_n = X,} F(X_1) \ldots F(X_n)$$

where the last sum is over all ways to decompose the set X into nonempty subsets $\{X_1, \ldots, X_n\}$, this time unordered!

We digress slightly to prove the following important result.

Lemma 4.1.3 (The $\mathbf{\Sigma\!\!\!\int}$ Lemma) *For F, a measurable function on $\times^p$ Power$(\mathfrak{X})$, we have the identity*

$$\int F(X_1, \ldots, X_p) \, dX_1 \ldots dX_p \equiv \int \sum_{X_1 + \cdots + X_p = X} F(X_1, \ldots, X_p) \, dX.$$

$$(4.8)$$

Proof If we write both of these expressions out longhand, then the left-hand side picks up the factors $\frac{1}{n_1! \ldots n_p!}$, where $\#X_k = n_k$. On the right-hand side, we get $\frac{1}{n!}$, where $\#X = n$. Both expressions are multiple integrals with respect to either $dX_1 \ldots dX_p$ or dX with $X_1 + \cdots + X_p = X$; however, on the right-hand side, we obtain an additional factor $\binom{n}{n_1, \ldots, n_p}$, giving the number of decompositions of X with n_k elements in the kth subset. This accounts precisely the combinatorial factors, so both sides are equal. $\square$

4.1.3 The Composition Formula

Let us consider a family of Green's functions F with $F(\emptyset) = 0$. We now wish to find the Green's functions $H(X)$ such that

$$Z_H[J] = h(Z_F[J]),$$

where h is an analytic function of complex numbers. If we have the Maclaurin series expansion $h(z) = \sum_n \frac{1}{n!} h_n z^n$, then

$$H \equiv \sum_n \frac{1}{n!} h_n F^{\diamond n},$$

and so

$$H(X) = \sum_n h_n \overset{\text{unordered}}{\underset{X_1 + \cdots + X_n = X}{\sum}} F(X_1) \ldots F(X_n).$$

As $F(\emptyset) = 0$, we effectively have the sum over all $\{X_1, \ldots, X_n\}$ consisting of nonempty disjoint subsets of X whose union is X – in other words, partitions! – for all possible n.

Let $\pi = \{X_1, \ldots, X_n\}$ be a partition of X, then we write

$$F^\pi \triangleq F(X_1) \ldots F(X_n).$$

The union of two disjoint subsets A and B is denoted as $A + B$. So we have

$$\overset{\text{unordered}}{\underset{X_1 + \cdots + X_n = X}{\sum}} F(X_1) \ldots F(X_n) = \sum_{\pi \in \text{Part}_n(X)} F^\pi.$$

The number of blocks making up the partition is denoted as $N(\pi)$.

We now arrive at the following result.

Theorem 4.1.4 *Let F be a measurable function on* $\text{Power}(\mathfrak{X})$ *with $F[\emptyset] = 0$, and h be a function with Maclaurin series $h(z) = \sum_n \frac{1}{n!} h_n z^n$, then the equation $Z_H[J] = h(Z_F[J])$ generates the Green's functions*

$$H(X) = \sum_{\pi \in \text{Part}(X)} h_{N(\pi)} F^\pi. \tag{4.9}$$

4.1.4 Green's Functions: Cumulants

In many practical applications, one wishes to characterize a probability distribution through its cumulants.

Let G be a system of Green's functions. The **cumulant Green's functions** K are defined by their generating function $W[J] = Z_K[J]$, where

$$W[J] = \ln Z_G[J]. \tag{4.10}$$

We may use a diagrams to describe the ordinary and cumulant Green's functions, as in Figure 4.3.

The series expansion for (4.10) can be then represented in diagram form now as in Figure 4.4.

Theorem 4.1.5 *The ordinary Green's functions can be expressed in terms of the cumulant Green's functions according to the rule*

$$G(X) = \sum_{\pi \in \text{Part}(X)} K^\pi. \tag{4.11}$$

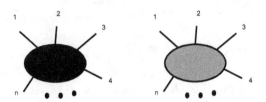

Figure 4.3 The diagram on the left (black center) represents an ordinary Green's function $G_{123...n}$, and the one on the right (gray center) represents a cumulant Green's function $K_{123...n}$.

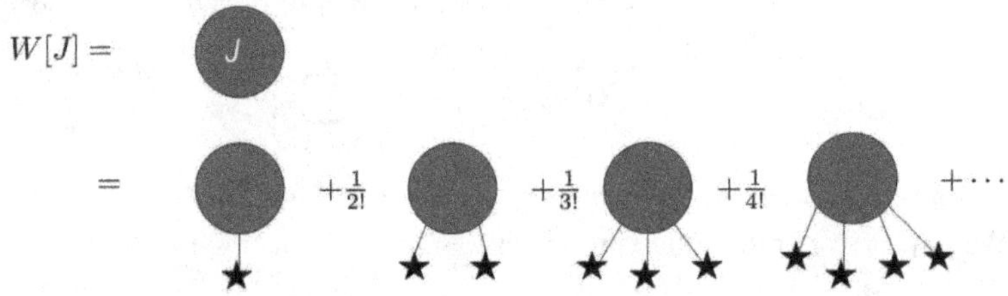

$$W[J] = \text{(diagram)}$$

$$= \text{(diagram)} + \tfrac{1}{2!}\,\text{(diagram)} + \tfrac{1}{3!}\,\text{(diagram)} + \tfrac{1}{4!}\,\text{(diagram)} + \cdots$$

Figure 4.4 The series expansion for $W[J] = \int K_X \, dX$. Note that $K_\emptyset = 0$, so there is no constant term in the expansion.

Conversely we have

$$K(X) = \sum_{\pi \in \text{Part}(X)} \mu(\pi)\, G^\pi, \qquad (4.12)$$

where $\mu(\pi) = (-1)^{N(\pi)-1}(N(\pi)-1)!$.

Proof For the first part, we simply use the preceding lemma with $h_n = 1$ so that $h(t) = e^t$ and take $F = K$ and $H = G$. Note that $K[\emptyset] \equiv 0$.

To invert the relationship, we now set F to be

$$\tilde{G}(X) = \begin{cases} G(X), & X \neq \emptyset; \\ 0, & X = \emptyset. \end{cases}$$

and $H = K$. In this case,

$$Z_K = \log Z_G = \log\left(1 + Z_{\tilde{G}}\right)$$

since $G(\emptyset) = 1$. We set $h(t) = \ln(1+t)$, and so $h_n = (-1)^{n-1}(n-1)!$ in this case. $\qquad \square$

Writing G_{123} as shorthand for $G(X)$ with $X = \{x_1, x_2, x_3\}$ and so on, we have

$$G_1 = K_1,$$
$$G_{12} = K_{12} + K_1 K_2,$$
$$G_{123} = K_{123} + K_{12} K_3 + K_{23} K_1 + K_{31} K_2 + K_1 K_2 K_3,$$
$$\vdots$$

and inversely

$$K_1 = G_1,$$
$$K_{12} = G_{12} - G_1 G_2,$$
$$K_{123} = G_{123} - G_{12} G_3 - G_{23} G_1 - G_{31} G_2 - 2 G_1 G_2 G_3,$$
$$\vdots$$

Let us examine the 4th order Green's function expanded in terms of its cumulants:

$$
\begin{aligned}
G_{1234} = {}& K_{1234} \\
&+ K_1 K_{234} + K_2 K_{134} + K_3 K_{124} + K_4 K_{123} + K_{12} K_{34} + K_{13} K_{24} + K_{14} K_{23} \\
&+ K_1 K_2 K_{34} + K_1 K_3 K_{24} + K_1 K_4 K_{23} + K_2 K_3 K_{14} + K_2 K_4 K_{13} + K_3 K_4 K_{12} \\
&+ K_1 K_2 K_3 K_4.
\end{aligned}
$$

This may be represented as the sum over the following diagrams; see Figure 4.5.

In general, an n-point Green's function $G_{12...n}$ will be a sum of B_n cumulant Green's functions, where B_n (the nth Bell number; see subsection 1.3.2) counts the number of ways to partition a set of n indices into nonempty subsets. More specifically, the Stirling numbers of the second kind, $\left\{ {n \atop m} \right\}$ count the number of

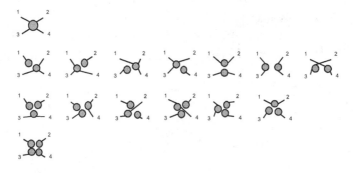

Figure 4.5 Diagrammatic expansion of G_{1234} into cumulants

ways to partition n items into exactly m subsets (blocks); see subsection 1.3.2: in the sum, there will be exactly $\left\{{n \atop m}\right\}$ terms, which are a product of m cumulants.

For the preceding case $n = 4$, we have $\left\{{4 \atop 1}\right\} = 1$, $\left\{{4 \atop 2}\right\} = 7$, $\left\{{4 \atop 3}\right\} = 6$, $\left\{{4 \atop 4}\right\} = 1$, and $B_4 = 15$.

4.1.5 Calculus for Fields

If a generating function $Z_G[J]$ is Fréchet differentiable about $J = 0$ to all orders, then we may work out the components $G(X)$ according to

$$G(X) = \left\{ \prod_{x \in X} \frac{\delta}{\delta J(x)} \right\} Z_G[J] \Bigg|_{J=0}. \tag{4.13}$$

A useful shorthand is to introduce the multiple Fréchet derivative

$$\frac{\delta}{\delta J^X} = \prod_{x \in X} \frac{\delta}{\delta J(x)}$$

along with the derivative $\dfrac{\delta Z_G[J]}{\delta J}$ defined by

$$\frac{\delta Z_G[J]}{\delta J} : X \mapsto \frac{\delta Z_G[J]}{\delta J^X}.$$

In particular, $\dfrac{\delta Z_G[J]}{\delta J}\Bigg|_{J=0} = G.$

Proposition 4.1.6 *We have* $\dfrac{\delta Z_G[J]}{\delta J} = \int G(\cdot + Y) J^Y dY$, *that is,*

$$\frac{\delta Z_G[J]}{\delta J^X} = \int G(X + Y) J^Y dY.$$

Proof   To see this, note that

$$\frac{\delta Z_G[J]}{\delta J(x)} = \frac{\delta}{\delta J(x)} \sum_{n=0}^{\infty} \frac{1}{n!} \int G_n(x_1, \ldots, x_n) J(x_1) \ldots J(x_n) \, dx_1 \ldots dx_n$$

$$= \sum_{n=1}^{\infty} \frac{1}{(n-1)!} \int G_n(x, x_2, \ldots, x_n) J(x_2) \ldots J(x_n) \, dx_2 \ldots dx_n$$

$$= \int G(\{x\} + Y) J^Y \, dY.$$

Therefore, more generally $\dfrac{\delta Z_G[J]}{\delta J} = \int G(\cdot + Y) J^Y dY.$ $\square$

Proposition 4.1.7 (The Leibniz Rule for Fields) *For analytic functional U and V, we have that*

$$\frac{\delta}{\delta J} U[J] V[J] = \frac{\delta U[J]}{\delta J} \diamond \frac{\delta V[J]}{\delta J}.$$

To see this, set $U = Z_F$ and $V = Z_G$, then we are required to show that

$$\frac{\delta}{\delta J^X} Z_F[J] Z_G[J] = \sum_{X_1 + X_2 = X} \frac{\delta Z_F[J]}{\delta J^{X_1}} \frac{\delta Z_G[J]}{\delta J^{X_2}}.$$

The proof follows by elementary induction. For several terms, we just find multiple Wick products.

We note that if $W = W[J]$ is a given functional, then

$$\frac{\delta h(W)}{\delta J(x_1)} = h'(W) \frac{\delta W}{\delta J(x_1)},$$

$$\frac{\delta^2 h(W)}{\delta J(x_1)\delta J(x_2)} = h''(W) \frac{\delta W}{\delta J(x_1)} \frac{\delta W}{\delta J(x_2)} + h'(W) \frac{\delta^2 W}{\delta J(x_1)\delta J(x_2)}.$$

$$\vdots$$

The pattern is easy to spot, and we establish it in the next lemma.

Lemma 4.1.8 (The Chain Rule for Fields) *Let $W = W[J]$ possess Fréchet derivatives to all orders and let h be smooth, then*

$$\frac{\delta}{\delta J^X} h(W[J]) = \sum_{\pi \in \text{Part}(X)} h^{(N(\pi))}(W[J]) \left(\frac{\delta W}{\delta J}\right)^\pi,$$

where $h^{(n)}(t)$ is the nth derivative of h at t.

Proof This is easily seen by induction. As $\dfrac{\delta h(W)}{\delta J^x} = h'(W) \dfrac{\delta W}{\delta J(x)}$, the identity is trivially true for $n = 1$. Now assume that it is true for n, and let $|X| = n$, then

$$\frac{\delta}{\delta J(x)} \frac{\delta h(W)}{\delta J^X} = \frac{\delta}{\delta J(x)} \sum_{\pi \in \text{Part}(X)} h^{(N(\pi))}(W[J]) \left(\frac{\delta W}{\delta J}\right)^\pi$$

$$= \sum_{\pi \in \text{Part}(X)} h^{(N(\pi)+1)}(W[J]) \frac{\delta W}{\delta J(x)} \left(\frac{\delta W}{\delta J}\right)^\pi$$

$$+ \sum_{\pi \in \text{Part}(X)} h^{(N(\pi))}(W[J]) \frac{\delta}{\delta J(x)} \left(\frac{\delta W}{\delta J}\right)^\pi.$$

However, the first term on the right-hand side is a sum over all parts of $X + \{x\}$ having x occurring as a singleton, while the second term, when differentiated with respect to $J(x)$, will be a sum over all parts of $X + \{x\}$ having x in some part containing at least one element of X. Thus we may write the preceding as

$$\frac{\delta h(W)}{\delta J^{X+\{x\}}} = \sum_{\pi \in \text{Part}(X+\{x\})} h^{(N(\pi))}(W[J]) \left(\frac{\delta W}{\delta J}\right)^\pi .$$

The identity then follows by induction.   □

The chain rule is in fact a generalization of the Faà di Bruno formula. We note that for $h(t) = e^t$, we get the equation

$$\frac{\delta}{\delta J^X} e^{W[J]} = e^{W[J]} \sum_{\pi \in \text{Part}(X)} \left(\frac{\delta W}{\delta J}\right)^\pi . \tag{4.14}$$

4.1.6 Zero-Dimensional Fields

We now take our space $\mathfrak{X}$ to consist of a single point denoted as $*$. In this case, our "fields" reduce to a single random variable ϕ, the field at $*$. The Green's functions G_n are then the moments $\mu_n = \mathbf{E}[\phi^n]$ and the generating functions $Z_G[t]$ are just the exponential generating series $g(t) = \sum_{n \geq 0} \frac{1}{n!} G_n t^n$. (Technically we have abandoned our no coincidence rule, but this is not a problem here.)

The chain rule formula then becomes

$$\frac{d^n}{dt^n} h(g(t)) = \sum_{\pi \in \text{Part}_n} h^{(N(\pi))}(g(t)) \prod_{A \in \pi} g^{(\#A)}(t) ,$$

where $h^{(n)}$ and $g^{(n)}$ are the nth derivatives of h and g. The sum is over partitions of n objects. We could rewrite this in terms of occupation numbers of partitions:

$$\frac{d^n}{dt^n} h(g(t)) = \sum_{\mathbf{n} \in \mathbb{N}_+^{\mathbb{N}}}^{E(\mathbf{n})=n} \nu_{\text{Part}}(\mathbf{n}) \, h^{(N(\mathbf{n}))}(g(t)) \prod_{k=1}^{\infty} \left(g^{(k)}(t)\right)^{n_k}$$

$$= \sum_{\mathbf{n} \in \mathbb{N}_+^{\mathbb{N}}}^{E(\mathbf{n})=n} \frac{1}{(1!)^{n_1}(2!)^{n_2}\ldots n_1! n_2! \ldots} \frac{n!}{} h^{(N(\mathbf{n}))}(g(t)) \prod_{k=1}^{\infty} \left(g^{(k)}(t)\right)^{n_k} .$$

This, of course, is just the Faà di Bruno formula. It may also be expressed using Bell polynomials as

$$\frac{d^n}{dt^n} h(g(t)) = \sum_{m=1}^{n} h^{(m)}(g(t)) \, \text{Bell}\left(n, m; g^{(1)}(t), g^{(2)}(t), \ldots\right) .$$

The relationship between moments and cumulants is now simply

$$\mu_n = \sum_{\pi \in Part_n} \prod_{A \in \pi} \kappa_{\#A} \equiv \sum_{m=1}^{n} \text{Bell}\,(n, m; \kappa_1, \kappa_2, \ldots),$$

which likewise recovers the earlier results.

4.1.7 Nonsymmetric Green's Functions

We will later encounter Green's functions of the form

$$G(x_1, \ldots, x_n) = \langle \Phi(x_1) \ldots \Phi(x_n) \rangle,$$

where $\langle \cdot \rangle$ is a quantum expectation (state) and the fields $\Phi(x)$ are operators that need not commute among themselves. In such cases, we have to deal with Green's functions where the order of the arguments are important.

To this end, we denote by $\mathbf{X} = (x_1, \ldots, x_n)$ the sequence and write accordingly $G = G(\mathbf{X})$. Despite this lack of complete symmetry, many of the ideas already encountered still apply. For instance, the generating function Z_G may still defined as in (4.1), though it only contains information about the *symmetrized* Green's function

$$\frac{1}{n!} \sum_{\sigma \in \mathfrak{S}_n} G(x_{\sigma(1)}, \ldots, x_{\sigma(n)}),$$

and this is all we would obtain when differentiating the generating function.

We define a subsequence of $\mathbf{X} = (x_1, \ldots, x_n)$ specifically to be a subsequence $(x_{i(1)}, \ldots, x_{i(p)})$, which respects the original order, that is, $i(1) < i(2) < \cdots < i(p)$. We include the possibility of an empty sequence ($p = 0$), and $\mathbf{X}$ itself ($p = n$). Two subsequences of $\mathbf{X}$ are disjoint if they have no common element. In this way, we may define a Wick product of nonsymmetric Green's functions as

$$F \diamond G\,(\mathbf{X}) \triangleq \sum_{\mathbf{X}_1 + \mathbf{X}_2 = \mathbf{X}} F(\mathbf{X}_1)\,G(\mathbf{X}_2), \tag{4.15}$$

where now the sum is over all pairs of disjoint subsequences of $\mathbf{X}$ whose union is $\mathbf{X}$.

A partition of $\mathbf{X}$ is understood as a set of nonempty pairwise disjoint subsequences of $\mathbf{X}$ whose union is $\mathbf{X}$. Their enumeration is exactly the same as that for sets. This is a consequence of the fact that, having fixed the sequence $\mathbf{X}$, we only deal with subsequences that respect this order, and so these are in one-to-one correspondence with subsets of the set $X = \{x_1, \ldots, x_n\}$.

We may therefore define cumulants of nonsymmetric Green's functions in the obvious way:

$$G(\mathbf{X}) = \sum K(\mathbf{X}_1) \ldots K(\mathbf{X}_m), \qquad (4.16)$$

where we now have a sum over all partitions $\{\mathbf{X}_1, \ldots, \mathbf{X}_m\}$ of $\mathbf{X}$ with $1 \leq m \leq n$. This is almost identical to the situation in Theorem 4.1.4, and the corresponding inverse relation likewise holds. Indeed, the expansions in (4.13) and (4.13) hold with the only caveat that moments are all order dependent, e.g. G_{123} is shorthand for $G(x_1, x_2, x_3)$, and so on, but this is not an issue as all terms in the development are displayed respecting this order.

4.2 A First Look at Boson Fock Space

When describing several quantum particles, it is important to take into account whether they are identical or not. The wave function of n particles will be a square-integrable function $\psi_n(x_1, \ldots, x_n)$ with the modulus-squared $\rho_n(x_1, \ldots, x_n) = |\psi_n(x_1, \ldots, x_n)|^2$ interpreted as the probability density to find the particles at the position $(x_1, \ldots, x_n)$. If the particles are indistinguishable, then quantum theory imposes the symmetry

$$\rho_n(x_{\sigma(1)}, \ldots, x_{\sigma(n)}) = \rho_n(x_1, \ldots, x_n)$$

for every permutation σ. That is, the probability density is completely symmetric under interchange of the particle labels. In Figure 4.6, as the identical particles are truly indistinguishable, only the locations are relevant. In other

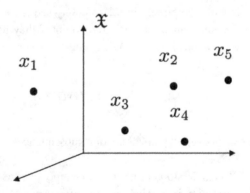

Figure 4.6 Quantum particles at positions $x_1, \ldots, x_5$.

words, the Gibbs $1/n!$ correction introduced in classical statistical physics is a physical feature in the quantum case!

While there may be several mathematical schemes that lead to the densities ρ_n being completely symmetric, only two mechanisms are encountered in Nature:

- **Bosons**: The wavefunctions are themselves completely symmetric:

$$\psi_n(x_{\sigma(1)}, \ldots, x_{\sigma(n)}) = \psi_n(x_1, \ldots, x_n),$$

- **Fermions**: The wavefunctions are completely antisymmetric:

$$\psi_n(x_{\sigma(1)}, \ldots, x_{\sigma(n)}) = (-1)^\sigma \psi(x_1, \ldots, x_n),$$

Here $(-1)^\sigma$ denotes the sign of the permutation.

More generally, we may have a state that has an indefinite number of particles. That is, we have a sequence of wave functions $(\psi_n)_{n \in \mathbb{N}_+}$ with the complex values $\psi_n(x_1, \ldots, x_n)$ being the probability amplitude for n particles. This includes the case $n = 0$, where ψ_0 is a complex scalar. We then have

$$p_n = \int |\psi_n(x_1, \ldots, x_n)|^2 dx_1 \ldots dx_n$$

as the probability that there will be exactly n particles, and the state normalization is $\sum_{n=0}^\infty p_n = 1$.

In the Boson case, we obtain a function on the power set of $\mathfrak{X}$ by introducing

$$\Psi \colon \mathrm{Power}(\mathfrak{X}) \mapsto \mathbb{C}$$
$$\colon \{x_1, \ldots, x_n\} \mapsto \sqrt{n!}\, \psi(x_1, \ldots, x_n). \tag{4.17}$$

Normalization then may be written in Guichardet shorthand as

$$\int |\Psi(X)|^2 \, dX = 1.$$

We have gone from the Hilbert space $L^2(\mathfrak{X}, dx)$ for a single particle, and we are now dealing with an indefinite number of indistinguishable Boson particles with state $\Psi \in L^2(\mathrm{Power}(\mathfrak{X}), dX)$, which is the **Boson Fock space** over $L^2(\mathfrak{X}, dx)$. We in fact have a mapping

$$\exp \colon L^2(\mathfrak{X}, dx) \mapsto L^2(\mathrm{Power}(\mathfrak{X}), dX)$$
$$\colon f \mapsto \exp(f). \tag{4.18}$$

This mapping is analytic and, as we shall prove in the next chapter, will map from a dense subset of $L^2(\mathfrak{X}, dx)$ to a dense subset of $\Psi \in L^2(\mathrm{Power}(\mathfrak{X}), dX)$. Moreover, we have

$$\langle \exp(f) | \exp(g) \rangle = e^{\langle f | g \rangle}.$$

If we consider again the degenerate situation of a zero-dimensional config-uration space $\mathfrak{X}$ consisting of a single point $*$, then $L^2(*) \equiv \mathbb{C}$. We recover the situation of the previous chapter with the Bose Fock space becoming the Hilbert space for the canonical commutation relations (CCR), and $\exp(\beta)$ being the Bargmann state with test "function" β.

5

Combinatorial Species

At this juncture, it is appropriate to mention a relatively recent branch of combinatorics that offers a powerful way to arrive at enumerations, and which is closely related to some of the ideas introduced in the previous chapters. This is the theory of combinatorial species, and we will outline its basic feature in this chapter, and rederive some of our basic enumeration problems (permutations, partitions, hierarchies, etc.) in this setting with relative (indeed astonishing) ease.

Several of these basic species – notably partitions (Part) and permutations (Perm) – have been encountered already, and some of the key constructions involved have already appeared when we were discussing the Wick product and the composition formula.

Let us consider a bijection ϕ on the configuration space $\mathfrak{X}$. This will induce a map ϕ between sets according to $\phi: A = \{x_1, \ldots, x_n\} \mapsto A' = \{\phi(x_1), \ldots, \phi(x_n)\}$, which in turn induces a map between partitions, that is, $\phi: \pi = \{A, B, C, \ldots\} \mapsto \{A', B', C', \ldots\}$; see Figure 5.1. the key feature of a bijection is that the induced maps preserve cardinality.

From a combinatorial point of view, nothing much has actually gone on. If we are interested in combinatorial structures such as partitions on a set of n items, then all we really need to know is that there really are n items to work with. The bijection ϕ leaves this invariant. We could focus on the number N_n (Part) of partitions of a set of n items, which we have seen is given by the Bell number B_n. We may then introduce the exponential generating function (EGF) for Partitions, which we take to be

$$\text{Part}(z) \triangleq \sum_{n \geq 0} \frac{1}{n!} N_n(\text{Part}) z^n \equiv e^{e^z - 1}.$$

In a similar fashion, we could consider any permutation σ of the elements of a subset $X = \{x_1, \ldots, x_n\}$. This leads to a permutation $\phi(\sigma)$ on which is

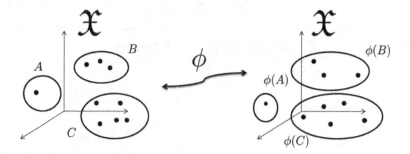

Figure 5.1 The combinatorial features of a partition of a set of items into subsets $\{A, B, C\}$ remain unchanged under the action of a bijection ϕ of the items.

the permutation on the image set $\phi(X)$ taking $\phi(x_k)$ to $\phi\left(x_{\sigma(k)}\right)$. As there are $N_n(\text{Perm}) = n!$ permutations on a set of n items, we get the EGF for permutations to be $\text{Perm}(z) = \frac{1}{1-z}$.

The idea can be extended to any combinatorial structure $\mathscr{S}$ that is built over a system on n items, where n is allowed take on values in $\mathbb{N}_+$. We could set $\mathscr{S}(X)$ to be the set of those structures built on items forming a given finite set X. (We have assumed implicitly that the items are labeled!) The EGF for $\mathscr{S}$ is then defined to be

$$\mathscr{S}(z) \triangleq \sum_{n \geq 0} \frac{1}{n!} N_n(\mathscr{S}) z^n.$$

As we have seen, the combinatorial features of $\mathscr{S}(X)$ should only depend on the set X up to bijections. (In fact, we have been far too restrictive in imposing that the ϕ's come about as maps from a fixed space $\mathfrak{X}$ to itself – all we really need is the bijective property.) The induced action of a bijection ϕ on the combinatorial structures is called a **transport** of the structure, and the transportation properties of structures are the key to the definition of combinatorial species introduced by Joyal (1981). A bijection $\phi: X \mapsto X$ is just a permutation of the vertices, which we can view alternatively as a passive relabeling of the vertices. The concept of structure should be independent of the labeling used, however.

A **species of structures** is a rule $\mathscr{S}$ that associates to each finite set X a finite set $\mathscr{S}(X)$, and to each bijection $\phi: X \mapsto Y$ a function $\mathscr{S}(\phi): \mathscr{S}(X) \mapsto \mathscr{S}(Y)$, such that

- $\mathscr{S}(id_X) = id_{\mathscr{S}(X)}$.
- Whenever we have bijections $X \overset{\phi}{\mapsto} Y \overset{\psi}{\mapsto} Z$, then $\mathscr{S}[\psi \circ \varphi] = \mathscr{S}[\psi] \circ \mathscr{S}[\varphi]$.

An element $s \in \mathscr{S}(X)$ is called an $\mathscr{S}$-structure on X, and the map $\mathscr{S}[\varphi]$ is called the *transport* of $\mathscr{S}$-structures along ϕ. The EGF of the species $\mathscr{S}$ is then

$$\mathscr{S}(z) \triangleq \sum_{n \geq 0} \frac{1}{n!} N(\mathscr{S}, n) z^n,$$

where $N(\mathscr{S}, n)$ gives the number of $\mathscr{S}$-structures over a set of n elements.

Some more examples are tabulated in the following:

	$\mathscr{S}$	$N(\mathscr{S}, n)$	$\mathscr{S}(z)$
Permutations	Perm	$n!$	$(1-z)^{-1}$
Linear orders	Lin	$n!$	$(1-z)^{-1}$
Cycles	Cyc	$(n-1)!$	$-\ln(1-z)$
Sets	Set	1	e^z
Power Sets	Power	2^n	e^{2z}

The species Perm, for instance, gives the permutations over finite sets. We have $n!$ permutations on a set of n elements. A **linear order** on a set X is a sequence of all the elements with no repetition. There is a one-to-one correspondence between the permutations and the linear orders over a given set; however, the most natural way to construct a bijection between the two is to fix a linear order, say $(x_1, \ldots, x_n)$, and any other linear order $(x_{\sigma(1)}, \ldots, x_{\sigma(n)})$ corresponds to the permutation: $x_i \mapsto x_{\sigma(i)}$. Let ϕ be a bijection on X, and therefore an element of Perm(X), then Perm(ϕ) transports permutations on X to themselves. Similarly, Lin(ϕ) will transport linear orders on X to themselves, but not in the same way. Permutations and linear orders are different species despite having the same enumeration.

A **cycle** is a permutation that acts transitively on X, that is, the orbit $\{\sigma^n(x) : n = 0, 1, 2, \ldots\}$ of any element $x \in X$ is all of X. The number of cycles on a set of n elements is $(n-1)!$.

The **set** species is given by Set$(X) = \{X\}$ and is (at first sight) rather trivial. It associates the one set $\{X\}$ with any set X. The power set of X, written as Power(X), is the set of all subsets of X as before.

5.1 Operations on Species

Let $\mathscr{S}$ and $\mathscr{R}$ be species, then their sum $\mathscr{S} + \mathscr{R}$ is defined by saying that an $(\mathscr{S} + \mathscr{R})$ structure on a finite set X is either an $\mathscr{S}$-structure or a $\mathscr{R}$-structure.

That is, $(\mathscr{S} + \mathscr{R})(X)$ is the disjoint union of $\mathscr{S}(X)$ and $\mathscr{R}(X)$. We may define $\mathscr{S} + \mathscr{S}$ by artificially attaching a color to each of the species, so that we could have, for instance, $\mathscr{S}_{\text{red}} + \mathscr{S}_{\text{blue}}$. We may use several colors for multiple sums of the same basic species.

The EGF of a sum of species is then just the sum of the respective series:

$$(\mathscr{S} + \mathscr{R})(z) = \mathscr{S}(z) + \mathscr{R}(z).$$

Let $\mathscr{S}$ be a species and $n \geq 0$ an integer. We define $\mathscr{S}_n$ by

$$\mathscr{S}_n(X) = \begin{cases} \mathscr{S}(X), & \text{if } |X| = n; \\ \emptyset, & \text{otherwise.} \end{cases}$$

In particular, every species may be (canonically) decomposed as

$$\mathscr{S} = \sum_{n \geq 0} \mathscr{S}_n$$

and we may also define

$$\mathscr{S}_+ = \sum_{n \geq 1} \mathscr{S}_n,$$

$$\mathscr{S}_{\text{even}} = \sum_{n \text{ even}} \mathscr{S}_n,$$

$$\mathscr{S}_{\text{odd}} = \sum_{n \text{ odd}} \mathscr{S}_n.$$

Let $\mathscr{S}$ and $\mathscr{R}$ be species, then their product $\mathscr{S} \diamond \mathscr{R}$ is defined by saying that an $\mathscr{S} \diamond \mathscr{R}$ structure on a finite set X is an $\mathscr{S}$-structure on some subset Y of X and a $\mathscr{R}$-structure on the complement X/Y.

As the notion suggests, this is just the Wick product in a different guise; indeed, we have

$$\mathscr{S} \diamond \mathscr{R}(X) \triangleq \sum_{X_1 + X_2 = X} \mathscr{S}(X_1) \mathscr{R}(X_2).$$

The EGF of a product of species is then naturally defined as the product of the respective series (just as in subsection 4.1.2):

$$(\mathscr{S} \diamond \mathscr{R})(z) = \mathscr{S}(z)\mathscr{R}(z).$$

Just as the notion of product of species abstracts the Wick product, so too can we abstract the concept of composition of species from that of composition of moment generating functions.

Let $\mathscr{F}$ and $\mathscr{G}$ be species with $\mathscr{G}[\emptyset] = \emptyset$. We define the composition $\mathscr{F} \circ \mathscr{G}$ to be the species

$$\mathscr{F} \circ \mathscr{G}(X) \triangleq \sum_{\pi \in \text{Part}(X)} \mathscr{F}[\pi] \prod_{Y \in \pi} \mathscr{G}[Y]. \tag{5.1}$$

Theorem 5.1.1 *Let $\mathscr{F}$ and $\mathscr{G}$ be two species of structures and suppose that $\mathscr{G}[\emptyset] = \emptyset$. Then the series associated to the species $\mathscr{F} \circ \mathscr{G}$ has EGF $\mathscr{F} \circ \mathscr{G}(z) = \mathscr{F}(\mathscr{G}(z))$.*

5.1.1 Examples of Species

Derangements

As an example, the number of derangements (permutations without fixed points) is denoted d_n. In a classic problem of combinatorial probability, we have that $p(n)d_n/n!$ is the probability that if n men leave their top hats in the cloak room before an opera, and the hats are then returned at random, that no one gets his original hat back.

We now show how to derive this using species. Let Der be the species of derangements, then Perm = Set $\diamond$ Der, which basically says that a permutation consists of fixed points (a set) and cycles of length 2 or more (derangements). We then get

$$\mathrm{Der}(z) = \frac{\mathrm{Perm}(z)}{\mathrm{Set}(z)} = e^{-z}/(1-z),$$

from which we read off $N_n(\mathrm{Der}) = d_n$ to be

$$d_n = n! \sum_{k=0}^{n} \frac{(-1)^k}{k!}.$$

Therefore, $p(n) = \sum_{k=0}^{n} \frac{(-1)^k}{k!}$, which is approximately $e^{-1} = 0.36788\ldots$.

Sets

We specify an element of the power set X by simultaneously the set itself and its complement in X. Therefore, Power = Set $\diamond$ Set, and so $\mathrm{Power}(z) = \mathrm{Set}(z)^2$.

Partitions

From the simple observation that a partition is a set of nonempty disjoint sets, we get the following characterization of partitions as a species. A partition is a set of nonempty subsets. The species of partitions therefore satisfies the relation

$$\mathrm{Part} = \mathrm{Set} \circ \mathrm{Set}_+.$$

We see that

$$\mathrm{Part}(z) = \exp(e^z - 1) \equiv \sum_n \frac{1}{n!} B_n z^n.$$

Pairs

The species with pairings as the structure will be denoted as $\text{Pair}(X)$. These are sets of subsets of size 2, and so

$$\text{Pair} = \text{Set} \circ \text{Set}_2.$$

It follows immediately that

$$\text{Pair}\,(z) = \exp\left\{\frac{1}{2}z^2\right\}$$

from which we read off that the number of ways to pair off n elements is

$$N_n(\text{Pair}) = \begin{cases} \frac{(2k)!}{2^k k!}, & n = 2k; \\ 0, & n \text{ odd}. \end{cases}$$

Permutations

Likewise, a permutation is a set of disjoint cycles, so we similarly have

$$\text{Perm} = \text{Set} \circ \text{Cycle}.$$

We deduce that $\text{Perm}\,(z) = \exp\{\text{Cycle}\,(z)\}$, and so we recover $\text{Cycle}\,(z) = \ln(1-z)^{-1}$.

5.2 Graphs

Let $\mathscr{G}$ denote the species of graphs. Note that

$$N_n\,(\mathscr{G}) = 2^{\binom{n}{2}},$$

since if we have a set X of size n giving the possible vertices, there will be $\binom{n}{2}$ possible edges that may or may not be present in the graph. The EGF lacks a closed form, but does have a radius of convergence equal to $\frac{1}{2}$.

We consider the species of connected graphs denoted by $\mathscr{G}_{\text{con}}$.

Proposition 5.2.1 *The species of connected graphs* $\mathscr{G}_{con}$ *generates all graphs by the identity* $\mathscr{G} = \text{Set} \circ \mathscr{G}_{con}$. *In particular,*

$$\mathscr{G}_{con}\,(z) = \ln\mathscr{G}\,(z).$$

Proof Every graph can be considered as a set of connected subgraphs, and so as species we have $\mathscr{G} = \text{Set} \circ \mathscr{G}_{\text{con}}$. For the EGFs, we then have $\mathscr{G}(z) = \exp\mathscr{G}_{\text{con}}(z)$. $\qquad\square$

From this point of view, we may define the species of connected structures in a species $\mathscr{S}$ by the relation $\mathscr{S} = \text{Set} \circ \mathscr{S}_{\text{con}}$.

5.3 Weighted Species

A **weight** w on a species $\mathscr{S}$ is a family of functions $w \colon \mathscr{S}(X) \mapsto \mathbb{C}$ for each finite set X. We shall be interested in weights that are *preserved under transport*; that is, for any bijection $\phi \colon X \mapsto Y$, we have $w(\phi S) = w(S)$ for every $S \in \mathscr{S}(X)$. In other words, the weight attached to a structure is independent of the underlying labeling. A pair $(\mathscr{S}, w)$ consisting of a species and a weight preserved under the transport is termed a **weighted species**.

We may then define a weighted EGF to be

$$\mathscr{S}^{(w)}(z) \triangleq \sum_{n=0}^{\infty} \frac{1}{n!} \sum_{S \in \mathscr{S}_n} w(S)\, z^n.$$

For the uniform weight $w(S) = 1$, we recover the usual EGF. Most of what we have done up to now carries over to weighted species.

The sum of $(\mathscr{S}, w)$ and $(\mathscr{R}, v)$ is $(\mathscr{S} + \mathscr{R}, u)$, where

$$u(S) = \begin{cases} w(S), & S \in \mathscr{S}(X); \\ v(S), & S \in \mathscr{R}(X). \end{cases}$$

The product of $(\mathscr{S}, w)$ and $(\mathscr{R}, v)$ is $(\mathscr{S} \diamond \mathscr{R}, u)$, where a typical product structure (S, R) with $S \in (X_1)$ and $R \in \mathscr{R}(X_2)$ will have the weight

$$u(S, R) = w(S)\, v(R).$$

The composition of $(\mathscr{S}, w)$ and $(\mathscr{R}, v)$ is $(\mathscr{S} \circ \mathscr{R}, u)$, where a typical structure $(S, R_1, \ldots, R_n)$ with $\pi = \{R_1, \ldots, R_n\}$, $S \in \mathscr{S}(\pi)$, and $R_i \in \mathscr{R}(X_i)$ will have the weight

$$u(S, R_1, \ldots, R_n) = w(S)\, v(R_1) \ldots v(R_n).$$

The rules for weighted EGFs are then

$$(\mathscr{S} + \mathscr{R})^u(z) = \mathscr{S}^w(z) + \mathscr{R}^v(z)$$
$$(\mathscr{S} \diamond \mathscr{R})^u(z) = \mathscr{S}^w(z)\, \mathscr{R}^v(z)$$
$$(\mathscr{S} \circ \mathscr{R})^u(z) = \mathscr{S}^w\left(\mathscr{R}^v(z)\right)$$

with the appropriate choice of u in each case.

5.3.1 Enumerating Permutations

A weight on the species of permutations is given by

$$w(\sigma) = \lambda^{N(\sigma)},$$

where $\lambda > 0$ and $N(\sigma)$ counts the number of cycles in the permutation σ. As permutations are just sets of cycles, we have

$$(\text{Perm}, w) = (\text{Set}, 1) \circ (\text{Cycle}_+, \lambda),$$

where we attached weight λ to each cycle. Therefore,

$$\text{Perm}^w(z) = \exp\{-\lambda \ln(1-z)\} = \frac{1}{(1-z)^\lambda}.$$

At this stage, however, we recognize the moment generating function for the Gamma distribution! Therefore,

$$\text{Perm}^w(z) = \sum_{n=1}^{\infty} \frac{1}{n!} \mu_n^{\text{Perm}}(\lambda) z^n,$$

where $\mu_n^{\text{Perm}}(\lambda)$ are the Gamma distribution moments

$$\mu_n^{\text{Perm}}(\lambda) = \lambda(\lambda+1)\cdots(\lambda+n-1) \equiv \lambda^{\bar{n}}.$$

We may expand the rising factorial power $\lambda^{\bar{n}}$ as[1]

$$\lambda^{\bar{n}} \equiv \sum_m \begin{bmatrix} n \\ m \end{bmatrix} \lambda^m.$$

5.3.2 Enumerating Partitions

A weight on the partitions is given by

$$w(\pi) = \lambda^{N(\pi)},$$

where $\lambda > 0$ and $N(\pi)$ is the number of blocks making up a partition π. We then have

$$(\text{Part}, w) = (\text{Set}, 1) \circ (\text{Set}_+, \lambda).$$

Therefore,

$$\text{Part}^w(z) = e^{\lambda(e^z-1)}.$$

We write as a power series in z

$$\text{Part}^w(z) = \sum_{n=1}^{\infty} \frac{1}{n!} \mu_n^{\text{Part}}(\lambda) z^n$$

---

[1] We now see why the Stirling numbers of the first kind ($\begin{bmatrix} n \\ m \end{bmatrix}$; see subsection 1.3.2), count the number of permutations of a set of n elements having exactly m cycles. Indeed, we have

$$\sum_{n=0}^{\infty} \frac{1}{n!} \sum_m \begin{bmatrix} n \\ m \end{bmatrix} \lambda^m z^n = \frac{1}{(1-z)^\lambda}.$$

but this time we recognize the moment generating function for the Poisson distribution! We now have that[2]

$$\mu_n^{\text{Part}} = \sum_m \begin{Bmatrix} n \\ m \end{Bmatrix} \lambda^m.$$

5.3.3 Enumerating Hierarchies

Proposition 5.4.1 (Enumerating Hierarchies) *The EGF for hierarchies will be denoted as* $\text{Hier}(z) = \sum_n \frac{1}{n!} h_n z^n$, *and satisfies*

$$e^{\text{Hier}(z)} = 2\text{Hier}(z) - z + 1.$$

Proof A hierarchy is a tree with a root, and from these grow at least two subhierarchies:

$$\text{Hier} = \text{Set}_1 + \text{Set}_{\geq 2}(\text{Hier})$$

or

$$\text{Hier}(z) = z + \left(e^{\text{Hier}(z)} - 1 - \text{Hier}(z) \right).$$

Rearranging gives the result. $\qquad\qquad\qquad\qquad\qquad\qquad\qquad\square$

5.4 Differentiation of Species

Let $\mathscr{S}$ be a species, then define the **species derivative** $\mathscr{S}'$ by

$$\mathscr{S}'(X) = \mathscr{S}(X + \{x_*\}),$$

where x_* is new element outside of X.

We have $N(\mathscr{S}', n) = N(\mathscr{S}, n + 1)$, which means that the EGFs are

$$\mathscr{S}'(z) = \frac{d}{dz} \mathscr{S}(z).$$

Properties of the differentiation operation are

- $(\mathscr{S} + \mathscr{R})' = \mathscr{S}' + \mathscr{R}'$
- $(\mathscr{S} \diamond \mathscr{R})' = \mathscr{S}' \diamond \mathscr{R}'$
- $(\mathscr{S} \circ \mathscr{R})' = \mathscr{S}' \circ \mathscr{R}'$

---

[2] And so we see once more that the Stirling numbers of the second kind $\begin{Bmatrix} n \\ m \end{Bmatrix}$ count the number of ways to partition a set of n elements into exactly m blocks. We have the identity

$$\sum_{n=0}^{\infty} \frac{1}{n!} \sum_m \begin{Bmatrix} n \\ m \end{Bmatrix} \lambda^m z^n = e^{\lambda(e^z - 1)}.$$

As an example, let us consider Cycle$'$. A cycle with a distinguished element can be understood as a linearly ordered list of the remaining elements of the cycle starting with the one next to the distinguished element, therefore Cycle$'$=Lin. We could argue that Lin=Set$_0$ +Set$_1$$\diamond$Lin and so Lin$(z) = 1 + zLin(z)$, or lin$(z) = (1 - z)^{-1}$, which we have already seen! Then Cycle$(z) = -\log(1 - z)$, by integrating and noting that Cycle$(0) = 0$.

We also have Lin$'$=Lin $\diamond$ Lin, and we could deduce Lin(z) from this differential equation with the condition Lin$(0) = 1$. Note that derivative of the species Tree gives the species Forest.

Finally, we remark that differentiation carries over immediately to weighted species since the derivative inherits the same weight.

6

Combinatorial Aspects of Quantum Fields: Feynman Diagrams

We now apply the Guichardet shorthand notation more extensively in field theory, and in particular, as complementary to the usual diagrammatic approach. We will deal with algebraic features, relegating the rigorous analysis to a later stage. Essentially what we are going to do is to use the Guichardet notation as an alternative to Feynman diagrams. (For other applications of combinatorial species to describing Feynman diagrams, see Faris, 2009/11.)

To be specific, what we will study in this section is the combinatorics behind euclidean quantum field theory or, equivalently, the Schwinger functions. These will be properly introduced in the next chapter, but for the time being we content ourselves with giving a rudimentary introduction.

6.1 Basic Concepts

Let $\mathfrak{X}$ be a finite set, then a classical field over $\mathfrak{X}$ is a function on $\mathfrak{X}$ taking real values: we will call such a function $\varphi = \{\varphi_x : x \in \mathfrak{X}\}$, a **field realization**. The space of all field realizations can be denoted as Φ, and here it is $\mathbb{R}^{\#\mathfrak{X}}$. A real-valued function on Φ may be called a functional and we suppose the existence of a functional $S[\cdot]$, called the action, such that $\Xi = \int_{\Phi} \exp\{S[\varphi]\}\,\mathcal{D}\varphi < \infty$, where $\mathcal{D}\varphi = \prod_{x \in \mathfrak{X}} d\varphi_x$ denotes a standard Lebesgue measure. We then arrive at a well-defined probability measure $\mathbf{P}$ on Φ determined by

$$d\mathbf{P}(\varphi) = \frac{1}{\Xi} e^{S[\varphi]}\,\mathcal{D}\varphi. \tag{6.1}$$

The expectation of a (typically nonlinear) functional $F = F[\cdot]$ of the field is the integral over Φ:

$$\langle F[\phi] \rangle = \int_{\Phi} F[\varphi]\,d\mathbf{P}(\varphi). \tag{6.2}$$

Here we adopt the convention that ϕ is a random variable, with specific realizations denoted by φ.

The label $x \in \mathfrak{X}$ is assumed to give all relevant information, which may be position on a finite lattice, spin, or such. So long as $\mathfrak{X}$ is finite, the mathematical treatment is straightforward. Otherwise, we find ourselves having to resort to infinite dimensional analysis. We may introduce a dual space $\mathfrak{J}$ of fields $J = \{J^x : x \in \Lambda\}$, which are referred to as the **source fields**. Our convention will be that the source fields carry a "contravariant" index, while the field, and its realizations, carry a "covariant" index. The duality between fields and sources will be written as $\langle \varphi, J \rangle = \sum_{x \in \mathfrak{X}} \varphi_x J^x$.

In statistical mechanics, we would then typically consider a sequence $(\mathfrak{X}_n)_n$ of finite subsets of a fixed infinite lattice with each $\mathfrak{X}_n$ contained within $\mathfrak{X}_{n+1}$ and $\cup_{n \geq 1} \mathfrak{X}_n$ giving the entire lattice. This bulk limit is studied, for instance, as the thermodynamic limit.

The general situation that we are really interested in, however, is when $\mathfrak{X}$ is continuous. If we want $\mathfrak{X} = \mathbb{R}^d$, then we should take $\mathfrak{J}$ to be the space of Schwartz functions on $\mathbb{R}^d$ and Φ to be the tempered distributions. (The field realizations are more singular than the sources!)

$$S\left(\mathbb{R}^d\right) = \mathfrak{J} \subset L^2\left(\mathbb{R}^d\right) \subset \Phi = S'\left(\mathbb{R}^d\right) \tag{6.3}$$

Here, the duality is denoted by $\langle \varphi, J \rangle = \int_{\mathfrak{X}} \varphi_x J^x dx$. In keeping with the idea that sources carry a contravariant index and fields a covariant one, we introduce an Einstein summation convention that repeated indices are to be summed/integrated over. So, for example,

$$\langle \varphi, J \rangle \equiv \varphi_x J^x. \tag{6.4}$$

The appropriate way to consider randomizing the field in the infinite dimensional case will then be to consider probability measures on the Borel sets of Φ, that is, on the σ-algebra generated by the weak topology on the space of test functions. We shall nevertheless assume that we may fix a probability measure $\mathbf{P}$ on Φ and compute expectations $\langle F[\phi] \rangle = \int_\Phi F[\varphi]\, d\mathbf{P}(\varphi)$ for suitable functionals F. (Here we denote the random variable corresponding to the field by ϕ.) In particular, we shall assume that the expectation

$$Z[J] = \mathbf{E}\left[e^{\langle \phi, J \rangle}\right] \equiv \int_\Phi e^{\langle \varphi, J \rangle}\, d\mathbf{P}(\varphi) \tag{6.5}$$

exists. This will act as a moment generating functional for the measure $\mathbf{P}$ and we shall work formally in the following, effectively assuming the existence of moments to all orders, leaving analytic considerations until later. The Green's functions $G(x_1, \ldots, x_n) = \mathbf{E}\left[\phi_{x_1} \ldots \phi_{x_n}\right]$, and we may write this as

$$G_X = \mathbf{E}[\phi_X] = \left.\frac{\delta Z[J]}{\delta J^X}\right|_{J=0}, \tag{6.6}$$

where for $X = \{x_1, \ldots, x_n\}$ we write $G(x_1, \ldots, x_n)$ as G_X (a completely symmetric tensor with covariant rank $\#X = n$) and introduce the shorthand

$$\phi_X = \prod_{x\in X} \phi_x \tag{6.7}$$

as well as

$$J^X = \prod_{x\in X} J^x, \qquad \frac{\delta^{\#X}}{\delta J^X} = \prod_{x\in X} \frac{\delta}{\delta J^x}. \tag{6.8}$$

Note that we need only the set X – the ordering of the elements is irrelevant! We of course have the formal expression $\frac{\delta}{\delta J^x}(J^y) = \delta_x^y$. Similarly, we shall write $\frac{\delta}{\delta \varphi_X} = \prod_{x\in X} \frac{\delta}{\delta \varphi_x}$.

A multilinear map $T: \times^n \mathfrak{J} \mapsto \mathbb{C}$ is called a tensor of covariant rank n and it will be determined by the components $T_{x_1 \ldots x_n}$ such that $T(J_{(1)}, \ldots, J_{(n)}) = T_{x_1 \ldots x_n} J_{(1)}^{x_1} \ldots J_{(n)}^{x_n}$. Likewise, we refer to a multilinear map from $\times^n \Phi$ to the complex numbers as a tensor of contravariant rank n.

We shall say that a functional $F = F[J]$ is analytic in J if it admits a series expansion $F[J] = \sum_{n\geq 0} \frac{1}{n!} f_{x_1 \ldots x_n} J^{x_1} \ldots J^{x_n}$, where $f_{x_1 \ldots x_n}$ are the components of a completely symmetric covariant tensor and as usual the repeated dummy indices are summed/integrated over. A more compact notation is to write the series expansion in terms of the Guichardet notation[1] as

$$F[J] = \int f_X J^X dX.$$

Note that $\frac{\delta}{\delta J^{y_1}} \ldots \frac{\delta}{\delta J^{y_m}} F[J] = \sum_{n\geq 0} \frac{1}{n!} f_{y_1 \ldots y_m x_1 \ldots x_n} J^{x_1} \ldots J^{x_n}$, and this now reads as

$$\frac{\delta^{\#Y}}{\delta J^Y}\left(\int f_X J^X dX\right) = \int f_{Y+X} J^X dX.$$

We recall the Leibniz rule

$$\frac{\delta}{\delta J^X}(F_1 \ldots F_m) = \sum_{X_1 + \cdots + X_m = X} \frac{\delta F_1}{\delta J^{X_1}} \cdots \frac{\delta F_m}{\delta J^{X_m}},$$

and the exponential formula $\frac{\delta}{\delta J^X} e^W = e^W \sum_{\pi \in \text{Part}(X)} \prod_{A\in\pi} \frac{\delta W}{\delta J^A}$ from subsection 4.14. Evaluating at $J = 0$ recovers $G_X = \sum_{\pi \in \text{Part}(X)} \prod_{A\in\pi} K_A$, as in (4.11).

---

[1] We could extend the Einstein summation to Guichardet notation and write, for instance, $f_X J^X$, where the repeated index X now implies the integral-sum $\int f_X J^X dX$ over the power set. While attractive, we avoid this notation as it is perhaps being a shorthand too far at this stage!

6.1.1 Presence of Sources

Let J be a source field. Given a probability measure $\mathbf{P}^0$, we may introduce a modified probability measure $\mathbf{P}^J$, absolutely continuous with respect to $\mathbf{P}^0$, and having Radon–Nikodym derivative

$$\frac{d\mathbf{P}^J}{d\mathbf{P}^0}(\varphi) = \frac{1}{Z_0[J]} \exp\{\langle \varphi, J \rangle\},$$

where $Z_0[J] = \mathbf{E}_0\left[e^{\langle \phi, J \rangle}\right]$ is the moment generating function for $\mathbf{P}^0$. The corresponding state is referred to as the state modified by the presence of a source field J, and we denote its expectation by $\mathbf{E}_J$.

Evidently, we just recover the reference measure $\mathbf{P}^0$ when we put $J = 0$. The Laplace transform of the modified state will be $Z_J[J'] = \langle \exp\{\langle \phi, J' \rangle\}\rangle_J$ and it is readily seen that this reduces to

$$Z_J[J'] = \frac{Z_0[J + J']}{Z_0[J]}.$$

In particular, the cumulants are obtained through $W_J[\cdot] = W_0[J + \cdot] - W_0[J]$ and we find

$$K_X^J = \frac{\delta W_J[J']}{\delta J'^X}\bigg|_{J'=0} = \frac{\delta W_0[J' + J]}{\delta J'^X}\bigg|_{J'=0}.$$

It is, however, considerably simpler to treat J as a free parameter and just consider $\mathbf{P}$ as being the parametrized family $\{\mathbf{P}^J : J \in \mathfrak{J}\}$. In these terms, we have

$$K_X^J = \frac{\delta W_0[J]}{\delta J^X} = \int K_{X+Y} J^Y \, dY, \tag{6.9}$$

where K_X are the zero-source cumulants. We point out that it is likewise more convenient to write the moments in the presence of a field J as

$$G_X^J = \frac{1}{Z[J]} \mathbf{E}_0\left[\phi_X e^{\langle \phi, J \rangle}\right] = \frac{1}{Z[J]} \frac{\delta Z[J]}{\delta J^X}.$$

The **mean field in the presence of the source**, $\bar{\phi}[J]$, is defined to be $\bar{\phi}_x[J] = \langle \phi_x \rangle_J$ and is given by the expression[2]

$$\bar{\phi}_x[J] = \frac{\delta W_0[J]}{\delta J^x} = \int K_{x+Y} J^Y \, dY$$

$$= \sum_{n \geq 0} \frac{1}{n!} \mathbf{E}_0\left[\phi_x \phi_{x_1} \ldots \phi_{x_n}\right] J^{x_1} \ldots J^{x_n} \tag{6.10}$$

and, of course, reduces to $\mathbf{E}_0[\phi_x]$ when $J = 0$.

---

[2] Summation convention in place!

6.2 Functional Integrals

We shall in the following make the assumption that we may write the probability measure in the functional integral form

$$\mathbf{P}[d\varphi] = \frac{1}{\Xi} e^{S[\varphi]} \mathcal{D}\varphi$$

for some analytic functional S, called the action. The normalization is then $\Xi = \int_\Phi e^{S[\varphi]} \mathcal{D}\varphi$, and we have the moment generator

$$Z[J] = \frac{1}{\Xi} \int_\Phi e^{\langle\varphi,J\rangle + S[\varphi]} \mathcal{D}\varphi.$$

The definition is of course problematic, but we justify this by fixing a Gaussian reference measure as a functional integral.

6.2.1 Gaussian States

We construct an Gaussian state explicitly in the finite dimensional case where $\#\mathfrak{X} < \infty$ by the following argument.

Let g^{-1} be a linear, symmetric operator on Φ with well-defined inverse g. We shall write g^{xy} for the components of g^{-1} and g_{xy} for the components of g. That is, the equation $g^{-1}\varphi = J$, or $g^{xy}\varphi_y = J^x$, will have unique solution $\varphi = gJ$, or $\varphi_x = g_{xy}J^y$. We shall assume that g is positive definite and so can be used as a metric. A Gaussian measure is then given by

$$\mathbf{P}_g(d\varphi) = \frac{1}{\sqrt{(2\pi)^{\#\mathfrak{X}} \det g}} \exp\left\{-\frac{1}{2} g^{xy}\varphi_x\varphi_y\right\} \prod_{x\in\mathfrak{X}} d\varphi_x, \qquad (6.11)$$

which we may say is determined from the quadratic action given by

$$S_g[\varphi] = -\frac{1}{2} g^{xy}\varphi_x\varphi_y \equiv -\frac{1}{2}\langle\varphi|g^{-1}|\varphi\rangle. \qquad (6.12)$$

The moment generating function is then given by

$$Z_g[J] = \exp\left\{\frac{1}{2} g_{xy}J^xJ^y\right\} \equiv e^{\frac{1}{2}\langle J|g|J\rangle}. \qquad (6.13)$$

In the infinite dimensional case, we may use (6.13) as the definition of the measure:

$$\mathbf{P}_g(d\varphi) = e^{-\frac{1}{2}\langle\varphi|g^{-1}|\varphi\rangle} \mathcal{D}\varphi.$$

The measure is completely characterized by the fact that the only nonvanishing cumulant is $K_{\{x,y\}} = g_{xy}$. We describe these using the following diagram in Figure 6.1.

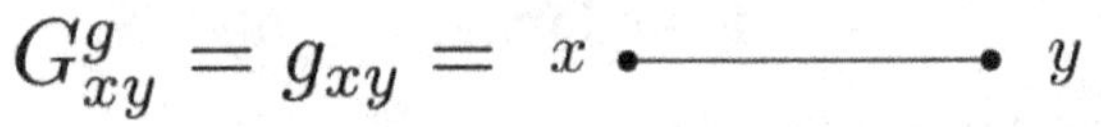

$$G^g_{xy} = g_{xy} = \quad x \; \bullet\!\!\!\!-\!\!\!-\!\!\!-\!\!\!-\!\!\!-\!\!\!-\!\!\!-\!\!\!\bullet \; y$$

Figure 6.1 The two-point function g_{xy} for a Gaussian distribution.

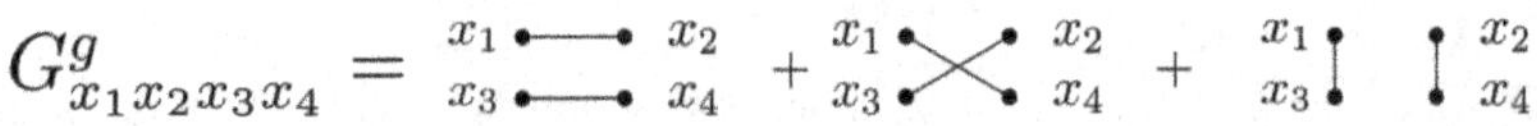

Figure 6.2 The fourth-order moments are $G^g_{x_1x_2x_3x_4} = g_{x_1x_2}g_{x_3x_4} + g_{x_1x_4}g_{x_2x_3} + g_{x_1x_3}g_{x_2x_4}$.

If we now use (4.11) to construct the Green's functions, we see that all odd moments vanish while

$$\mathbf{E}_g\left[\phi_{x(1)} \cdots \phi_{x(2k)}\right] = \sum_{\text{Pair}(2k)} g_{x(p_1)x(q_1)} \cdots g_{x(p_k)x(q_k)}, \qquad (6.14)$$

where the sum is over all pair partitions of $\{1,\ldots,2k\}$. The right-hand side will of course consist of $\frac{(2k)!}{2^k k!}$ terms. To this end, we introduce some notation. Let $\mathcal{P} \in \text{Pair}(X)$ be a given pair partition of a subset X, say $\mathcal{P} = \left\{x_{p(1)}, x_{q(1)}\right), \ldots, \left(x_{p(k)}, x_{q(k)}\right)\}$, then we write

$$g_{\mathcal{P}} = g_{x_{p(1)}x_{q(1)}} \cdots g_{x_{p(k)}x_{q(k)}}$$

in which case the Gaussian moments are

$$G^g_X \equiv \sum_{\mathcal{P}\in\text{Pair}(X)} g_{\mathcal{P}}. \qquad (6.15)$$

For instance, the $2k$th-order moments are then a sum of $\frac{(2k)!}{2^k k!}$ terms, and we describe in diagrams the fourth-order term in Figure 6.2.

6.2.2 General States

Suppose we are given a reference Gaussian probability measure $\mathbf{P}_g$. and suppose that $V[\cdot]$ is some analytic functional on Φ, say $V[\varphi] = \sum_{n\geq 0}\frac{1}{n!}v^{y_1\ldots y_n}$ $\varphi_{y_1}\cdots\varphi_{y_n}$, or more compactly,

$$V[\varphi] = \int v^X \varphi_X \, dX. \qquad (6.16)$$

A probability measure, $\mathbf{P}$, absolutely continuous with respect to $\mathbf{P}_g$, is then prescribed by taking its Radon–Nikodym to be

$$\frac{d\mathbf{P}}{d\mathbf{P}_g}(\varphi) = \frac{1}{\Xi}\exp\{V[\varphi]\},$$

provided, of course, that the normalization

$$\Xi \equiv \mathbf{E}_g\left[e^{V[\phi]}\right] < \infty. \tag{6.17}$$

Formally, we have

$$\mathbf{P}[d\varphi] = \frac{1}{\Xi}e^{S[\varphi]}\mathcal{D}\varphi,$$

where the action is

$$S[\varphi] = S_g[\varphi] + V[\varphi]$$
$$= -\frac{1}{2}\langle\varphi|g^{-1}|\varphi\rangle + \int v^X \varphi_X\,dX.$$

Likewise, $\Xi \equiv \int e^{S[\varphi]}\mathcal{D}\varphi$.

6.2.3 Feynman Diagrams

Lemma 6.2.1 *Let* $\mathbf{E}[\cdot]$ *be an expectation with the corresponding Green's function* $G_X = \mathbf{E}[\phi_X]$. *For analytic functionals* $A[\phi] = \int A^X \phi_X dX$, $B[\phi] = \int B^X \phi_X dX$, *and so on, we have the formula*

$$\mathbf{E}[A[\phi]B[\phi]\ldots Z[\phi]] = \int (A \diamond B \diamond \ldots \diamond Z)^X\,dX. \tag{6.18}$$

Proof The expectation in (6.18) reads as

$$\int A^{X_a}B^{X_b}\ldots Z^{X_z}\mathbf{E}\left[\phi_{X_a}\phi_{X_b}\ldots\phi_{X_z}\right]dX_a dX_b\ldots dX_z$$
$$= \int A^{X_a}B^{X_b}\ldots Z^{X_z}G_{X_a+X_b+\cdots+X_z}\,dX_a dX_b\ldots dX_z$$
$$\equiv \int\left(\sum_{X_a+X_b+\cdots+X_z=X}A^{X_a}B^{X_b}\ldots Z^{X_z}\right)G_X dX$$

where we use the Σ lemma, Lemma 4.1.3. □

Corollary 6.2.2 *Given an analytic functional* $V[\phi] = \int v^X \phi_X dX$, *where we assume that* $V[0] = 0$, *then*

$$\mathbf{E}\left[e^{V[\phi]}\right] = \int\left(\sum_{\pi\in\text{Part}(X)}v^\pi\right)G_X dX, \tag{6.19}$$

where $v^\pi \triangleq \prod_{A\in\pi}v^A$.

Proof   We have $v^\emptyset = 0$, and so we use (6.18) to get

$$\mathbf{E}\left[V[\phi]^n\right] = \int \left(v^{\diamond n}\right)^X G_X dX$$

$$\equiv n! \int \left(\sum_{\pi \in \mathrm{Part}_n(X)} v^\pi\right) G_X dX.$$

The relation (6.19) then follows by summing the exponential series.   □

The expression (6.19) applies to a general state. If we wish to specify to a Gaussian state $\mathbf{E}_g$ then we get the following expression.

Theorem 6.2.3 *Let* $\Xi = \mathbf{E}_g\left[e^{V[\phi]}\right]$, *as in (6.17), then*

$$\Xi \equiv \int \sum_{\pi \in \mathrm{Part}(X)} \sum_{\mathcal{P} \in \mathrm{Pair}(X)} v^\pi g_{\mathcal{P}} \, dX. \tag{6.20}$$

The proof is then just a simple substitution of the explicit form (6.15) for the Gaussian moments into (6.19). To understand this expression, let us look at a typical term appearing on the right-hand side. Let us fix a set X, say $X = \{x_1, \ldots, x_{10}\}$. There must be an even number of elements; otherwise, the contribution vanishes! We fix a partition $\pi = \{A, B, C\}$ of X, say $A = \{x_1, x_2, x_3, x_4\}$, $B = \{x_5, x_6, x_7\}$, and $C = \{x_8, x_9, x_{10}\}$, and a pair partition $\mathcal{P}$ consisting of the pairs $(x_1, x_2), (x_3, x_5), (x_4, x_6), (x_7, x_9), (x_8, x_{10})$. The contribution is the

$$v^{x_1 x_2 x_3 x_4} v^{x_5 x_6 x_7} v^{x_8 x_9 x_{10}} g_{x_1 x_2} g_{x_3 x_5} g_{x_4 x_6} g_{x_7 x_9} g_{x_8 x_{10}},$$

where we have an implied integration over repeated dummy indices. This can be described diagrammatically as follows (see Figure 6.3): for each of the

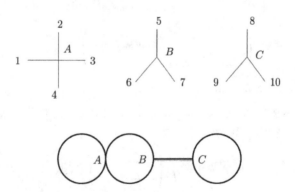

Figure 6.3 A Feynman diagram; see text.

elements of the partition π, we draw a vertex with the corresponding number of legs, in this case, a four-vertices for $v^A = v^{x_1 x_2 x_3 x_4}$ and a pair of three-vertices for $v^B = v^{x_5 x_6 x_7}$, $v^C = v^{x_8 x_9 x_{10}}$. We then connect the vertices pairwise according to the partition $\mathcal{P}$, with each connected pair (x_p, x_q) picking up a factor $g_{x_p x_q}$; the final diagram should have all lines contracted.

Note that the diagram at the end is presented in a simple form (the labels being redundant) and corresponds to a scalar. We sum over all diagrams that can be generated this way. However, there may be several choices that lead to topologically identical diagrams.

More generally, we have the following.

Theorem 6.2.4 *Let* $\mathbf{P}[d\varphi] = \frac{1}{\Xi} e^{S[\varphi]} \mathcal{D}\varphi$, *with*

$$S[\varphi] = -\frac{1}{2}\langle \varphi | g^{-1} | \varphi \rangle + \int v^X \varphi_X \, dX, \qquad (6.21)$$

then moments of the state are given by

$$G_X \equiv \frac{1}{\Xi} \int \sum_{\pi \in \mathrm{Part}(Y)} \sum_{\mathcal{P} \in \mathrm{Pair}(X+Y)} v^\pi g_{\mathcal{P}} \, dY. \qquad (6.22)$$

Here the rules are as follows: choose all possible subsets Y, all possible partitions π of Y, and all possible pair partitions $\mathcal{P}$ of $X + Y$; draw an m-vertex for each part of π of size m, label all the edges at each vertex by the corresponding elements of Y, and connect up all elements of $X + Y$ according the pair partition. We integrate over all Y's, and sum over all π's and $\mathcal{P}$'s.

The formula (6.22) is not yet a perturbation expansion in powers of the potential v since the normalization factor Ξ^{-1} is also a functional of v. To obtain such an expansion, we introduce the notion of a **linked** Feynman diagram. A connected Feynman diagram in the expansion of the integral in (6.22) is linked to the (external) set X, if it contains at least one element $x \in X$. Given two disjoint sets X and Y with $\#Y = n$, we choose a pair $(\pi, \mathcal{P}) \in \mathrm{Part}(Y) \times \mathrm{Pair}(X + Y)$. Then $\int v^\pi g_{\mathcal{P}} \, dy_1 \dots dy_n$ is a product of Feynman integrals, each integral being depicted by a connected Feynman diagram.[3] We say that $(\pi, \mathcal{P})$ is linked to X if all these integrals contain at least one external argument $x \in X$. In the pictorial description, that means all connected Feynman diagrams obtained from $\int v^\pi g_{\mathcal{P}} \, dy_1 \dots dy_n$ are linked to X. For an analytic formulation of this property, it is convenient to define the function

$$\Theta_X(\pi, \mathcal{P}) \triangleq \begin{cases} 1 \text{ if } (\pi, \mathcal{P}) \text{ is linked to } X, \\ 0 \text{ otherwise.} \end{cases} \qquad (6.23)$$

---

[3] The product may include factors such as $g_{x_1 x_2}$ without y-integration (0-dim. integration).

In general, a pair $(\pi, \mathcal{P}) \in \mathrm{Part}\,(Y) \times \mathrm{Pair}\,(X + Y)$ is not linked to X. But it is possible to split the set Y into two subsets $Y = Y_1 + Y_2$ such that the following conditions are satisfied:

(i) The function $v^\pi g_\mathcal{P}$ is factorized into $v^\pi g_\mathcal{P} = v^{\pi_1} g_{\mathcal{P}_1} v^{\pi_2} g_{\mathcal{P}_2}$ with $\pi_1 \in$ Part (Y_1), $\mathcal{P}_1 \in$ Pair (Y_1) and $\pi_2 \in$ Part (Y_2), $\mathcal{P}_2 \in$ Pair $(X + Y_2)$.
(ii) All factors of $v^{\pi_2} g_{\mathcal{P}_2}$ are linked to X.

That means $v^\pi g_\mathcal{P}$ can be separated into a factor without a link and into a factor, which is depicted by linked diagrams. This observation leads to the following

Lemma 6.2.5 *The integrand in (6.22) has the representation*

$$\sum_{\pi \in \mathrm{Part}(Y)} \sum_{\mathcal{P} \in \mathrm{Pair}(X+Y)} v^\pi g_\mathcal{P} = \sum_{Y_1 + Y_2 = Y} f(Y_1) g(Y_2, X). \tag{6.24}$$

with the functions

$$f(Y_1) = \sum_{\pi \in \mathrm{Part}(Y_1)} \sum_{\mathcal{P} \in \mathrm{Pair}(Y_1)} v^\pi g_\mathcal{P} \quad \text{and}$$

$$g(Y_2, X) = \sum_{\pi \in \mathrm{Part}(Y_2)} \sum_{\mathcal{P} \in \mathrm{Pair}(Y_2 + X)} \Theta_X \, (\pi, \mathcal{P}) \, v^\pi g_\mathcal{P}.$$

The integral in (6.22) is then evaluated with $\not\Sigma$ Lemma.

$$\int \sum_{\pi \in \mathrm{Part}(Y)} \sum_{\mathcal{P} \in \mathrm{Pair}(X+Y)} v^\pi g_\mathcal{P} \, dY = \int f(Y_1) dY_1 \int g(Y_2, X) dY_2$$

$$\overset{(6.20)}{=} \Xi \int \sum_{\pi \in \mathrm{Part}(Y_2)} \sum_{\mathcal{P} \in \mathrm{Pair}(Y_2 + X)} \Theta_X \, (\pi, \mathcal{P}) \, v^\pi g_\mathcal{P} \, dY_2 \, .$$

Inserting this result into (6.22), the normalization Ξ drops out, and as final result we obtain the **linked cluster expansion** for the moments

$$G_X = \int \sum_{\pi \in \mathrm{Part}(Y)} \sum_{\mathcal{P} \in \mathrm{Pair}(Y+X)} \Theta_X \, (\pi, \mathcal{P}) \, v^\pi g_\mathcal{P} \, dY, \tag{6.25}$$

which contains only terms linked to X.

6.2.4 The Dyson–Schwinger Equation

Theorem 6.2.6 (Wick Expansion) *We have the moment generating function* $Z[J]$ *for the measure* $\mathbf{P}[d\varphi] = \frac{1}{\Xi}e^{S[\varphi]}\,\mathcal{D}\varphi$ *given by*

$$Z[J] = \frac{1}{\Xi}\mathbf{E}_g\left[e^{\langle\varphi,J\rangle+V[\varphi]}\right] = \frac{1}{\Xi}\exp\left\{V\left[\frac{\delta}{\delta J}\right]\right\}Z_g[J]. \qquad (6.26)$$

Here we understand

$$\exp\left\{V\left[\frac{\delta}{\delta J}\right]\right\} \equiv \exp\left\{\int dX\, v^X \frac{\delta}{\delta J^X}\right\}$$

and the proof rests on the identity for suitable analytic functionals F:

$$F\left[\frac{\delta}{\delta J}\right]Z_g[J] = F\left[\frac{\delta}{\delta J}\right]\mathbf{E}_g\left[e^{\langle\varphi,J\rangle}\right] = \mathbf{E}_g\left[F[\varphi]\,e^{\langle\varphi,J\rangle}\right].$$

We now derive a functional differential equation for the generating function.

Lemma 6.2.7 *The Gaussian generating functional* Z_g *satisfies the differential equations*

$$\left\{F_g^x\left[\frac{\delta}{\delta J}\right] + J^x\right\}Z_g[J] = 0, \qquad (6.27)$$

where $F_g^x[\varphi] = \frac{\delta}{\delta\varphi_x}S_g[\varphi] \equiv -g^{xy}\varphi_x$.

Proof Explicitly, we have $Z_g = \exp\left\{\frac{1}{2}J^x g_{xy}J^y\right\}$ so that $\frac{\delta}{\delta J^x}Z_g = g_{xy}J^y Z_g$, which can be rearranged as

$$\left(-g^{xy}\frac{\delta}{\delta J^y} + J^x\right)Z_g = 0.$$

$\square$

Lemma 6.2.8 *For the moment generating function* $Z[J]$ *for the measure* $\mathbf{P}[d\varphi] = \frac{1}{\Xi}e^{S[\varphi]}\,\mathcal{D}\varphi$, *we have*

$$\left\{F_I^x\left[\frac{\delta}{\delta J}\right] - g^{xy}\frac{\delta}{\delta J^y} + J^x\right\}Z[J] = 0, \qquad (6.28)$$

where $F_I^x[\varphi] = \frac{\delta}{\delta\varphi_x}V[\varphi]$.

Proof We observe that from the Wick expansion identity (6.30) we have

$$J^x Z[J] = \frac{1}{\Xi}J^x \exp\left\{V\left[\frac{\delta}{\delta J}\right]\right\}Z_g[J]$$

and using the commutation identity

$$\left[J^x, \exp\left\{V\left[\frac{\delta}{\delta J}\right]\right\}\right] = -F_I^x\left[\frac{\delta}{\delta J}\right]\exp\left\{V\left[\frac{\delta}{\delta J}\right]\right\}$$

we find

$$
\begin{aligned}
J^x Z[J] &= \frac{1}{\Xi}\exp\left\{V\left[\frac{\delta}{\delta J}\right]\right\} J^x Z_g[J] \\
&\quad - \frac{1}{\Xi}F_I^x\left[\frac{\delta}{\delta J}\right]\exp\left\{V\left[\frac{\delta}{\delta J}\right]\right\} Z_g[J] \\
&= g^{xy}\frac{\delta}{\delta J^y}Z[J] - F_I^x\left[\frac{\delta}{\delta J}\right]Z[J],
\end{aligned}
$$

which gives the result. □

Putting these two lemmas together, we obtain the following result.

Theorem 6.2.9 (The Dyson–Schwinger Equation) *The generating functional* $Z[J]$ *for a probability measure* $\mathbf{P}[d\varphi] = \frac{1}{\Xi}e^{S[\varphi]}D\varphi$ *satisfies the differential equation*

$$\left\{F^x\left[\frac{\delta}{\delta J}\right] + J^x\right\}Z[J] = 0, \tag{6.29}$$

where $F^x[\varphi] = \frac{\delta S[\varphi]}{\delta \varphi_x} = -\frac{1}{2}g^{xy}\varphi_y + F_I^x[\varphi]$.

Corollary 6.2.10 *Under the conditions of the preceding theorem, if the perturbation is* $V[\varphi] = v^X\varphi_X dX$, *then the ordinary moments of* $\mathbf{P}$ *satisfy the algebraic equations*

$$G_{X+x} = \sum_{x'\in X}g_{xx'}G_{X-x'} + g_{xy}\int dY\, v^{y+Y}G_{X+Y}. \tag{6.30}$$

Proof We have $F^x = -g^{xy}\varphi_y + \int v^{x+X}\varphi_X dX$. The Dyson–Schwinger equation then becomes

$$-g^{xy}\frac{\delta Z}{\delta J^y} + \int dY\, v^{x+Y}\frac{\delta Z}{\delta J^Y} + J^x Z = 0$$

and we consider applying the further differentiation $\frac{\delta^X}{\delta J^X}$ to obtain

$$-g^{xy}G_{y+X} + dY\, v^{x+Y}G_{X+Y} + \frac{\delta}{\delta J^X}\left(J^x Z\right) = 0.$$

The result follows from setting $J = 0$. □

The Dyson–Schwinger equation (6.30) can be expressed in diagrammatic form as in Figure 6.4.

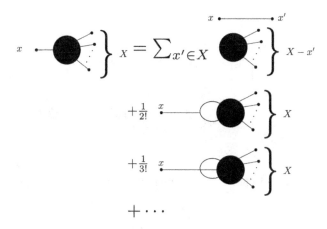

Figure 6.4 The Dyson–Schwinger equation (6.30) in diagrammatic form.

We may write (6.30) in index notation as

$$
\mathbf{E}\left[\phi_x\phi_{x_1}\ldots\phi_{x_m}\right] = \sum_{i=1}^{m} g_{xx_i}\mathbf{E}\left[\phi_{x_1}\ldots\widehat{\phi_{x_i}}\ldots\phi_{x_m}\right]
$$

$$
+g_{xy}\sum_{n\geq 0}\frac{1}{n!}v^{yy_1\ldots y_n}\mathbf{E}\left[\phi_{y_1}\ldots\phi_{y_n}\phi_{x_1}\ldots\phi_{x_m}\right],
$$

where the hat indicates an omission. This hierarchy of equations for the Green's functions is equivalent to the Dyson–Schwinger equation.

We remark that the first term on the right-hand side of (6.30) contains the moments $\mathbf{E}\left[\phi_{X-x'}\right]$, which are of order two smaller than the left-hand side $\mathbf{E}\left[\phi_{X+x}\right]$. The second term on the right-hand side of (6.30) contains the moments of higher order, and so we generally cannot use this equation recursively. In the Gaussian case, we have $\mathbf{E}_g\left[\phi_{X+x}\right] = \sum_{x'\in X} g_{xx'}\mathbf{E}_g\left[\phi_{X-x'}\right]$ $\mathbf{E}_g\left[\phi_{X-x'}\right]$, from which we can deduce (6.14) by just knowing the first and second moments, $\mathbf{E}_g\left[\phi_x\right] = 0$ and $\mathbf{E}_g\left[\phi_x\phi_y\right] = g_{xy}$.

The Dyson–Schwinger Equation for Cumulants

The Dyson–Schwinger equations may be alternatively stated for W. They are

$$
J^x - g^{xy}\frac{\delta W}{\delta J^y} + \int dX\, v^{x+X} \sum_{\pi\in\mathrm{Part}(X)}\prod_{A\in\pi}\frac{\delta W}{\delta J^A} = 0. \tag{6.31}
$$

The Dyson–Schwinger equation for cumulants may be rearranged as $\frac{\delta W}{\delta J^x} = g_{xy}J^y + g_{xy}\int dX\, v^{y+X}\sum_{\pi\in\mathrm{Part}(X)}\prod_{A\in\pi}\frac{\delta W}{\delta J^A}$, and the diagrammatic form is

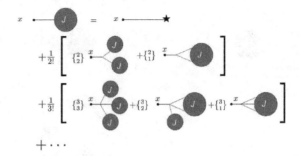

Figure 6.5 The Dyson–Schwinger equation based on $\frac{\delta W}{\delta J^x}$.

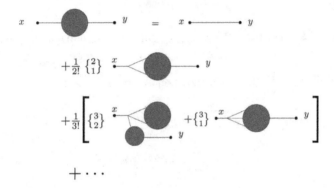

Figure 6.6 The Dyson–Schwinger equation for the second cumulant.

presented in Figure 6.5. We group together topologically identical diagrams and include the combinatorial coefficients.

By taking a further derivative $\frac{\delta}{\delta J^y}$ of the equation (6.31) and evaluating at $J = 0$, we obtain the diagrammatic expression for the second cumulant in Figure 6.6.

A useful way to see this is to note that we may define an operator D by

$$D_x = \frac{1}{Z[J]} \frac{\delta}{\delta J^x} (Z[J]\,\cdot) \equiv \frac{\delta}{\delta J^x} + \frac{\delta W}{\delta J^x}$$

and do the Dyson–Schwinger equation (6.29), $\left\{ F^x \left[\frac{\delta}{\delta J} \right] + J^x \right\} Z[J] = 0$, which becomes $\{ F^x[D] + J^x \} 1 = 0$, that is,

$$\left\{ F^x \left[\frac{\delta}{\delta J} + \frac{\delta W[J]}{\delta J} \right] + J^x \right\} 1 = 0. \tag{6.32}$$

6.3 Tree Expansions

We have introduced moment generating functions formally structured as

$$Z[J] = \frac{1}{\Xi} \int_{\Phi} e^{\langle \varphi, J \rangle + S[\varphi]} \, \mathcal{D}\varphi.$$

Following a well-worn principle that we do not try to justify here, one might reasonably expect that the greatest contribution to the integral comes from the field ψ for which the exponent $\langle \varphi, J \rangle + S[\varphi]$ is stationary, that is, we have the approximation

$$Z[J] \simeq e^{\langle \psi, J \rangle + S[\psi]},$$

where $\frac{\delta}{\delta \varphi_x} \left(\langle \varphi, J \rangle + S[\varphi] \right) \Big|_{\varphi = \psi} = 0$. We make the following assumptions: the stationary solution $\psi = \psi[J]$ exists and is unique for each fixed J. It will satisfy automatically the identity

$$J^x + \frac{\delta S[\varphi]}{\delta \varphi_x} \Big|_{\varphi = \psi[J]} = 0,$$

that is,

$$J^x - g^{xy} \psi_y + \int v^{x+X} \psi_X \, dX = 0$$

or rearranging gives[4]

$$\psi_x = J_x + \int v_x^X \psi_X \, dX. \tag{6.33}$$

We may rewrite (6.33) as $\psi = gJ + f(\psi)$, where $f(\psi)_x$ is the rather involved expression $\int v_x^X \psi_X \, dX$. However, we may in principle iterate to get

$$\psi[J] = gJ + f(gJ + f(J + f(gJ + \cdots))),$$

or in more detail

$$\psi_x = J_x + \int dX_0 \, v_x^{X_0} \prod_{x_1 \in X_0} \left(J_{x_1} + \int dX_1 \, v_{x_1}^{X_1} \prod_{x_2 \in X_1} \left(J_{x_2} + \cdots \right) \right)$$

$$= J_x + \int dX_0 \, v_x^{X_0} J_{X_0} + \int dX_0 dX_1 \, v_x^{X_0} \prod_{x_1 \in X_0} v_{x_1}^{X_1} J_{X_1} + \cdots.$$

---

[4] Here we lower contravariant indices using the metric g, i.e., $J_x = g_{xy} J^y \left(= \int g_{xy} J(y) \, dy \right)$ and $v_x^X = g_{xy} v^{y+X}$.

The preceding expression admits a remarkable diagrammatic interpretation. We use the following diagrams for the tensors ψ_x and J_x:

$$\psi_x = \quad \bullet\!\!-\!\!\bigcirc$$

$$J_x = \quad \bullet\!\!-\!\!\bigstar$$

For simplicity, we assume that v^X is nonzero only if $\#X = 3, 4$, that is, $\phi^3 + \phi^4$ theory. Then the equation (6.33) reads as

$$\psi_x = J_x + \frac{1}{2!}v_x^{y_1y_2}\psi_{y_1}\psi_{y_2} + \frac{1}{3!}v_x^{y_1y_2y_3}\psi_{y_1}\psi_{y_2}\psi_{y_3} \tag{6.34}$$

and this may be depicted by the diagram shown in Figure 6.7.

The process of iteration then leads to the following expansion in terms of the diagrams shown in Figure 6.8.

The general term here is a phylogenetic tree. In the expansion in Figure 6.8, we have lumped topologically equivalent trees together, hence the combinatorial factors appear. However, the sum here is over all phylogenetic trees – or equivalently all hierarchies! – which is why it is sometimes known as a tree expansion. In $\phi^3 + \phi^4$ theory, each node of the tree will have either two or three branches.

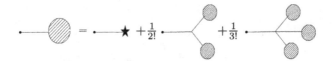

Figure 6.7 Use of diagrams to depict the equation (6.34).

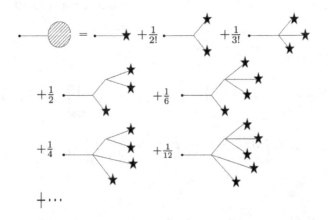

Figure 6.8 A tree expansion of the field ψ_x in $\phi^3 + \phi^4$ theory.

We may write the expansion in terms of hierarchies

$$\psi_x = \int dX \sum_{H \in \text{Hier}(X)} v_x(H) J^X, \qquad (6.35)$$

where $v_x(H)$ is the weight attached to a given hierarchy H. The weight is easily calculated by drawing out the tree diagram: for each node we have the tree section

$$B$$
$$A_1 A_2 \quad A_m$$

which gets a factor $v_{x_B}^{x_{A_1} \cdots x_{A_m}}$, where various labels $x_B, x_{A_1}, \ldots, x_{A_m}$ are dummy variables in $\mathfrak{X}$. Apart from the root ($x_B = x$), and the leaves, all these variables are contracted over in the product of such factors (the contractions corresponding to the branches between nodes!).

6.4 One-Particle Irreducibility

In this section, we derive algebraic rules for decomposing the cumulant (connected) Green's functions into further basic diagrams. These are known as the one-particle irreducible (1PI) diagrams, as they have the property that cutting a line corresponding to a propagator g does not split the diagram into disconnected parts; see, for instance, Rivers (1987). This decomposition has been very important in relation to later renormalization methods.

6.4.1 Legendre Transforms

Suppose that $W = W[J]$ is a convex analytic function. That is, W is a real-valued analytic functional with the property that

$$W[tJ_1 + (1-t)J_2] \leq tW[J_1] + (1-t)W[J_2] \qquad (6.36)$$

for all $0 < t < 1$ and $J_1, J_2 \in \mathfrak{J}$. The Legendre transform (more exactly, the Legendre-Fenchel transform) of W is then defined by

$$\Gamma[\varphi] = \inf_{J \in \mathfrak{J}} \{W[J] - \langle \varphi, J \rangle\}. \qquad (6.37)$$

$\Gamma[\varphi]$ will then be a concave (i.e., Γ is convex) analytic functional in φ and we may invert the formula as follows:

$$W[J] = \sup_{\varphi \in \Phi} \{\Gamma[\varphi] + \langle \varphi, J \rangle\}. \tag{6.38}$$

If the functional W is taken to be strictly convex, that is, if we have strict inequality in (6.36), then the infimum is attained at a unique source $\bar{J} = \bar{J}[\varphi]$ for each fixed φ, and so

$$\Gamma[\varphi] \equiv W[\bar{J}[\varphi]] - \langle \varphi, \bar{J}[\varphi] \rangle.$$

Moreover, we may invert $\bar{J}: \Phi \mapsto \mathfrak{J}$ to get a mapping $\bar{\phi}: \mathfrak{J} \mapsto \Phi$, and for fixed J the supremum is given by $\bar{\phi}[J]$, thus giving $W[J] = \bar{\Gamma}[J] + \langle \bar{\phi}[J], J \rangle$, where we take

$$\bar{\Gamma}[J] \triangleq \Gamma[\bar{\phi}[J]]. \tag{6.39}$$

The extremal conditions are then

$$\bar{J}^x \equiv -\frac{\delta \Gamma}{\delta \varphi_x}, \qquad \bar{\phi}_x \equiv \frac{\delta W}{\delta J^x}. \tag{6.40}$$

Let $W''[J]$ be the symmetric tensor with entries

$$W''_{xy}[J] = \frac{\delta^2 W[J]}{\delta J^x \delta J^y} \equiv \frac{\delta \bar{\phi}_x[J]}{\delta J^y}.$$

This will be positive definite – it will be interpreted in the following as the covariance of the field in the presence of the source J. Likewise, if we let $\Gamma''[\varphi]$ be the linear operator with entries $\frac{\delta^2 \Gamma}{\delta \varphi_x \delta \varphi_y}$, then we have

$$\Gamma''^{xy}[\varphi] = \frac{\delta^2 \Gamma[\varphi]}{\delta \varphi_x \delta \varphi_y} = -\frac{\delta \bar{J}^y[\varphi]}{\delta \varphi_x},$$

and so we conclude that $W''[J]$ and $-\bar{\Gamma}''[J] = \Gamma''[\bar{\phi}[J]]$ will be inverses for each other. In other words,

$$\frac{\delta^2 W[J]}{\delta J^x \delta J^y} \frac{\delta^2 \Gamma}{\delta \varphi_y \delta \varphi_z}\bigg|_{\varphi = \bar{\phi}[J]} = -\delta^z_x. \tag{6.41}$$

Lemma 6.4.1 *Let $F: \Phi \mapsto \mathbb{R}$ be a functional and define $\bar{F}: \mathfrak{J} \mapsto \mathbb{R}$ by $\bar{F}[J] \triangleq F[\bar{\phi}(J)]$, then*

$$\frac{\delta \bar{F}[J]}{\delta J^x} = \frac{\delta^2 W[J]}{\delta J^x \delta J^y} \frac{\delta F}{\delta \varphi}\bigg|_{\bar{\phi}[J]}.$$

Proof This is just the chain rule, as $\frac{\delta \bar{\phi}^y[J]}{\delta J^x} = \frac{\delta^2 W[J]}{\delta J^x \delta J^y}$. □

Let us introduce the tensor coefficients

$$\Theta^X [J] \triangleq \frac{\delta\Gamma}{\delta\varphi_X}\bigg|_{\bar\phi[J]}.$$ (6.42)

Note that $\Theta^{x,y}[J]$ are the components of $\bar\Gamma''[J]$.

Lemma 6.4.2 *We have the following recurrence relation*

$$\frac{\delta W}{\delta J^{X+y}} = \sum_{\pi\in\mathrm{Part}_<(X)} \frac{\delta^2 W}{\delta J^y\delta J^p}\,\Theta^{p+Z_\pi}\left(\prod_{A\in\pi}\frac{\delta W}{\delta J^{A+z_A}}\right),$$ (6.43)

where, again, each z_A is a dummy variable associated with each component part A and $Z_\pi = \{z_A : A\in\pi\}$.

Proof Taking $\frac{\delta}{\delta J^x}$ of (6.41) and using the multiderivative form of the Leibniz rule, we find

$$
\begin{aligned}
0 &= \frac{\delta}{\delta J^x}\left(\frac{\delta^2 W}{\delta J^x\delta J^y}\,\Theta^{\{y,z\}}\right)\\
&= \sum_{X_1+X_2=X}\frac{\delta W}{\delta J^{X_1+x+y}}\,\frac{\delta}{\delta J^{X_2}}\,\Theta^{\{y,z\}}\\
&= \sum_{X_1+X_2=X}\frac{\delta W}{\delta J^{X_1+x+y}}\sum_{\pi\in\mathrm{Part}(X_2)}\Theta^{y+z+Z_\pi}\left(\prod_{A\in\pi}\frac{\delta W}{\delta J^{A+z_A}}\right).
\end{aligned}
$$

Now the $X_1 = X, X_2 = \emptyset$ term in the last sum yields $\frac{\delta W}{\delta J^{X+x+y}}\Theta^{\{y,z\}}$, which will be the highest-order derivative appearing in the expression. We take this over to the left-hand side, and then multiply both sides by $-\frac{\delta^2 W}{\delta J^x\delta J^z}$, which is the inverse of $\Theta^{\{y,z\}}$. Finally, we get the expression

$$
\begin{aligned}
\frac{\delta W}{\delta J^{X+x+y}} = \sum_{X_1+X_2=X}\sum_{\pi\in\mathrm{Part}(X_2)}&\frac{\delta^2 W}{\delta J^y\delta J^p}\frac{\delta W}{\delta J^{X_1+x+q}}\\
&\times\left(\prod_{A\in\pi}\frac{\delta W}{\delta J^{A+z_A}}\right)\Theta^{\{p,q\}+Z_\pi}.
\end{aligned}
$$

We note that the sum is over all $X_1 + x$ (where $X_1 \subset X$, but not $X_1 = X$), and partitions of X_2 can be reconsidered as a sum over all partitions of $X + x$, excepting the coarsest one. Let us do this and set $X' = X + x$ and $Z'_\pi = Z_\pi + q$; dropping the primes then yields (6.43). □

6.4.2 Field–Source Relations

We now suppose that W in the preceding is the cumulant generating functional, that it is convex, and that the Legendre transform functional Γ admits an analytic expansion of the type

$$\Gamma[\varphi] = \sum_{n \geq 0} \frac{1}{n!} \Gamma^{y_1 \ldots y_n} \varphi_{y_1} \ldots \varphi_{y_n} = \int \Gamma^X \varphi_X \, dX \qquad (6.44)$$

for constant coefficients Γ^X – these are given a diagrammatic representation in Figure 6.9.

In this case, $\bar{\phi}_x = \frac{\delta W}{\delta J^x}$ will give the mean field (6.10) in the presence of source J. Inversely, the source field $\bar{J}$ will be then given by

$$\bar{J}^x[\varphi] = -\frac{\delta \Gamma}{\delta \varphi_x} = -\int \Gamma^{x+X} \varphi_X \, dX. \qquad (6.45)$$

In longhand, $\bar{J}^x[\varphi] = -\sum_{n \geq 0} \frac{1}{n!} \Gamma^{xy_1 \ldots y_n} \varphi_{y_1} \ldots \varphi_{y_n}$.

Gaussian States

For the Gaussian state, we have $W_g[J] = \frac{1}{2} g_{xy} J^x J^y$, and so here $\bar{\phi}_x = g_{xy} J^y$. The inverse map is then $\bar{J}^x[\varphi] = g^{xy} \varphi_y$, and so we obtain

$$\Gamma_g[\varphi] = -\frac{1}{2} g^{xy} \varphi_x \varphi_y.$$

Note that $\Gamma_g \equiv S_g$.

General States

For a more general state, we will have $\Gamma = \Gamma_g + \Gamma_I$, where

$$\Gamma_I[\varphi] = \Gamma^x \varphi_x + \frac{1}{2} \pi^{xy} \varphi_x \varphi_y + \sum_{n \geq 3} \frac{1}{n!} \Gamma^{x_1 \ldots x_n} \varphi_{x_1} \ldots \varphi_{x_n},$$

$$\Gamma^{x_1 \cdots x_n} = \quad$$

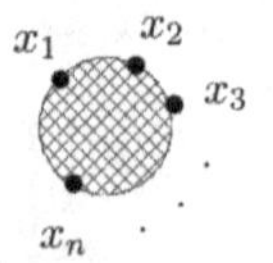

Figure 6.9 The coefficients Γ^X appearing in (6.44) are completely symmetric contravariant tensors, and we use this diagram to represent them – their contravariant nature is expressed by the fact that they have connection points rather than legs.

that is, $\Gamma^{xy} = -g^{xy} + \pi^{xy}$. It follows that

$$\bar{J}^x[\varphi] = -\Gamma^x + \left(g^{xy} - \pi^{xy}\right)\varphi_y - \frac{1}{2}\Gamma^{xyz}\varphi_y\varphi_z - \cdots \qquad (6.46)$$

and (substituting $\varphi = \bar{\phi}$) we may rearrange this to get

$$\bar{\phi}_x = g_{xy}\left(J^y + \Gamma^y + \pi^{yz}\bar{\phi}_z + \frac{1}{2}\Gamma^{yzw}\bar{\phi}_z\bar{\phi}_w + \cdots\right). \qquad (6.47)$$

Without loss of generality, we may take $\Gamma^x = 0$ as we could always absorb this term otherwise into J^x. With this choice, we have $\Gamma[\varphi = 0] = 0$ and $\bar{\phi}[J = 0] = 0$.

We note that $\bar{\phi}$ appears on the right-hand side of (6.47) in a generally non-linear manner: let us rewrite this as $\bar{\phi} = gJ + f\left(\bar{\phi}\right)$, where $f_x(\varphi) = g_{xy}\pi^{yz}\varphi_z + \sum_{n \geq 3} g_{xy}\Gamma^{yz_1\dots z_n}\varphi_{z_1}\dots\varphi_{z_n}$. We may reiterate (6.47) to get an expansion

$$\bar{\phi} = gJ + f\left(gJ + f\left(gJ + f\left(gJ + \cdots\right)\right)\right),$$

and we know that this expansion should be resummed to give the series expansion in terms of J as in (6.10).

Self-Energy

Now $\Gamma''^{xy}[\varphi] = -g^{xy} + \Gamma_I''^{xy}[\varphi]$, or $\Gamma''[\varphi] = -\left(1 - \Gamma_I''[\varphi]g\right)g^{-1}$, so we conclude that

$$W''[J] = -\Gamma''\left[\bar{\phi}[J]\right]^{-1} = g\frac{1}{1 - \bar{\Gamma}_I''[J]g} = \frac{1}{g^{-1} - \bar{\Gamma}_I''[J]}, \qquad (6.48)$$

where $\left(\bar{\Gamma}_I''[J]\right)^{xy} \equiv \pi^{xy} + \sum_{n \geq 1}\frac{1}{n!}\Gamma^{xyz_1\dots z_n}\bar{\phi}_{z_1}[J]\dots\bar{\phi}_{z_n}[J]$. This relation may be alternatively written as the series

$$W''[J] = g + g\bar{\Gamma}_I''[J]g + g\bar{\Gamma}_I''[J]g\bar{\Gamma}_I''[J]g + \cdots \qquad (6.49)$$

In particular, we note that $W''[J = 0] = K$ is the covariance matrix while $\bar{\Gamma}_I''[J = 0] \equiv \pi$, and we obtain the series expansion

$$K = g + g\pi g + g\pi g\pi g + \cdots \qquad (6.50)$$

We now wish to determine a formula relating the cumulants in the presence of the source J to the tensor coefficients $\Theta^X[J]$ defined in (6.42).

Theorem 6.4.3 *We have the following recurrence relation*

$$K_{X+y}^J = \sum_{\pi \in \text{Part}_<(X)} K_{yp}^J \Theta^{p+Z_\pi}[J]\left(\prod_{A \in \pi} K_{A+z_A}^J\right), \qquad (6.51)$$

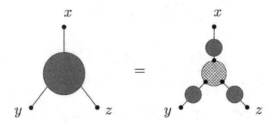

Figure 6.10 Third-order terms.

where, again, each z_A is a dummy variable associated with each component part A and $Z_\pi = \{z_A : A \in \pi\}$.

This, of course, just follows straight from (6.43). The crucial thing about (6.51) is that the right-hand side contains lower-order cumulants and so can be used recursively. Let us iterate once:

$$K^J_{X+y} = K^J_{yp} \sum_{\pi \in \text{Part}_<(X)} \Theta^{p+Z_\pi}[J] \prod_{A \in \pi} K^J_{A+z_A} =$$

$$K^J_{yp} \sum_{\pi \in \text{Part}_<(X)} \Theta^{p+Z_\pi}_J[J] \prod_{A \in \pi} K^J_{qz_A} \sum_{\pi' \in \text{Part}_<(A)} \Theta^{q+Z_{\pi'}}[J] \prod_{B \in \pi'} K^J_{B+z_B}.$$

What happens is that each part A of the first partition gets properly divided up into subparts, and we continue until we eventually break down X into its singleton parts. However, this is just a top-down description of a hierarchy on X. Proceeding in this way, we should obtain a sum over all hierarchies of X. At the root of the tree, we have the factor K^J_{yp}, and for each node/part A, appearing anywhere in the sequence, labeled by dummy index z_A, say, we will break it into a proper partition $\pi' \in \text{Part}_<(A)$ and with multiplicative factor $K^J_{qz_A} \sum_{\pi' \in \text{Part}_<(A)} \Theta^{q+Z_{\pi'}}[J]$, where $Z_{\pi'}$ will be the set of labels for each part $B \in \pi'$.

If we set $X = \{x_1, x_2\}$, then we have only one hierarchy to sum over and we find $K^J_{yx_1x_2} = K^J_{yp}K^J_{x_1z_1}K^J_{x_2z_2}\Theta^{qz_1z_2}[J]$; see Figure 6.10 for a graphical representation.

The four-vertex term K^J_{xyzw} is likewise developed in Figure 6.11.

It is useful to use the bottom-up description to give the general result. First we introduce some new coefficients defined by

$$\Upsilon^Y_{x_1 \ldots x_n}[J] \triangleq K^J_{x_1z_1} \ldots K^J_{x_nz_n} \Theta^{\{z_1, \ldots, z_n\}+Y}[J].$$

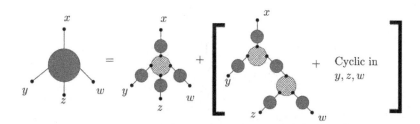

Figure 6.11 Fourth-order terms.

with the exceptional case $\Upsilon_{xy} \triangleq K_{xy}^J$. Then we find that

$$
K_{X+y}^J = \sum_{\mathcal{H}=\{\pi^{(1)},\ldots,\pi^{(m)}\}\in\mathfrak{H}(X)} \Upsilon_{y+Z_{\pi^{(m)}}}[J] \prod_{A^{(m)}\in\pi^{(m)}} \Upsilon_{Z_{\pi^{(m-1)}}}^{z_{A^{(m)}}}[J]
$$

$$
\times \prod_{A^{(m-1)}\in\pi^{(m-1)}} \Upsilon_{Z_{\pi^{(m-2)}}}^{z_{A^{(m-1)}}}[J] \ldots \prod_{A^{(1)}\in\mathcal{A}^{(1)}} \Upsilon_{Z_{\pi^{(2)}}}[J]. \tag{6.52}
$$

For instance, we have the following expansions for the lowest cumulants:

$$
K_{yx_1x_2}^J = \Upsilon_{yx_1x_2},
$$

$$
K_{yx_1x_2x_3}^J = \Upsilon_{yx_1x_2x_3} + \left(\Upsilon_{yx_1}{}^r\Upsilon_{rx_2x_3} + \cdots\right),
$$

$$
K_{yx_1x_2x_3x_4}^J = \Upsilon_{yx_1x_2x_3x_4}
$$

$$
+ \left(\Upsilon_{yx_1}{}^r\Upsilon_{rx_2x_3x_4} + \cdots\right) + \left(\Upsilon_{yx_1x_2}{}^r\Upsilon_{rx_3x_4} + \cdots\right)
$$

$$
+ \left(\Upsilon_{yx_1}{}^r\Upsilon_{rx_2}{}^q\Upsilon_{qx_3x_4} + \cdots\right) + \left(\Upsilon_y{}^{rq}\Upsilon_{rx_1x_2}\Upsilon_{qx_3x_4} + \cdots\right)
$$

The terms in round brackets involve permutations of the x_j indices leading to distinct terms. Thus, there are $1 + \frac{1}{2}\binom{4}{2} = 4$ terms making up the right-hand side for the fourth-order cumulant: the first term in round brackets corresponds to the hierarchy $\{\{\{x_1\}\},\{\{x_2\},\{x_3\}\}\}$ and there are three such second-order hierarchies. There are $1 + \binom{5}{3} + \frac{1}{2}\binom{5}{1}\binom{4}{2} = 26$ terms making up the right-hand side for the fifth-order cumulant.

7

Entropy, Large Deviations, and Legendre Transforms

In this chapter, we go back and consider the mathematical definition of entropy as a measure of statistical uncertainty as introduced by Shannon. We relate this to large deviation theory – a generalization of Laplace asymptotics for sequences of probability distributions – and see how the cumulant moments and rate functions (entropic distance) of limit probability distributions are related by a Legendre–Fenchel transform. Indeed, there is an analogy between entropic functions and the one-particle irreducible generating functions. We describe briefly the Freidlin–Wentzell theory as an example of large deviations for stochastic processes.

7.1 Entropy and Information

7.1.1 Shannon Entropy

We begin by trying to quantify the amount of surprise associated by the occurrence of a particular probabilist outcome or, more generally, event. Intuitively, the surprise of an event should depend only on the probability of its occurrence, so we seek a function

$$s: [0, 1] \mapsto [0, \infty),$$
$$: p \mapsto s(p),$$

with $s(p)$ measuring our surprise at seeing an event of probability p occur. If an event is certain to occur, then there is no surprise if it happens so we require that $s(1) = 0$. This should be the least surprising event! It is convenient to define the unlikeliness of the event to be $\frac{1}{p}$, that is, inversely proportional to the probability p. Another desirable feature is that the greater the unlikeliness is, then the more we are surprised, and this translates to s being monotone decreasing: $s(p) > s(q)$ whenever $p < q$. We would also desire that events that have similar levels of unlikeliness lead to levels of surprise that are also close: mathematical s must be a continuous function.

This still leaves the exact form of s wide open, but we impose one more condition coming from probability theory. If two events are independent, then the surprise at seeing both occur is no more and no less than the sum of the surprises of both. That is,

$$s(pq) = s(p) + s(q).$$

This now restricts us to the logarithms:

$$s(p) \triangleq -\log p. \tag{7.1}$$

The only freedom we have left is to choose the base of the logarithm, and we will fix base 2. The sequel log will always be understood as $\log_2$.

The surprise is therefore the logarithm of the unlikeliness, or equivalently $p = 2^{-s}$.

Now let us consider a probability vector $\pi = (p_1, \ldots, p_n)$. The vector π belongs to the simplex $\Sigma_n = \{(p_1, \ldots, p_n) : p_k \geq 0, \sum_k p_k = 1\}$. We may think of a discrete random variable X, which is allowed to take on at most n values, say $x_1, \ldots, x_n$, with probabilities $p_k = \Pr\{X = x_k\}$: in this way π gives its probability distribution. If we measure X, then our surprise is finding the value x_k should then be $-\log p_k$. The average surprise is then

$$h(X) = -\sum_k p_k \ln p_k. \tag{7.2}$$

We adopt the standard convention that $-x \ln x$ takes the limit value

$$-\lim_{x \to 0^+} x \ln x = 0$$

for $x = 0$ by continuity. In other words, while events that have extremely high levels of unlikeliness lead to a high level of surprise, they make negligible contribution to $h(X)$ due to their low probability to occur. This is a consequence of the fact that the surprise only grows logarithmically.

It turns out that $h(X)$ plays an important role in information theory. This emerged from the work of Shannon on coding theory and, recognizing the same mathematical form as the entropy of an ensemble in statistical mechanics – Boltzmann's H-function (1.3) – Shannon used the term entropy for $h(X)$.

Strictly speaking, the entropy h is a function of the probability distribution of X, and not X itself. In other words, $h(X)$ depends only on the probabilities for X to take on its various variables – it doesn't matter what these values are, so long as they are distinct, let alone what probability space we choose to represent X on.

Relative Entropy

Let π and π' be probability vectors on Σ_n, then we define the relative entropy of π' to π as

$$D\left(\pi'||\pi\right) \triangleq \sum_{k=1}^{n} p'_k \log \left(\frac{p'_k}{p_k}\right).$$

It is easy to see that $D\left(\pi'||\pi\right) \geq 0$ with equality if and only if $\pi' = \pi$. (This is Gibbs' inequality, and we will prove it more generally later.) We note that the relative entropy can be written as

$$D\left(\pi'||\pi\right) = \sum_{k=1}^{n} p'_k \left(-\log p_k + \log p'_k\right) = \mathbb{E}'\left[S - S'\right],$$

where S is the surprise of π (i.e, the random vector with values $s_k = -\log p_k$) and S' is the corresponding surprise of π'. This can be further broken down as

$$D\left(\pi'||\pi\right) = \mathbb{E}'\left[S\right] - H\left(\pi'\right).$$

7.1.2 Differential Entropy

We generalize the Shannon entropy to continuous random variables X possessing a probability distribution function ρ_X. For definiteness, we fix on $\mathbb{R}^n$-valued random vectors X and define the **differential entropy** to be

$$H\left(X\right) \triangleq -\int \rho_X \ln \rho_X.$$

Again the notation associates the entropy with the continuous random variable X as a shorthand for the actual dependence, which is on the distribution ρ_X. However, there are a number of subtle differences between the discrete case.

First there is the fact the probability density ρ_X sets a scale that is in turn inherited by the entropy. The integral is $-\int_{\mathbb{R}^n} \rho_X\left(x\right) \ln \rho_X\left(x\right) dx$, and this assumes that we have already replaced any physical units required to describe the variables with dimensionless ones. If X was originally a distance parameter measured in meters, then ρ would have units m^{-n}, and so the logarithm would not make sense.

Second, unlike the discrete case where the probabilities p_k must lie in the range 0 to 1, the values $\rho_X\left(x\right)$ need only be positive. We therefore typically encounter the function $y = -x \ln x$ for values $x > 1$ as well, leading to possible negative terms appearing in the integrand; see Figure 7.1. Technically, the integral may be restricted to the support of probability density if desired.

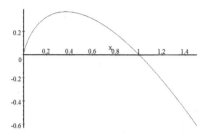

Figure 7.1 The graph of $y = -x \ln x$.

We note that the entropy may be considered as the expectation

$$H(X) \triangleq -\mathbb{E}[\ln \rho_X].$$

Indeed, $\ln \rho_X$ may be called the surprise function of the random variable.

One issue that presents itself straight away is the question of physical units for the differential entropy. The discrete entropy is naturally dimensionless as it involves probabilities; however, H deals with probability densities. Even after making X dimensionless, we could apply an affine linear rescaling $X \to AX + c$ with A invertible. Here we find

$$H(AX + c) = H(X) + \ln |A|.$$

We see that the differential entropy is translation invariant, but is only determined up to an additive constant dependent on the choice of units to render the random variable dimensionless.

Entropy of Gaussian Variables

Let X be Gaussian random variable ($n = 1$) with mean μ and standard deviation σ. Then $\rho_X(x) = \frac{1}{\sqrt{2\pi}\sigma} e^{-\frac{1}{2\sigma^2}(x-\mu)^2}$

$$H_{\text{Gaussian}}(X) = -\mathbb{E}\left[-\frac{1}{2\sigma^2}(X - \mu)^2 - \ln \sqrt{2\pi}\sigma \right]$$
$$= \frac{1}{2} + \frac{1}{2}\ln(2\pi) + \ln \sigma,$$

or $\sigma = \frac{1}{\sqrt{2\pi e}} e^{H_{\text{Gaussian}}(X)}$. Perhaps not surprisingly, the entropy increases as the uncertainty σ does. The result generalizes to the case of an n-dimensional Gaussian with covariance matrix C:

$$H_{\text{Gaussian}}(X) = \frac{n}{2} + \frac{n}{2}\ln(2\pi) + \frac{1}{2}\ln |C|.$$

Rearranging gives

$$|C|^{1/n} = \frac{1}{2\pi e} e^{\frac{2}{n} H_{\text{Gaussian}}(X)}.$$

Relative Entropy

Let ρ_1 and ρ_2 be probability density functions on $\mathbb{R}^n$, then the **relative entropy** of ρ_1 with respect to ρ_2 is defined to be

$$D(\rho_1 \| \rho_2) \triangleq \int \rho_1 \ln \frac{\rho_1}{\rho_2}.$$

This is also known as the **Kullback–Leiber divergence,** the **entropy of discrimination,** and the **information distance.** More exactly, let P_1 and P_2 be probability measures on $\mathbb{R}^n$, then we define their relative entropy as $\int dP_1 \ln \frac{dP_1}{dP_2}$ if P_1 is absolutely continuous with respect to P_2 (with $\frac{dP_1}{dP_2}$ being the Radon–Nikodym derivative), but taken as $-\infty$ otherwise. We have the following property.

Proposition 7.1.1 (Gibbs' Inequality) *The relative entropy satisfies*

$$D(\rho_1 \| \rho_2) \geq 0.$$

Furthermore, $D(\rho_1 \| \rho_2) = 0$ if and only if $\rho_1 = \rho_2$.

Proof We may write $D(\rho_1 \| \rho_2) = \int_A \rho_1 \phi(y)$, where $y = \frac{\rho_2}{\rho_1}$, $\phi(y) = -\ln y$ and A is the support of ρ_1. As ϕ is convex, we may apply Jensen's inequality to get $\int \rho_1 \phi(y) \geq \phi(\int \rho_1 y)$ but $\int \rho_1 y = 1$ and $\phi(1) = 0$. As ϕ is strictly convex, we get the equality only if $y \equiv 1$ almost everywhere. □

As a measure of the distance from ρ_2 to ρ_1, $D(\rho_1 \| \rho_2)$ has the desirable property of being positive and vanishing only if ρ_1 and ρ_2 are the same. It is not a metric, however, since it is not symmetric!

Let ρ_k be both n-dimensional Gaussian variables with mean vectors μ_k and covariance matrices C_k respectively for $k = 1, 2$. Then

$$D(\rho_1 \| \rho_2) = \frac{1}{2} (\mu_1 - \mu_2)^\top C_2^{-1} (\mu_1 - \mu_2) + \frac{1}{2} \text{tr} \left\{ C_1 C_2^{-1} \right\}$$
$$- \frac{1}{2} \ln |C_1 C_2^{-1}| - \frac{1}{2} n.$$

7.2 Law of Large Numbers and Large Deviations

Let X be a random variable, say taking values in $\mathbb{R}^d$, and suppose that it has a well-defined cumulant generating function

$$W_X(t) = \ln \int e^{t \cdot x} \mathbb{K}_X[dx]$$

for $t \in \mathbb{R}^d$, where $\mathbb{K}_X$ is its probability distribution. (Note that $t.x$ is shorthand here for $\sum_{j=1}^d t_j x_j$.) The law of large numbers is the principle that if we generate independent copies of X, say random variable $X_1, X_2, \ldots$, then their geometric means $\bar{X}_n = \frac{1}{n} \sum_{k=1}^n X_k$ ought to have a distribution that is increasingly concentrated about the mean value $\mu = \int x \mathbb{K}_X[dx]$ as n becomes large.

Indeed, we see that

$$W_{\bar{X}_n}(t) = \ln \mathbb{E}\left[e^{\frac{1}{n}t.(X_1 + \cdots + X_n)}\right] = n \ln \mathbb{E}\left[e^{\frac{1}{n}t.X}\right] = n \ln W_X\left(\frac{t}{n}\right).$$

and assuming that the cumulant function is differentiable about $t = 0$, we get $\lim_{n \to \infty} W_{\bar{X}_n}(t) = W_X'(0) = \mu$.

Alternatively, we may say that the probability distribution $\mathbb{K}_n$ of $\bar{X}_n$ converges to δ_μ. On the one hand, we have the conclusion that

$$\lim_{n \to \infty} \Pr\left\{\bar{X}_n \in A\right\} = 1$$

for any (say open) set $A \subset \mathbb{R}^d$ containing the mean μ. On the other, we would get $\Pr\{\bar{X}_n \notin A\}$ should vanish if A does not contain the mean μ. Clearly the recording of a value for the geometric mean $\bar{X}_n$ away from μ, known as a large deviation, is an increasingly rare event for large n, and the next question is how to quantify how small these probabilities tend to get. We first look at a concrete example.

As an example, we take X to be a Bernoulli variable: taking value 1 with probability p and 0 with probability $1 - p$. This has the cumulant generating function

$$W_X(t) = \ln\left(1 - p + e^t p\right)$$

and, of course, the mean value $\mu = p$. For integer m between 0 and n, we have

$$\Pr\left\{\bar{X}_n = \frac{m}{n}\right\} = \binom{m}{n} p^m (1 - p)^{n-m}.$$

We now assume that both n and m are large with $\frac{m}{n} = q \in (0, 1)$. Using Stirling's approximation $n! \approx \sqrt{2\pi n} e^{-n} n^n$, we find that

$$\Pr\left\{\bar{X}_n = \frac{m}{n}\right\} \approx e^{-nI(q) + O(\ln n)},$$

where

$$I(q) = q \ln \frac{q}{p} + (1 - q) \ln \frac{1 - q}{1 - p}.$$

We see that the probability of getting a large deviation (that is, a mean value q different from p from a sample of size n) is exponentially small. Another way of saying this is that our surprise at recording a sample average q different from p grows roughly proportional to the sample size n – specifically the surprise is to leading orders $nI(q)$. In particular,

$$\lim_{n \to \infty} -\frac{1}{n} \ln \Pr\left\{\bar{X}_n = q\right\} = I(q).$$

The result can actually be significantly strengthened. Not only is the probability for $\bar{X}_n$ to equal q decay exponentially with rate $I(q)$, but so too does the probability for $\bar{X}_n$ to be equal to q or further away! The function $I(q)$ is called the rate function and is sketched later in this chapter. The main features are that it is convex with a minimum of zero at $q = p$.

There are a number of things going on in this example, but let's pull out some key points. The probability of a collection of large deviation events is a sum over small probabilities that are decaying exponentially fast with the sample size n. Mathematically, the event with the least rate of decay is the one that should dominate – it may be exponentially small but the others are vanishing faster! Put another way, the decay of a collection of large deviation events with sample size t is determined by the least surprising of these events. The rate function takes a specific form in this example and in fact it is just the relative entropy $D\left(\pi_q \| \pi_p\right)$ of the Bernoulli probability vector $\pi_q = (q, 1 - q)$ to $\pi_p = (p, 1 - p)$. Evidently, the rate function ought to be determined by the common distribution $\mathbb{K}_X$ of the random variables that we're sampling and indeed it does depend on the parameter p that fixes the Bernoulli distribution in this case. However, we remark that the rate function and the cumulant generating function are Legendre transforms of each other:

$$I(q) = \min_t \left\{tq - W_X(t)\right\}.$$

In this case, the minimizer $t_*(q)$ is then seen to be $\ln\left(\frac{q}{p}\frac{1-p}{1-q}\right)$ and one readily sees that $I(q) = t_*(q)\, q - W_X(t_*(q))$. Conversely, the rate function determines the distribution as we can invert to get $W_X(t) = \min_q \left\{tq - I(q)\right\}$.

Going beyond the Bernoulli case, it turns out that the straightforward generalization holds for the distribution of the sample mean for random variables on $\mathbb{R}$, and this is known as Cramér's Theorem. Effectively we have the law of large numbers, which is telling us that the distribution $\mathbb{K}_n$ of the average, $\bar{X}_n$, of a sample of size n converges to the degenerate distribution δ_μ. The large deviation principle tells us a bit more, roughly that $\mathbb{K}_n[dx]$ behaves asymptotically as $e^{-nI(x)}$ "dx," where "dx" is a proxy for some background measure that we don't care about. Here $I(x)$ is called the rate function, and it should be strictly positive except at $x = \mu$, where it vanishes. This is indeed very rough, but gives the basic intuition needed. In order to state these precisely, it is convenient to make some formal definitions.

7.2.1 Large Deviation Property

The following is a mathematical definition.

Let $(\mathbb{K}_n)_n$ be a sequence of probability measures on a Polish space $\mathfrak{X}$, and $I: \mathfrak{X} \mapsto [0, \infty]$, a function with compact level sets, that is, $\{x \in \mathfrak{X} : I(x) \le c\}$ is compact for each $c \ge 0$. We say that $(\mathbb{K}_n)_n$ satisfies a **large deviation property** with rate function I if for every open set A

$$\limsup_{n \to \infty} -\frac{1}{n} \ln \mathbb{K}_n [A] \le \inf_{x \in A} I(x),$$

and for every closed subset C, we have

$$\liminf_{n \to \infty} -\frac{1}{n} \ln \mathbb{K}_n [A] \ge \inf_{x \in A} I(x).$$

This means that if A is any Borel subset, then

$$\inf_{x \in \bar{A}} I(x) \le \liminf_{n \to \infty} -\frac{1}{n} \ln \mathbb{K}_n [A] \le \limsup_{n \to \infty} -\frac{1}{n} \ln \mathbb{K}_n [A] \le \inf_{x \in A^o} I(x),$$

where A^o is the interior of A and $\bar{A}$ is its closure. In particular, if the large deviation property holds, then for any set with $\inf_{x \in \bar{A}} I(x) = \inf_{x \in A^o} I(x)$ we have that the following limit exists:

$$\lim_{n \to \infty} -\frac{1}{n} \ln \mathbb{K}_n [A] \equiv \inf_{x \in A} I(x).$$

We say that a pair of sequences $(a_n)_n$ and $(b_n)_n$ are **asymptotically logarithmically equivalent**, written $a_n \asymp b_n$, if

$$\lim_{n \to \infty} \frac{1}{n} \ln a_n = \lim_{n \to \infty} \frac{1}{n} \ln b_n.$$

If a sequence of measures $(\mathbb{K}_n)_n$ satisfies a large deviation property with rate function I, then we denote this by

$$\mathbb{K}_n [dx] \asymp e^{-nI(x)} \text{``}dx\text{''}.$$

Let us stress that the large deviation property is a mathematical definition. The measures $(\mathbb{K}_n)_n$ are not necessarily supposed to arise as distributions of sample mean variables S_n, though they usually relate to some form of distributions for rare events. The following result is useful at this level of generality.

Theorem 7.2.1 (Gärtner–Ellis) *Let* $(\mathbb{K}_n)_n$ *be a sequence of probability measures on* $\mathfrak{X} \equiv \mathbb{R}^d$. *Suppose that limit*

$$W(t) = \lim_{n \to \infty} \frac{1}{n} \ln \int e^{nt.x} \mathbb{K}_n [dx]$$

exists for each $t \in \mathbb{R}^d$, *and defines a differentiable function, then* $(\mathbb{K}_n)_n$ *satisfies a large deviation property with rate function* I *given by*

$$I(x) = \sup_t \{t.x - W(t)\}.$$

The result may be justified on the following grounds: if $(\mathbb{K}_n)_n$ did satisfy a large deviation property, then

$$\int e^{nt.x} \mathbb{K}_n [dx] \asymp \int e^{n[t.x - I(x)]} \text{``}dx\text{''} \asymp e^{n \sup_x \{t.x - I(x)\}},$$

where in the last part one argues that it is the largest rate that dominates; this would imply that $W(t) \equiv \sup_x \{t.x - I(x)\}$, which may be inverted to get $I(x) = \sup_t \{t.x - W(t)\}$ provided $W(\cdot)$ is differentiable.

A related result is Varadhan's Theorem.

Theorem 7.2.2 (Varadhan) *Let* $(\mathbb{K}_n)_n$ *be a sequence of probability measures on a Polish space* $\mathfrak{X}$, *and* $I: \mathfrak{X} \mapsto [0, \infty]$ *a rate function with compact level sets. Then the large deviation property is equivalent to the condition that*

$$\lim_{n \to \infty} \frac{1}{n} \ln \int e^{n\Phi(x)} \mathbb{K}_n [dx] = \sup_x \{\Phi(x) - I(x)\}$$

for every continuous bounded real function Φ.

The intuition justifying this is similar to before:

$$\int e^{n\Phi(x)} \mathbb{K}_n [dx] \asymp \int e^{n\Phi(x)} e^{-nI(x)} \text{``}dx\text{''} \asymp e^{n \sup_x \{\Phi(x) - I(x)\}},$$

where again we use the rule of thumb that the largest exponential rate dominates. This motivates the following guess that

$$\lim_{n \to \infty} \frac{1}{n} \ln \int e^{n\Phi(x)} \mathbb{K}_n [dx] = \sup_x \{\Phi(x) - I(x)\}.$$

Note that if we take $\mathfrak{X} \equiv \mathbb{R}^d$ in Varadhan's Theorem, then we cannot automatically take $\Phi(x)$ to be $t.x$ as this is not bounded. The following two results are very useful in practical applications.

Theorem 7.2.3 (The Contraction Principle) *Let* $\phi: \mathfrak{X} \mapsto \mathfrak{X}'$ *be continuous and consider the pullback probabilities*

$$\mathbb{K}'_n = \mathbb{K}_n \circ \phi^{-1}.$$

Then if $(\mathbb{K}_n)_n$ *satisfies a large deviation principle with rate function I, it follows that* $(\mathbb{K}'_n)_n$ *satisfies a large deviation principle with rate function I' given by*

$$I'\left(x'\right) = \inf\left\{I\left(x\right) : \phi\left(x\right) = x'\right\}.$$

Theorem 7.2.4 (Deformation Principle) *Let* Φ *be a bounded continuous function, then define probabilities*

$$\tilde{\mathbb{K}}_n\left[dx\right] = \frac{1}{\Xi_n}e^{n\Phi(x)}\,\mathbb{K}_n\left[dx\right],$$

where the normalization is $\Xi_n = \int e^{n\Phi(x)}\mathbb{K}_n\left[dx\right]$. *If* $(\mathbb{K}_n)_n$ *satisfies a large deviation property with rate function I, then* $\left(\tilde{\mathbb{K}}_n\right)_n$ *satisfies a large deviation property with rate function*

$$\tilde{I}\left(x\right) = I\left(x\right) - \Phi\left(x\right) - \inf\left\{I - \Phi\right\}.$$

7.2.2 Large Deviation Results

As we have mentioned, the definition of the large deviation property applies to more general situations than sampling a random variable. However, we now return to this situation and state the classical result, which is now a corollary to the Gärtner–Ellis Theorem.

Theorem 7.2.5 (Cramér) *Let X be an* $\mathbb{R}^d$ *with distribution* $\mathbb{K}_X$, *with cumulant generating function* $W_X\left(t\right) = \ln\int e^{tx}\mathbb{K}_X\left[dx\right]$ *well defined for all* $t \in \mathbb{R}^d$, *and let* $\mathbb{K}_n$ *be the distribution for the geometric mean of n independent samples of X. Then* $(\mathbb{K}_n)_n$ *satisfies a large deviation property with the rate function*

$$I\left(x\right) = \sup_t\left\{t.x - W_X\left(t\right)\right\}.$$

Here we arrive W_X as the form of the scaled cumulant function limit W since

$$\frac{1}{n}\ln\int e^{nt.x}\mathbb{K}_n\left[dx\right] = \frac{1}{n}\ln\mathbb{E}\left[e^{t.(X_1+\cdots+X_n)}\right] = \ln\mathbb{E}\left[e^{t.X}\right] = W_X\left(t\right).$$

We note that W_X is automatically analytic, so the differentiability condition comes for free.

Empirical Measures

Next let $\mathfrak{X}$ be a Polish space and $\mathcal{M}\left(\mathfrak{X}\right)$ be the space of Radon probability measures on $\mathfrak{X}$, equipped with the topology of weak convergence. (In fact, $\mathcal{M}\left(\mathfrak{X}\right)$ can be turned into a Polish space with the Wasserstein metric.) Given an i.i.d. sequence $X_1, X_2, \ldots$, we may define the empirical measures

$$M_n\left(\omega; dx\right) = \frac{1}{n} \sum_{k=1}^{n} \delta_{X_k(\omega)}[dx].$$

So we may think of $M_n\left(\cdot, dx\right)$ as a random measure, that is, a random variable taking values in $\mathcal{M}\left(\mathfrak{X}\right)$.

A sequence of probability measures on $\mathcal{M}(\mathfrak{X})$ is then given by

$$\mathbb{P}_n\left[A\right] = \Pr\left\{M_n\left(\cdot, dx\right) \in A\right\}.$$

Intuitively one expects that the empirical measures should converge in some sense as n tends to infinity to the common distribution $\mathbb{K}_X$ for the terms in the i.i.d. sequence. This can be elegantly formulated as a large deviation property.

Theorem 7.2.6 (Sanov) *The family $\left(\mathbb{P}_n\right)_n$ of distributions for the empirical measures obtained by sampling a random variable X satisfy a large deviation property with the rate function*

$$I\left(K'\right) = D\left(K'||K_X\right).$$

To complete the picture, we should remark that relative entropy has a natural variational expression:

$$D\left(K'||K_X\right) = \sup_f\left\{\int_{\mathfrak{X}} f dK' - W_X\left(f\right)\right\}$$

where

$$W_X\left(f\right) = \ln\int_{\mathfrak{X}} e^{f(x)}\, K_X[dx].$$

The supremum is taken over all continuous real Borel-measurable functions on $\mathfrak{X}$ (and we can even restrict to bounded functions!).

Actually, Sanov's Theorem is a higher-level version of Cramér's Theorem, and we may deduce the latter from the form using the contraction principle with the map $\phi\colon \mathcal{M}\left(\mathfrak{X}\right) \mapsto \mathbb{R}^d$:

$$\phi\left(\mathbb{K}\right) \triangleq \int_{\mathfrak{X}} x\, \mathbb{K}[dx].$$

Proof of Sanov's Theorem When $\mathfrak{X}$ Is Finite

It is instructive to look at Sanov's Theorem in the case where $\mathfrak{X}$ is finite, say $\sharp\mathfrak{X} = d$. In this case, we may introduce random variables $N_n\left(\omega, x\right)$ for each $x \in \mathfrak{X}$ that count the number of times x occurs in the sequence $\left(X_1\left(\omega\right), \ldots, X_n\left(\omega\right)\right)$. Indeed, $N_n\left(\cdot, x\right)$ has a $\text{Bin}(n, p\left(x\right))$ distribution where $p\left(x\right) = \Pr\left\{X = x\right\}$. The empirical measure simplifies to

$$M_n (\cdot, dx) = \sum_{x \in \mathfrak{X}} p_n (\cdot, x) \, \delta_x [dx],$$

where $p_n (\omega, x) = \frac{N_n(\omega, x)}{n}$.

We may now view M_n as an $\mathbb{R}^d$ valued random variable – constrained to be a probability vector. Specifically, we may think of M_n as the random vector $(p_n (\cdot, x))_{x \in \mathfrak{X}}$. In this context, we can apply Gärtner–Ellis, and here the scaled cumulant function $W (t)$ is

$$\frac{1}{n} \ln \mathbb{E} \left[e^{n \sum_x t(x) p_n(\cdot, x)} \right] = \frac{1}{n} \ln \mathbb{E} \left[e^{\sum_x t(x) N_n(\cdot, x)} \right] \equiv \left(\sum_x e^{t(x)} p (x) \right)$$

where we use the fact that the occupation numbers $\{N_n (\cdot, x)\}_{x \in \mathfrak{X}}$ have a multinomial distribution $Bin (n, \pi)$ where $\pi = \{p (x)\}_{x \in \mathfrak{X}}$. The Gärtner–Ellis Theorem then states that the distributions of the M_n satisfy a large deviation property with the rate function

$$I \left(\pi' \right) = \sup_t \left\{ \sum_x t (x) p' (x) - \sum_x e^{t(x)} p (x) \right\}$$

where $\pi' \equiv \{p' (x)\}_{x \in \mathfrak{X}}$. The supremum is attained, and we have the conditions (from differentiating wrt. $t (x)$ for fixed x)

$$0 = p' (x) - \frac{e^{t(x)} p (x)}{e^{W(t)}}$$

or

$$t (x) = \ln \frac{p' (x)}{p (x)} + W (t).$$

Substituting back in gives

$$I \left(\pi' \right) = \sum_x p' (x) \ln \frac{p' (x)}{p (x)} \equiv D \left(\pi' || \pi \right).$$

7.3 Large Deviations and Stochastic Processes

7.3.1 Path Integrals and the Wiener Process

Recall that the Wiener process $\{W(t) : t \geq 0\}$ is a family of real-valued random variables parameterized by time $t \geq 0$, such that

$$\mathbb{E}[e^{\int_0^\infty f(t) dW(t)}] = e^{\frac{1}{2} \int_0^\infty f(t)^2 dt} \tag{7.3}$$

for any real-valued square-integrable function f. The process is a Markov process, and indeed, the probability density associated with it taking values $q_1, \ldots, q_n$ at times $t_1 < \cdots < t_n$ is

$$\rho_n(q_1, t_1, \ldots, q_n, t_n) = \left\{ \prod_{k=1}^{n} T(q_k, t_k | q_{k-1}, t_{k-1}) \right\} \rho(q_1, t_1), \qquad (7.4)$$

where the initial probability density is $\rho(q, t) = \frac{1}{\sqrt{2\pi t}} e^{-\frac{q^2}{2t}}$ and (for $t > t'$) the transition mechanism is

$$T(q, t | q', t') = \frac{1}{\sqrt{2\pi (t - t')}} e^{-\frac{(q-q')^2}{2(t-t')}}. \qquad (7.5)$$

From the probability densities ρ_n, we can use Kolmogorov's Reconstruction Theorem to construct a probability space $(\Omega_W, \mathcal{F}_W, \mathbb{P}_W)$ for the process. In particular, Wiener was able to give a more explicit construction where we take sample space to be the set of all continuous paths parameterized by time $t \geq 0$ (trajectories) starting at the origin,

$$\Omega_W = \mathcal{C}_0[0, \infty), \qquad (7.6)$$

such that if $\mathbf{q} = \{q(t) : t \geq 0\}$ is an outcome, then $W(t)$ takes on the value $q(t)$. The sigma-algebra, $\mathcal{F}_W$, in this case is the one generated by the cylinder sets: that is, the sets of all trajectories that are required at a finite number of certain times to be in certain fixed interval regions. We refer to this as the canonical Wiener process, and from now on take it as the default probability space $(\Omega_W, \mathcal{F}_W, \mathbb{P}_W)$ for the Wiener process. We have

$$\mathbb{E}[e^{\int_0^\infty f(t)dW(t)}] = \int_{\mathcal{C}_0[0,\infty)} e^{\int_0^\infty f(t)dq(t)} \mathbb{P}_W[d\mathbf{q}]. \qquad (7.7)$$

However, it is sometime convenient to think of a path integration, or functional integral, formulation of Wiener averages. Let $t > 0$ and choose $0 \leq t_1 < \cdots < t_n \leq t$ for fixed n. Then we have

$$\mathbb{E}[F_t[W(t_1), \ldots, W(t_n)]] = \int F_t[q_1, \ldots, q_n] \mathcal{N}_n e^{-\frac{1}{2} \sum_{k=1}^{n} \frac{(q_k - q_{k-1})^2}{t_k - t_{k-1}}} dq_1 \ldots dq_n,$$

$$(7.8)$$

with $q_0 = 0$, and $\mathcal{N}_n = \prod_k = 1^n (2\pi (t_k - t_{k-1}))^{-1}$. Taking a continuum limit $n \to \infty$ with $\max(t_k - t_{k-1}) \to 0$, we may formally consider the limit functional $F_t[\mathbf{W}]$ of the Wiener process over the time interval $[0, t]$ with expectation

$$\mathbb{E}[F_t[\mathbf{W}]] = \int_{\mathcal{C}_0[0,t]} F_t[\mathbf{q}] e^{-\frac{1}{2} \int_0^t \dot{q}(\tau)^2 d\tau} \mathscr{D}\mathbf{q}, \qquad (7.9)$$

where $\mathscr{D}\mathbf{q}$ is the formal limit of $\mathcal{N}_n \, dq_1 \ldots dq_n$, and $\int_0^t \dot{q}(\tau)^2 d\tau$ the limit of $\sum_{k=1}^n \frac{(q_k - q_{k-1})^2}{t_k - t_{k-1}}$. This would be then normalized (setting $F_t \equiv 1$) by

$$\int_{\mathcal{C}_0[0,t)} e^{-\frac{1}{2}\int_0^t \dot{q}(\tau)^2 d\tau} \, \mathscr{D}\mathbf{q} = 1. \tag{7.10}$$

While this is appealing, sadly neither of the limits makes sense separately: a well-known theorem of André Weil shows that there is no translationally invariant measure on the trajectories, so $\mathscr{D}\mathbf{q}$ is meaningless; a theory of Wiener shows that while the noncontinuous trajectories for a set of $\mathbf{P}_W$-measure zero, so too do the differentiable trajectories, and so $\int_0^t \dot{q}(\tau)^2 d\tau$ is a well-defined integral for almost no trajectory.

Nevertheless, the combination

$$\mathbb{P}_t[d\mathbf{q}] \equiv e^{-\frac{1}{2}\int_0^t \dot{q}(\tau)^2 d\tau} \, \mathscr{D}\mathbf{q} \tag{7.11}$$

can be used formally to great effect. For instance, we have

$$\mathbb{E}[e^{\int_0^\infty f(t) dW(t)}] = \int_{\mathcal{C}_0[0,t]} e^{\int_0^t f(t) dq(t)} \, e^{-\frac{1}{2}\int_0^t \dot{q}(\tau)^2 d\tau} \, \mathscr{D}\mathbf{q}$$

$$= e^{\frac{1}{2}\int_0^t f(\tau)^2 d\tau} \int_{\mathcal{C}_0[0,\infty)} e^{-\frac{1}{2}\int_0^t [\dot{q}(\tau) - f(\tau)]^2 d\tau} \, \mathscr{D}\mathbf{q}$$

$$= e^{\frac{1}{2}\int_0^t f(\tau)^2 d\tau},$$

where we translate $q(t) \mapsto q(t) + \int_0^t f(\tau) d\tau$.

Now let us state a simple result about the rescaled Wiener process.

Theorem 7.3.1 (Schilder) *Let $\{W(t)\}_{0 \le t \le T}$ be the canonical Wiener process on $\mathcal{C}_0[0, T]$, and let $\mathbb{K}_n$ be the law for $\frac{1}{\sqrt{n}} W(\cdot)$. Then the family $\{\mathbb{K}_n\}_n$ satisfies a large deviation property with the rate function*

$$I(\mathbf{q}) = \frac{1}{2} \int_0^T |\dot{q}(t)|^2 \, dt, \tag{7.12}$$

with $I(\mathbf{q}) = \infty$ if the path is not absolutely continuous.

A proof of this may be found in Dembo and Zeitouni (1998); however, we can see the germ of the idea by observing that

$$\mathbb{K}_n[d\mathbf{q}] \equiv e^{-\frac{1}{2}n \int_0^T \dot{q}(\tau)^2 d\tau} \, \mathscr{D}\mathbf{q}. \tag{7.13}$$

Indeed, by Gärtner–Ellis, we have

$$W(f) = \lim_{n \to \infty} \frac{1}{n} \ln \int_{C_0[0,T]} e^{\int_0^T f(t)\dot{q}(t)dt} \mathbb{K}_n[d\mathbf{q}]$$

$$= \lim_{n \to \infty} \frac{1}{n} \ln \int_{C_0[0,T]} e^{n \int_0^T f(t)\dot{q}(t)dt} \mathbb{P}_W[d\mathbf{q}]$$

$$= \frac{1}{2} \int_0^T |f(t)|^2 dt,$$

with the Legendre–Fenchel transform

$$I(\mathbf{q}) = \inf_f \left\{ \int_0^T f(t)\dot{q}(t)\, dt - W(f) \right\}$$

$$= \frac{1}{2} \int_0^T |\dot{q}(t)|^2\, dt - \frac{1}{2} \inf_f \int_0^T |f(t) - \dot{q}(t)|^2\, dt$$

$$= \frac{1}{2} \int_0^T |\dot{q}(t)|^2\, dt.$$

7.3.2 Large Deviation for Diffusions

A diffusion process in $\mathbb{R}^n$ driven by m independent Wiener processes $\{W_\alpha(t)\}$ is described in component form by

$$dX^i(t) = v^i(X(t))\, dt + \sum_{\alpha=1}^m b^i_\alpha(X(t))dW_\alpha(t), \qquad (7.14)$$

$(i = 1, \ldots, n)$, with initial conditions $X(0) = x_0 \in \mathbb{R}^n$. Here the differentials are understood in the Itō sense, that is, $dX(t)$ means $X(t+dt) - X(t)$ for positive increments dt. We therefore have that $\mathbb{E}[dX^i(t)]\, dt = \mathbb{E}[v^i(X(t))]\, dt$, so v is the average velocity vector field.

To work out a path integral formulation, let us consider the 1-d case $dX(t) = v(X(t))dt + b(X(t))dW(t)$. Roughly speaking, for $t_2 > t_1$ with $t_2 - t_1$ small, the variable $[X(t_2) - X(t_1) - v(X(t_1)(t_2 - t_1)]/b(X(t_1))$ is approximately a standard Gaussian and so the transition probability is

$$T(x_2, t_2|x_1 t_1) \approx \frac{1}{\sqrt{2\pi(t_2 - t_1)}} e^{-\frac{\left(x_2 - x_1 - v(x_1)(t_2 - t_1)\right)^2}{2(t_2 - t_1)b(x_1)^2}}$$

$$\times \left(1 + \nabla v(x_1)(t_2 - t_1)\right)^{-1/2}, \qquad (7.15)$$

the final term arising as a Jacobian. We can guess the path integral form at this stage. The associated probability measure on the space of sample trajectories, $C_{x_0}[0, t]$ in the multidimensional case, is

$$\mathbb{P}_{X,t}(d\mathbf{x}) = e^{-\int_0^t \mathscr{L}_X(x,\dot{x})} \mathscr{D}\mathbf{x}, \tag{7.16}$$

where the action functional now comes from the associated Lagrangian

$$\mathscr{L}_X(x,\dot{x}) = \frac{1}{2}[\dot{x} - v(x)]^\top \Sigma_{XX}^{-1}(x) [\dot{x} - v(x)] + \frac{1}{2}\nabla v(x), \tag{7.17}$$

where $\Sigma_{XX}(x)$ is the $n \times n$ diffusion matrix with entries $\sum_{\alpha=1}^m b_\alpha^i(x) b_\alpha^j(x)$. We assume that the diffusion matrix is invertible at each $x \in \mathbb{R}^n$.

Theorem 7.3.2 (Freidlin–Wentzell) *Let* $\{X_n(t)\}_{0 \le t \le T}$ *be the Itō diffusion process satisfying the stochastic differential equation (SDE)*

$$dX_n(t) = v(X_n(t)) dt + \frac{1}{\sqrt{n}} \sum_\alpha b_\alpha(X_n(t)) dW_n(t) \tag{7.18}$$

with $X_n(0) = 0$, *where the* $B_\alpha(\cdot)$ *are independent canonical Wiener processes and the drift vector field* $v(\cdot)$ *is uniformly Lipschitz continuous. We make the assumption that* Σ_X *is strictly positive definite (so that the generator of the diffusion is elliptic). Then the family* $\{\mathbb{K}_n\}_n$ *given by* $\mathbb{K}_n = \mathbb{P} \circ X_n^{-1}$ *satisfies a large deviation property with the rate function*

$$I(\mathbf{q}) = \frac{1}{2} \int_0^T [\dot{q} - v(q)]^\top \Sigma_{XX}^{-1}(q) [\dot{q} - v(q)] dt, \tag{7.19}$$

with $I(\mathbf{q}) = \infty$ *if the path is not in the Sobolev space* $H^1([0,t], \mathbb{R}^d)$ *of* $\mathbb{R}^d$*-valued* L^2 *functions on* $[0, T]$ *with* L^2 *derivative.*

A proof of this may be found in Dembo and Zeitouni (1998) and Freidlin and Wentzell (1998), but it is not too difficult to motivate the result formally by the same type of argument given after Schilder's Theorem.

There is an intriguing connection between the rate functions, I, appearing here, and the one-particle irreducible generating functions, Γ, considered in the previous chapter. Both occur as Legendre–Fenchel transforms of the cumulants of various random variables, processes, or fields. This connection between probabilistic distributions and quantum field theory goes back to Jona-Lasinio (1983). Subsequently, a least effective action (Γ) principle for fluctuations in terms of stochastic processes was developed by Eyink (1996) along similar lines.

8

Introduction to Fock Spaces

Fock spaces were introduced by Vladimir Fock as the appropriate Hilbert space setting for quantum fields corresponding to Bosonic quanta. They have since emerged as a rich source of mathematical investigation, and are relevant to a large number of areas of mathematics and deserving independent investigation. In this chapter, we set up the basic definitions for Full (that is, distinguishable quanta), Boson, and Fermion cases.

8.1 Hilbert Spaces

8.1.1 Notations and General Statements

Linear spaces are spaces over the field of complex number $\mathbb{C}$ unless stated differently. Hilbert spaces are denoted as $\mathfrak{H}$ or $\mathfrak{h}$ with elements $f, g, \ldots$. The positive definite inner product $f \in \mathfrak{H}, g \in \mathfrak{H} \mapsto \langle f \mid g \rangle \in \mathbb{C}$ is linear in g and antilinear in f and satisfies the hermitean symmetry

$$\langle g \mid f \rangle = \langle f \mid g \rangle^*. \tag{8.1}$$

The inner product determines a norm on $\mathfrak{H}$ according to

$$\|f\| = \sqrt{\langle f \mid f \rangle} \geq 0. \tag{8.2}$$

Indeed, we note that $\|f\| = 0$ if and only if $f = 0$. The inner product may in fact be recovered from the norm by means of the polarization identity

$$4\langle f \mid g \rangle = \|f + g\|^2 - \|f - g\|^2 - i\|f + ig\|^2 + i\|f - ig\|^2. \tag{8.3}$$

All Hilbert spaces considered here are separable, that is, we shall always assume the existence of a finite or countable orthonormal basis.

The completeness axiom for Hilbert spaces is the requirement that every sequence of vectors that is Cauchy in the norm must converge in the norm to an element of the Hilbert space. In general, an infinite dimensional linear space $\mathfrak{E}$ with a positive definite inner product need not necessarily be complete, and in such cases we refer to the space as an inner product space, or pre-Hilbert, space. However, one can always define a (unique) completion $\widehat{\mathfrak{E}}$ of $\mathfrak{E}$, which is a Hilbert space, see e.g. Reed and Simon (1972).

Hilbert spaces can emerge in a more general setting. Let $\mathfrak{E}$ be a linear space with a positive hermitean form $f, g \in \mathfrak{E} \mapsto B(f, g) \in \mathbb{C}$, i.e. a form that satisfies $B(f, g) = B(g, f)^*$ and $B(f, f) \geq 0$ for all $f, g \in \mathfrak{E}$. The space $\mathfrak{E}_0 \triangleq \{f : B(f, f) = 0\}$ may be nontrivial. Then $B(f, g)$ induces a positive definite inner product into the quotient space $\mathfrak{E}/\mathfrak{E}_0$, and the completion of this quotient space is a Hilbert space. The space $\mathfrak{E}$ is called semi-Hilbert space. See e.g. Maurin (1968)[1].

The continuous linear operators $A, B, \ldots$ on an inner product space $\mathfrak{E}$ form the linear space $\mathcal{L}(\mathfrak{E})$, and a norm of this space is given by the operator norm

$$\|A\| = \sup_{f, \|f\|=1} \|Af\| = \sup_{f, g, \|f\|=1, \|g\|=1} |\langle f \mid Ag \rangle|$$
$$\equiv \sup_{f, g : \|f\|=1, \|g\|=1} \mathrm{Re}\langle f \mid Ag \rangle. \tag{8.4}$$

If the space $\mathfrak{E}$ is completed to the Hilbert space $\mathfrak{H}$, the operator A can be extended to a continuous operator on $\mathfrak{H}$ with the norm (8.4).

A mapping $f \in \mathfrak{H} \mapsto f^\star \in \mathfrak{H}$ is denoted as **conjugation** if it has the properties

$$f \mapsto f^\star \text{ is antilinear, i.e. } (\alpha f)^\star = \alpha^\star f^\star \text{ for } \alpha \in \mathbb{C} \text{ and } f \in \mathfrak{H},$$
$$f \mapsto f^\star \text{ is isometric, i.e. } \|f^\star\| = \|f\| \text{ for } f \in \mathfrak{H}, \tag{8.5}$$
$$f \mapsto f^\star \text{ is involutive, i.e. } f^{\star\star} = f \text{ for } f \in \mathfrak{H}.$$

Using the polarization identity, we derive from the isometry of this antilinear mapping

$$\langle f^\star \mid g^\star \rangle = \langle f \mid g \rangle^*. \tag{8.6}$$

Hence any conjugation is antiunitary.

Proposition 8.1.1 *If* $f \mapsto f^\star$ *is a conjugation, then*

$$(f \mid g) \triangleq \langle f^\star \mid g \rangle \tag{8.7}$$

is a bilinear symmetric nondegenerate form on $\mathfrak{H}$.

---

[1] In this reference, the semi-Hilbert-space is called a pre-Hilbert space. In most of the literature – and in the present book – the notion "pre-Hilbert space" is reserved to spaces with a positive definite form.

Proof The form is obviously bilinear. The identities (8.1) and (8.6) yield

$$\langle f \mid g \rangle = \langle f^\star | g \rangle \overset{(8.6)}{=} \langle f | g^\star \rangle^\star \overset{(8.1)}{=} \langle g^\star | f \rangle = \langle g | f \rangle \tag{8.8}$$

and (8.7) is symmetric. Assume $\langle g \mid f \rangle = 0$ for all $f \in \mathfrak{H}$, then $\langle g \mid f \rangle = \langle g^\star \mid f \rangle = 0$ for all $f \in \mathfrak{H}$. Take $f = g^\star$, then $\langle g^\star \mid g^\star \rangle = 0$ and consequently, $g^\star = 0$. Since the involution is isometric that is only possible for $g = 0$. $\quad\square$

The subset $\mathfrak{H}_{\mathbb{R}} \triangleq \{f = f^\star \mid f \in \mathfrak{H}\}$ is a linear space over the field $\mathbb{R}$, and it is a real Hilbert space with the inner product of $\mathfrak{H}$. Any $f \in \mathfrak{H}$ can be written as sum $f = f_1 + if_2$ with vectors $f_{1,2} \in \mathfrak{H}_{\mathbb{R}}$. These vectors are $f_1 = (f + f^\star)/2$ and $f_2 = -i(f - f^\star)/2$. Any orthonormal (ON) basis $\{e_k : k \in \mathbb{K}\}$ of the real space $\mathfrak{H}_{\mathbb{R}}$ is an ON basis of the complex space $\mathfrak{H}$.

8.1.2 Examples

Take the L^2-space $L^2(\mathbb{R}^n)$ with functions $\mathbb{R}^n \ni \mathbf{x} \mapsto \varphi(\mathbf{x}) \in \mathbb{C}$, then a simple conjugation is the usual complex conjugation of functions

$$\varphi(\mathbf{x}) \mapsto \left(\varphi^\star\right)(\mathbf{x}) \triangleq \varphi(\mathbf{x})^\star. \tag{8.9}$$

The functions that are "real" with respect to this involution, i.e. $\varphi^\star = \varphi$, are just the functions with values in $\mathbb{R}$. However, the Fourier transform $(\mathscr{F}\varphi)(\mathbf{k}) = f(\mathbf{k})$ maps the involution (8.9) onto the involution

$$f(\mathbf{k}) \mapsto (f^\star)(\mathbf{k}) = f(-\mathbf{k})^\star \tag{8.10}$$

for the functions $f(\mathbf{k}) \in L^2(\mathbb{R}^n)$. In this context, "real" now implies $f(\mathbf{k}) = f(-\mathbf{k})^\star$, that is, then $\left(\mathscr{F}^{-1}f\right)(\mathbf{x}) = \varphi(\mathbf{x})$ is a real-valued function.

8.2 Tensor Spaces

There is an extensive literature about tensor algebras. We only refer to Bourbaki (1974), Greub (1978), and to the appendix in Dieudonné (1972).

8.2.1 The Tensor Product

General Definitions

Let $\phi \colon \mathfrak{E}_1 \times \mathfrak{E}_2 \times \cdots \times \mathfrak{E}_n \mapsto \mathfrak{E}$ be an n-linear mapping of the pre-Hilbert spaces $\mathfrak{E}_k$, $k = 1, 2, \ldots, n$, into the pre-Hilbert space $\mathfrak{E}$ with the following properties:

(i) If $f_k, g_k \in \mathfrak{E}_k$, $k = 1, 2, \ldots, n$, then we may define the inner product of $f = \phi(f_1, \ldots, f_n) \in \mathfrak{E}$ and $g = \phi(g_1, \ldots, g_n) \in \mathfrak{E}$ to be

$$\langle f \mid g \rangle = \prod_{k=1}^{n} \langle f_k \mid g_k \rangle. \tag{8.11}$$

(ii) The linear hull of $\{\phi(f_1, \ldots, f_n) \mid f_1 \in \mathfrak{E}_1, \ldots, f_n \in \mathfrak{E}_n\}$ is dense in $\mathfrak{E}$.

This mapping is called the **tensor product** of the spaces $\mathfrak{E}_k$. The usual notations are $\mathfrak{E} = \mathfrak{E}_1 \underline{\otimes} \mathfrak{E}_2 \underline{\otimes} \cdots \underline{\otimes} \mathfrak{E}_n$ and $f = f_1 \underline{\otimes} \cdots \underline{\otimes} f_n$. Those elements of $\mathfrak{E}$, which have the representation $f = f_1 \underline{\otimes} \cdots \underline{\otimes} f_n$, are called **decomposable**. The tensor space $\mathfrak{E}$ is the linear hull of all decomposable tensors. The tensor product is associative:

$$\left(\mathfrak{E}_1 \underline{\otimes} \mathfrak{E}_2\right) \underline{\otimes} \mathfrak{E}_3 = \mathfrak{E}_1 \underline{\otimes} \left(\mathfrak{E}_2 \underline{\otimes} \mathfrak{E}_3\right) = \mathfrak{E}_1 \underline{\otimes} \mathfrak{E}_2 \underline{\otimes} \mathfrak{E}_3. \tag{8.12}$$

A trivial example is where all the factor spaces are one-dimensional, that is, $\mathfrak{E}_1 = \cdots = \mathfrak{E}_n = \mathbb{C}$, in which case $\mathbb{C} \underline{\otimes} \mathbb{C} \underline{\otimes} \cdots \underline{\otimes} \mathbb{C} = \mathbb{C}$.

An immediate consequence of the definition is that if the vectors $e_j^k \in \mathfrak{E}_k$, $j \in \mathbb{J}_k \subset \mathbb{N}$ form an orthonormal basis of the space $\mathfrak{E}_k, k = 1, 2, \ldots, n$, then the tensors

$$e_{j_1}^1 \underline{\otimes} e_{j_2}^2 \underline{\otimes} \cdots \underline{\otimes} e_{j_n}^n \in \mathfrak{E}, \qquad (j_1, \ldots, j_n) \in \mathbb{J}_1 \times \cdots \times \mathbb{J}_n \tag{8.13}$$

form an orthonormal basis of $\mathfrak{E}$. Hence any element $F \in \mathfrak{E}$ can be represented as

$$f = \sum_{(j_1, \ldots, j_n) \in \mathbb{J}_1 \times \cdots \times \mathbb{J}_n} c(j_1, \ldots, j_n) \, e_{j_1}^1 \underline{\otimes} e_{j_2}^2 \underline{\otimes} \cdots \underline{\otimes} e_{j_n}^n \tag{8.14}$$

with coefficients $c(j_1, \ldots, j_n) \in \mathbb{C}$ such that

$$\|f\|^2 = \sum_{(j_1, \ldots, j_n) \in \mathbb{J}_1 \times \cdots \times \mathbb{J}_n} |c(j_1, \ldots, j_n)|^2. \tag{8.15}$$

Our interest will be in the situation where the factors are Hilbert spaces $\mathfrak{H}_1$ and $\mathfrak{H}_2$, in which case the algebraic tensor product $\mathfrak{H}_1 \underline{\otimes} \mathfrak{H}_2$ consists of all decomposable tensors. We shall more often work with the completion of the linear hull of $\mathfrak{H}_1 \underline{\otimes} \mathfrak{H}_2$, as this is again a Hilbert space, which we denote by $\mathfrak{H}_1 \otimes \mathfrak{H}_2$.

For example, let $\mathfrak{H}_1 = L^2(\mathcal{M}_1, d\mu_1)$ and $\mathfrak{H}_2 = L^2(\mathcal{M}_2, d\mu_2)$ be two $\mathcal{L}^2$-spaces. Then the tensor space $\mathfrak{H}_1 \otimes \mathfrak{H}_2$ can be identified with $L^2(\mathcal{M}_1 \times \mathcal{M}_2, d\mu_1 \times d\mu_2)$, and the bilinear product is the numerical product

$$\begin{aligned} f(x_1) &\in L^2(\mathcal{M}_1, d\mu_1), \ g(x_2) \in L^2(\mathcal{M}_2, d\mu_2) \mapsto \\ h(x_1, x_2) &= f(x_1)g(x_2) \in L^2(\mathcal{M}_1 \times \mathcal{M}_2, d\mu_1 \times d\mu_2). \end{aligned} \tag{8.16}$$

The inner product obviously factorizes according to (8.11). If $\{\varphi_m(x_1)\}$ is an orthonormal basis of $L^2(\mathcal{M}_1, d\mu_1)$ and $\{\psi_n(x_2)\}$, an orthonormal basis of $L^2(\mathcal{M}_2, d\mu_2)$, then $\{\varphi_m(x_1)\psi_n(x_2)\}$ is an orthonormal basis of $L^2(\mathcal{M}_1 \times \mathcal{M}_2, d\mu_1 \times d\mu_2)$; see Reed and Simon (1972, Sect. II.4).

8.2.2 The Tensor Algebra

Let $\mathfrak{H}$ be a Hilbert or pre-Hilbert space with dimension $\dim \mathfrak{H} \geq 1$. Then the linear hull of all decomposable tensors of degree n of this Hilbert space is denoted by $\mathfrak{H}^{\otimes n} = \mathfrak{H} \otimes \mathfrak{H} \otimes \cdots \otimes \mathfrak{H}$, with n factors where $n = 1, 2, \ldots$. We also take $\mathfrak{H}^{\otimes 0}$ to be the one-dimensional space $\mathbb{C}$.

The **tensor algebra** over a (pre-)Hilbert space $\mathfrak{H}$ is the set $\underline{\Gamma}(\mathfrak{H})$ as the set of all sequences $F = (F_0, F_1, \ldots)$, with $F_m \in \mathfrak{H}^{\otimes m}$, $m = 0, 1, 2, \ldots$, and $F_n = 0$ if $n > N$ for some $N \in \mathbb{N}$.

We now show step-by-step that this is indeed an algebra. First, we see that $\underline{\Gamma}(\mathfrak{H})$ becomes a vector space with the following definitions: if $F = (F_0, F_1, \ldots)$ and $G = (G_0, G_1, \ldots)$ are elements of $\underline{\Gamma}(\mathfrak{H})$, then we may define their linear combination $H = \alpha F + \beta G$ for $\alpha, \beta \in \mathbb{C}$ in the obvious way as

$$H = (H_0, H_1, \ldots) \in \underline{\Gamma}(\mathfrak{H}), \qquad \text{with } H_n = \alpha F_n + \beta G_n \in \mathfrak{H}^{\otimes n}. \quad (8.17)$$

An inner product on the tensor algebra may be defined as follows: for $F, G \in \underline{\Gamma}(\mathfrak{H})$, we set

$$\langle F \mid G \rangle = \sum_{n=0}^{\infty} \langle F_n \mid G_n \rangle_n, \quad (8.18)$$

where $\langle F_n \mid G_n \rangle_n$ is the inner product of $\mathfrak{H}^{\otimes n}$. The norm is given as usual by

$$\|F\| = \sqrt{\langle F \mid F \rangle}. \quad (8.19)$$

With these definitions, $\underline{\Gamma}(\mathfrak{H})$ is a pre-Hilbert space. If we identify $F_n \in \mathfrak{H}^{\otimes n}$ with the corresponding sequence $(0, \ldots 0, F_n, 0, \ldots) \in \underline{\Gamma}(\mathfrak{H})$, then the spaces $\mathfrak{H}^{\otimes n}$ are orthogonal subspaces of $\underline{\Gamma}(\mathfrak{H})$. As an alternative to writing $F = (F_0, F_1, \ldots) \in \underline{\Gamma}(\mathfrak{H})$, we shall write in the sequel

$$F = \bigoplus_{n=0}^{\infty} F_n. \quad (8.20)$$

From (8.12) follows the simple identity $\mathfrak{H}^{\otimes m} \otimes \mathfrak{H}^{\otimes n} = \mathfrak{H}^{\otimes (m+n)}$ for the tensor product of the Hilbert spaces $\mathfrak{H}^{\otimes m}$ and $\mathfrak{H}^{\otimes n}$. It is straightforward to derive the identity

$$\langle F \otimes G \mid F' \otimes G' \rangle = \langle F \mid F' \rangle \langle G \mid G' \rangle \quad (8.21)$$

for tensors $F, F' \in \mathfrak{H}^{\otimes m}$ and $G, G' \in \mathfrak{H}^{\otimes n}$.

With the additional rule $\lambda \otimes F_n = F_n \otimes \lambda = \lambda F_n$ for $\lambda \in \mathbb{C} = \mathfrak{H}^{\underline{\otimes}0}$ and $F_n \in \mathfrak{H}^{\underline{\otimes}n}$, for $n \geq 0$, the tensor product can be extended to an associative product on the space $\underline{\Gamma}(\mathfrak{H})$. For $F = \bigoplus_{n=0}^{\infty} F_n$ and $G = \bigoplus_{n=0}^{\infty} G_n$ with $F_n, G_n \in \mathfrak{H}^{\underline{\otimes}n}$, we define $H = F \underline{\otimes} G$ by

$$H = \bigoplus_{n=0}^{\infty} H_n \text{ with } H_n = \sum_{k=0}^{n} F_k \underline{\otimes} G_{n-k}. \tag{8.22}$$

Since the sum terminates, there is no problem of convergence. With this definition, it is clear that the linear space $\underline{\Gamma}(\mathfrak{H})$ becomes an algebra, and the unit of this algebra is $1 \in \mathbb{C} = \mathfrak{H}^{\underline{\otimes}0} \subset \underline{\Gamma}(\mathfrak{H})$.

8.2.3 The Fock Space

The completion of the tensor space $\mathfrak{H}^{\underline{\otimes}n}$ will be denoted $\mathfrak{H}^{\otimes n}$, and the completion of the tensor algebra $\underline{\Gamma}(\mathfrak{H})$ is the direct orthogonal sum

$$\Gamma(\mathfrak{H}) = \bigoplus_{n=0}^{\infty} \mathfrak{H}^{\otimes n}. \tag{8.23}$$

This space consists of all (finite and infinite) series

$$F = \bigoplus_{n=0}^{\infty} F_n \text{ with } F_n \in \mathfrak{H}^n \text{ and } \sum_{n=0}^{\infty} \|F_n\|^2 = \|F\|^2 < \infty, \tag{8.24}$$

and is referred to as the **Fock space**.

From (8.21) follows the norm identity

$$\|F_m \underline{\otimes} G_n\| = \|F_m\| \, \|G_n\| \tag{8.25}$$

for $F_m \in \mathfrak{H}^{\underline{\otimes}m}$ and $G_n \in \mathfrak{H}^{\underline{\otimes}n}$. This implies that the tensor product of $\mathfrak{H}^{\underline{\otimes}m}$ and $\mathfrak{H}^{\underline{\otimes}n}$ is in fact continuous and may therefore be extended to the Hilbert spaces $\mathfrak{H}^{\otimes m}$ and $\mathfrak{H}^{\otimes n}$ by continuity. For arbitrary tensors $F_m \in \mathfrak{H}^{\otimes m}$ and $G_n \in \mathfrak{H}^{\otimes n}$, we then obtain $F_m \otimes G_n \in \mathfrak{H}^{\otimes(m+n)}$ with the norm identity (8.25). The tensor product is therefore defined on the algebraic sum of the completed spaces

$$\Gamma_{\text{fin}}(\mathfrak{H}) = \bigoplus_{n \geq 0} \mathfrak{H}^{\otimes n}, \tag{8.26}$$

and $\Gamma_{\text{fin}}(\mathfrak{H})$ is an algebra. Unfortunately the tensor product (8.22) is not continuous in the norm (8.19) on $\Gamma(\mathfrak{H})$, and it has no continuous extension onto the whole Fock space $\Gamma(\mathfrak{H})$.[2]

---

[2] It is, however, possible to define other Hilbert norms such that the tensor product is continuous on the whole Fock space. See Kupsch and Smolyanov (2000).

As an example, take the Hilbert space $\mathfrak{H} = L^2(\mathbb{R}, d\mu)$ where $d\mu$ is a positive measure on $\mathbb{R}$. Then the tensors in $\mathfrak{H}^{\otimes n}$, $n \in \mathbb{N}$, are functions $f_n(x_1, \ldots, x_n)$ on $\mathbb{R}^n$, which are square integrable with respect to the product measure $\underbrace{d^n\mu = d\mu \times \cdots \times d\mu}_{n}$, i.e. $f_n \in L^2(\mathbb{R}^n, d^n\mu)$. An element of the Fock space $\Gamma(\mathfrak{H})$ is a sequence $f = \{f_0, f_1, \ldots\}$ of tensors $f_0 \in \mathbb{C}$ and $f_n \in L^2(\mathbb{R}^n, d^n\mu)$ with a finite norm

$$\|f\|^2 = |f_0|^2 + \sum_{n=1}^{\infty} \int_{\mathbb{R}^n} |f_n(x_1, \ldots, x_n)|^2 \, d^n\mu. \tag{8.27}$$

8.3 Symmetric Tensors

8.3.1 The Fock Space of Symmetric Tensors

With $\mathfrak{S}_n$, we denote the set of all $n!$ permutations of the numbers $\{1, 2, \ldots, n\}$. For each $\sigma \in \mathfrak{S}_n$, the mapping

$$f_1 \otimes f_2 \otimes \cdots \otimes f_n \in \mathfrak{H}^{\otimes n} \mapsto f_{\sigma(1)} \otimes f_{\sigma(2)} \otimes \cdots \otimes f_{\sigma(n)} \in \mathfrak{H}^{\otimes n}$$

is linear in each factor (in fact, it is unitary!). The symmetrization prescription

$$P_+^{(n)} f_1 \otimes f_2 \otimes \cdots \otimes f_n \triangleq \frac{1}{n!} \sum_{\sigma \in \mathfrak{S}_n} f_{\sigma(1)} \otimes f_{\sigma(2)} \otimes \cdots \otimes f_{\sigma(n)} \tag{8.28}$$

defines a linear operator on $\mathfrak{H}^{\otimes n}$, $n \geq 2$. (For $n = 0, 1$, we just take P_+^n to be the identity.) If $\tau \in \mathfrak{S}_n$ is a fixed element of $\mathfrak{S}_n$, the set $\{\sigma\tau \mid \sigma \in \mathfrak{S}_n\}$ is again $\mathfrak{S}_n$: the set of all permutations on $\{1, \ldots, n\}$ being the same as the set of all permutations on $\{\tau(1), \ldots, \tau\}$ for any permutation τ. The operator (8.28) therefore satisfies

$$P_+^{(n)} f_1 \otimes f_2 \otimes \cdots \otimes f_n = P_+^{(n)} f_{\tau(1)} \otimes f_{\tau(2)} \otimes \cdots \otimes f_{\tau(n)} \tag{8.29}$$

for any permutation $\tau \in \mathfrak{S}_n$.

Excercise Prove the identities

$$P_+^{(n)} \left(P_+^{(n)} f_1 \otimes f_2 \otimes \cdots \otimes f_n \right) = P_+^{(n)} f_1 \otimes f_2 \otimes \cdots \otimes f_n \tag{8.30}$$

and

$$\begin{aligned} &\langle P_+^{(n)} f_1 \otimes f_2 \otimes \cdots \otimes f_n \mid g_1 \otimes g_2 \otimes \cdots \otimes g_n \rangle \\ &= \langle f_1 \otimes f_2 \otimes \cdots \otimes f_n \mid P_+^{(n)} g_1 \otimes g_2 \otimes \cdots \otimes g_n \rangle \end{aligned} \tag{8.31}$$

for arbitrary $f_k, g_k \in \mathfrak{H}$, $k = 1, \ldots, n$.

By linearity these identities imply the operator identities

$$P_+^{(n)2} = P_+^{(n)} \text{ and } P_+^{(n)*} = P_+^{(n)} \tag{8.32}$$

on $\mathfrak{H}^{\otimes n}$. Hence $P_+^{(n)}$ is a continuous[3] projection operator on $\mathfrak{H}^{\underline{\otimes} n}$ and can be extended to a continuous orthogonal projection operator on $\mathfrak{H}^{\otimes n}$. The image

$$\Gamma_n^+(\mathfrak{H}) \triangleq P_+^{(n)} \mathfrak{H}^{\otimes n} \subset \mathfrak{H}^{\otimes n} \tag{8.33}$$

is a closed subspace of $\mathfrak{H}^{\otimes n}$ and the elements $F \in \Gamma_n^+(\mathfrak{H})$ are the **symmetric tensors** of degree n.

With $P_+^{(0)} \lambda \triangleq \lambda \in \mathbb{C} = \mathfrak{H}^{\otimes 0}$ and $P_+^{(1)} f \triangleq f \in \mathfrak{H} = \mathfrak{H}^{\otimes 1}$, the operators $P_+^{(n)}$ are defined on all the spaces $\mathfrak{H}^{\otimes n}$, $n = 0, 1, 2, \ldots$. We now define a projection operator P_+ on the Fock space $\Gamma(\mathfrak{H})$ by

$$F = \bigoplus_{n=0}^{\infty} F_n \mapsto P_+ F \triangleq \bigoplus_{n=0}^{\infty} P_+^{(n)} F_n \text{ if } F_n \in \mathfrak{H}^{\otimes n}. \tag{8.34}$$

The restriction of this operator to $\mathfrak{H}^{\otimes n}$ agrees with $P_+^{(n)}$.

The closed linear subspace

$$\Gamma^+(\mathfrak{H}) \triangleq P_+ \Gamma(\mathfrak{H}) = \bigoplus_{n=0}^{\infty} \Gamma_n^+(\mathfrak{H}) \tag{8.35}$$

is the **Fock space of symmetric tensors** or the **Boson Fock space**.

As example, take the Hilbert space $\mathfrak{H} = L^2(\mathbb{R}, d\mu)$ as in the Fock example. Then the tensors in $\Gamma_n^+(\mathfrak{H})$, $n \in \mathbb{N}$, are functions $f_n^+(x_1, \ldots, x_n)$ in $L^2(\mathbb{R}^n, d^n\mu)$, which are symmetric with respect to the exchange of the variables $x_j \leftrightarrow x_k$. An element of the Fock space $\Gamma^+(\mathfrak{H})$ is a sequence $f = \{f_0, f_1, \ldots\}$ of tensors $f_0 \in \mathbb{C}$ and symmetric functions $f_n \in L^2(\mathbb{R}^n, d^n\mu)$ with a finite norm (8.27).

8.3.2 The Algebra of Symmetric Tensors

The algebraic subspace of symmetric tensors of degree n is denoted by $\mathcal{A}_n^+(\mathfrak{H})$, that is,

$$\mathcal{A}_n^+(\mathfrak{H}) = P_+^{(n)} \underline{\mathfrak{H}}^{\otimes n} = \underline{\Gamma}(\mathfrak{H}) \cap \Gamma_n^+(\mathfrak{H}) \subset \mathfrak{H}^{\underline{\otimes} n}.$$

The tensor product (8.22) can be defined for arbitrary tensors in the space

$$\mathcal{A}^+(\mathfrak{H}) \triangleq P_+ \underline{\Gamma}(\mathfrak{H}) = \bigoplus_{n \geq 0} \mathcal{A}_n^+(\mathfrak{H}) = \underline{\Gamma}(\mathfrak{H}) \cap \Gamma^+(\mathfrak{H}). \tag{8.36}$$

---

[3] The continuity follows from the observation that $\|P_s F\|^2 = (P_s F \mid P_s F) = \left(P_s^2 F \mid F\right) = (P_s F \mid F) \leq \|P_s F\| \, \|F\|$. Hence $\|P_s F\| \leq \|F\|$ for all $F \in \mathcal{H}^{\underline{\otimes} n}$.

Since $F \otimes G$ is in general not an element of $\mathcal{A}^+(\mathfrak{H})$, for arbitrary $F, G \in \mathcal{A}^+(\mathfrak{H})$, we must define

$$F, G \in \mathcal{A}(\mathfrak{H}) \mapsto P_+ (F \otimes G) \in \mathcal{A}(\mathfrak{H}) \tag{8.37}$$

to get a bilinear product within $\mathcal{A}^+(\mathfrak{H})$, which – as a consequence of (8.29) – is symmetric

$$P_+ (F \otimes G) \equiv P_+ (G \otimes F). \tag{8.38}$$

Excercise Derive the identity

$$P_+ (P_+ F \otimes P_+ G) = P_+ (F \otimes G) \tag{8.39}$$

for tensors $F \in \mathfrak{H}^{\underline{\otimes} m}$ and $G \in \mathfrak{H}^{\underline{\otimes} n}$ and arbitrary numbers $m, n \in \mathbb{N}$.

As a consequence of (8.39) the product (8.37) is associative. The standard definition of the **symmetric tensor product** is $F \vee G$, defined by

$$F \in \mathcal{A}_m^+, \, G \in \mathcal{A}_n^+ \mapsto F \vee G = \sqrt{\frac{(m+n)!}{m! \, n!}} P_+ (F \otimes G) \in \mathcal{A}_{m+n}^+. \tag{8.40}$$

There is a good reason for the combinatorial coefficient, which we now explain. In principle, we could take any sequence of nonzero complex numbers $(c(n))_n$ and then define a product for symmetric tensors of degree m and n by

$$F \in \mathcal{A}_m^+, \, G \in \mathcal{A}_n^+ \mapsto F \circ G = \frac{c(m+n)}{c(m)c(n)} P_+ (F \otimes G) \in \mathcal{A}_{m+n}^+, \tag{8.41}$$

and extend this definition by linearity to a product on $\mathcal{A}^+(\mathfrak{H})$. Let us first prove the following general result.

Proposition 8.3.1 *The bilinear extension of the product (8.41) is an associative symmetric product on $\mathcal{A}^+(\mathfrak{H})$.*

Proof We give an indication of the proof. Take $F_m \in \mathcal{A}_m^+(\mathfrak{H})$, $G_n \in \mathcal{A}_n^+(\mathfrak{H})$, and $H_k \in \mathcal{A}_k^+(\mathfrak{H})$, then

$$(F \circ G) \circ H = \frac{c(m+n+k)}{c(m+n)c(k)} \frac{c(m+n)}{c(m)c(n)} P_+ (F \otimes G \otimes H) \tag{8.42}$$

$$= \frac{c(m+n+k)}{c(m)c(n)c(k)} P_+ (F \otimes G \otimes H) \tag{8.43}$$

and

$$F \circ (G \circ H) = \frac{c(m+n+k)}{c(m)c(n+k)} \frac{c(n+k)}{c(n)c(k)} P_+ (F \otimes G \otimes H) \tag{8.44}$$

$$= \frac{c(m+n+k)}{c(m)c(n)c(k)} P_+ (F \otimes G \otimes H). \tag{8.45}$$

Hence the product is associative for tensors of fixed degree. The statement for general tensors follows by linearity. The symmetry follows from (8.38). □

With each of these definitions, the product (8.41) is symmetric and $\mathcal{A}^+(\mathfrak{H})$ is an associative algebra that is generated by vectors $f \in \mathfrak{H}$.

Our specification $c(n) = \sqrt{n!}$, however, leads to the following exceptional property relating to orthogonal subspaces.

Lemma 8.3.2 *If $\mathfrak{H}_1$ and $\mathfrak{H}_2$ are two orthogonal subspaces of $\mathfrak{H}$, then the inner product of tensors $F_1 \vee F_2$ and $G_1 \vee G_2$ with $F_k, G_k \in \mathcal{A}^+(\mathfrak{H}_k)$, $k = 1, 2$, factorizes into*

$$\langle F_1 \vee F_2 \mid G_1 \vee G_2 \rangle = \langle F_1 \mid G_1 \rangle \langle F_2 \mid G_2 \rangle. \tag{8.46}$$

Proof The proof is first given for tensors of fixed degree $F_{km} \in \mathcal{A}_m^+(\mathfrak{H}_k)$, $G_{kn} \in \mathcal{A}_n^+(\mathfrak{H}_k)$. The inner product (8.18) of these tensors is

$$\langle F_{1m} \vee F_{2n} \mid G_{1p} \vee G_{2q} \rangle = \sqrt{\frac{(m+n)!}{m!\,n!}} \sqrt{\frac{(p+q)!}{p!\,q!}} \langle F_{1m} \otimes F_{2n} \mid P_+(G_{1p} \otimes G_{2q}) \rangle. \tag{8.47}$$

Thereby, we have used the definition (8.40) of the product and the properties $P_+^* = P_+ = P_+^2$ of the projection operator. Since $F_{1m} \vee F_{2n} \in \mathcal{A}_{m+n}^+(\mathfrak{H})$ and $G_{1p} \vee G_{2q} \in \mathcal{A}_{p+q}^+(\mathfrak{H})$, the inner product vanishes unless $m + n = p + q$. Due to linearity in the right factor, it is sufficient to continue the calculation with decomposable tensors $G_{1p} = P_+^{(p)}(g_1 \otimes \cdots \otimes g_p)$, $g_j \in \mathfrak{H}_1$, and $G_{2q} = P_+^{(q)}(g_{p+1} \otimes \cdots \otimes g_{p+q})$, $g_{p+j} \in \mathfrak{H}_2$. Then we can use the explicit formula (8.28) for the projection operators. The factorization (8.21) yields that the inner product (8.47) vanishes unless both conditions $m = p$ and $n = q$ are satisfied. Moreover, we obtain the identities

$$\langle F_{1m} \vee F_{2n} \mid G_{1m} \vee G_{2n} \rangle$$
$$= \frac{(m+n)!}{m!\,n!} \langle F_{1m} \otimes F_{2n} \mid P_+^{(m+n)}(g_1 \otimes \cdots \otimes g_{m+n}) \rangle$$
$$= \frac{1}{m!\,n!} \sum_{\sigma \in \mathfrak{S}_{m+n}} \langle F_{1m} \otimes F_{2n} \mid g_{\sigma(1)} \otimes \cdots \otimes g_{\sigma(m+n)} \rangle$$
$$= \frac{1}{m!\,n!} \langle F_{1m} \mid \sum_{\sigma \in \mathfrak{S}_m} \left(g_{\sigma(1)} \otimes \cdots \otimes g_{\sigma(m)} \right) \rangle$$
$$\times \langle F_{2n} \mid \sum_{\tau \in \mathfrak{S}_n} \left(g_{m+\tau(1)} \otimes \cdots \otimes g_{m+\tau(n)} \right) \rangle$$

$$= \langle F_{1m} \mid P_+^{(m)} (g_1 \otimes \cdots \otimes g_m) \rangle \langle F_{2n} \mid P_+^{(n)} (g_{m+1} \otimes \cdots \otimes g_{m+n}) \rangle$$

$$= \langle F_{1m} \mid G_{1m} \rangle \langle F_{2n} \mid G_{2n} \rangle. \tag{8.48}$$

Hence (8.46) is true for tensors of fixed degree.

For general elements of $\mathcal{A}(\mathfrak{H}_k)$ we have $F_k = \bigoplus_m F_{km}$ and $G_k = \bigoplus_n G_{kn}$ with $F_{km} \in \mathcal{A}_m(\mathfrak{H}_k)$ and $G_{kn} \in \mathcal{A}_n^+(\mathfrak{H}_k)$, and the left side of (8.46) is

$$
\begin{aligned}
\langle F_1 \vee F_2 \mid G_1 \vee G_2 \rangle &= \sum_{mnpq} \langle F_{1m} \vee F_{2n} \mid G_{1p} \vee G_{2q} \rangle \\
&= \sum_{mn} \langle F_{1m} \vee F_{2n} \mid G_{1m} \vee G_{2n} \rangle \\
&= \sum_{mn} \langle F_{1m} \mid G_{1m} \rangle \langle F_{2n} \mid G_{2n} \rangle. \tag{8.49}
\end{aligned}
$$

But this result agrees with

$$\langle F_1 \mid G_1 \rangle \langle F_2 \mid G_2 \rangle = \sum_m \langle F_{1m} \mid G_{1m} \rangle \sum_n \langle F_{2n} \mid G_{2n} \rangle$$

and (8.46) is true for all tensors.   □

If we start with the general definition (8.41) of the product, the result (8.48) can only be derived if

$$\frac{c(m+n)}{c(m)c(n)} = \sqrt{\frac{(m+n)!}{m!\, n!}}$$

is satisfied. The product (8.40) is therefore uniquely determined by the factorization property. From (8.40), we obtain $1 \otimes F = F \otimes 1 = F$ for all $F \in \mathcal{A}^+(\mathfrak{H})$. Hence $\mathcal{A}^+(\mathfrak{H})$ is an algebra with a unit, and the element $1 \in \mathbb{C} \subset \mathcal{A}^+(\mathfrak{H})$ is the unit of the algebra. This unit is often called the vacuum vector and written as Ω or $|0\rangle$.

From the definition (8.40) and the norm identity (8.25), we obtain

$$\|F_m \vee G_n\| \leq \sqrt{\frac{(m+n)!}{m!\, n!}} \|F_m\| \|G_n\| \leq 2^{\frac{m+n}{2}} \|F_m\| \|G_n\| \tag{8.50}$$

for $F \in \mathcal{A}_m^+(\mathfrak{H})$ and $G \in \mathcal{A}_n^+(\mathfrak{H})$. Hence the tensor product of $\mathcal{A}_m^+(\mathfrak{H})$ and $\mathcal{A}_n^+(\mathfrak{H})$ can be extended to the Hilbert spaces $\Gamma_m^+(\mathfrak{H})$ and $\Gamma_n^+(\mathfrak{H})$ by continuity. For arbitrary tensors $F_m \in \Gamma_m^+(\mathfrak{H})$ and $G_n \in \Gamma_n^+(\mathfrak{H})$, we then obtain a tensor $F_m \vee G_n \in \Gamma_{m+n}^+(\mathfrak{H})$ with the norm estimate (8.50). The space $\Gamma_{fin}^+(\mathfrak{H}) = \Gamma^+(\mathfrak{H}) \cap \Gamma_{fin}(\mathfrak{H})$ is therefore an algebra. But the symmetric tensor product is not continuous on $\mathcal{A}^+(\mathfrak{H})$, and it has no continuous extension onto the whole Bose Fock space $\Gamma^+(\mathfrak{H})$.

We now derive some useful results for decomposable tensors. The tensors $f_1 \vee \cdots \vee f_m = \sqrt{m!}P_+ f_1 \otimes \cdots \otimes f_m$ and $g_1 \vee \cdots \vee g_n = \sqrt{n!}P_+ g_1 \otimes \cdots \otimes g_n$ with $f_j, g_k \in \mathfrak{H}$ have the inner product

$$
\begin{aligned}
\langle f_1 \vee \cdots \vee f_m \mid g_1 \vee \cdots \vee g_m \rangle &= m! \, \langle f_1 \otimes \cdots \otimes f_m \mid P_+ g_1 \otimes \cdots \otimes g_m \rangle \\
&= \delta_{n,m} \sum_{\sigma \in \mathfrak{S}_m} \langle f_1 \otimes \cdots \otimes f_m \mid g_{\sigma(1)} \otimes \cdots \otimes g_{\sigma(m)} \rangle \\
&= \delta_{n,m} \sum_{\sigma \in \mathfrak{S}_m} \prod_{j=1}^{m} \langle f_j \mid g_{\sigma(j)} \rangle.
\end{aligned}
$$

Hence the general rule is

$$
\langle f_1 \vee \cdots \vee f_m \mid g_1 \vee \cdots \vee g_n \rangle = \begin{cases} 0 & \text{if } m \neq n \\ \operatorname{per}\left(\langle f_j \mid g_k \rangle \right) & \text{if } m = n \end{cases} \tag{8.51}
$$

where we have used the *permanent* of a matrix $A = \left(\alpha_{jk} \right)_{j,k=1,\ldots,n}$

$$
\operatorname{per} A \triangleq \sum_{\sigma \in \mathfrak{S}_n} \prod_{j=1}^{n} \alpha_{j\sigma(j)}. \tag{8.52}
$$

A simple consequence of (8.51) is

$$
\left\| f^{\vee n} \right\|^2 = n! \, \|f\|^{2n}. \tag{8.53}
$$

We now return to the original occurrence of Fock space in chapter 4 as the Guichardet space $L^2 \left(\text{Power}\,(\mathfrak{X}), dX \right)$ over a one-particle space $L^2 (\mathfrak{X}, dx)$. In fact, we have the straightforward identification

$$
\Gamma^+ \left(L^2 (\mathfrak{X}, dx) \right) \cong L^2 \left(\text{Power}\,(\mathfrak{X}), dX \right),
$$

which is given by the unitary map

$$
F = \bigoplus_n F_n \in \Gamma^+ \left(L^2 (\mathfrak{X}, dx) \right) \mapsto \tilde{F} \in L^2 \left(\text{Power}\,(\mathfrak{X}), dX \right),
$$

where

$$
\tilde{F} : \{x_1, \ldots, x_n\} = \sqrt{n!} \, F_n (x_1, \ldots, x_n).
$$

As the reader may suspect, the $\sqrt{n!}$ factors play a combinatorial role. Indeed, the unitary map has the following action:

$$
F \vee G \mapsto \tilde{F} \diamond \tilde{G}.
$$

The previous lemma then can be alternatively proved as follows: assume that $F_k, G_k \in \mathcal{A}^+ (\mathfrak{H}_k)$ for $k = 1, 2$ with $\mathfrak{H}_1$ and $\mathfrak{H}_2$ orthogonal subspaces, then

$$\langle \tilde{F}_1 \diamond \tilde{F}_2 | \tilde{G}_1 \diamond \tilde{G}_2 \rangle = \int \sum_{X_1 + X_2 = X} \tilde{F}_1 (X_1)^* \, \tilde{F}_2 (X_2)^* \sum_{Y_1 + Y_2 = X} \tilde{G}_1 (Y_1) \, \tilde{G}_2 (Y_2) \, dX$$

$$= \int \sum_{X_1 + X_2 = X} \tilde{F}_1 (X_1)^* \, \tilde{G}_1 (X_1) \, \tilde{F}_2 (X_2)^* \, \tilde{G}_2 (X_2) \, dX,$$

where we used the orthogonality property to show that two decompositions of X had to coincide. We can then use the $\sum$ formula, Lemma 4.1.3, to show that the last term equals

$$\int \tilde{F}_1 (X_1)^* \, \tilde{G}_1 (X_1) \, \tilde{F}_2 (X_2)^* \, \tilde{G}_2 (X_2) \, dX_1 dX_2 = \langle \tilde{F}_1 | \tilde{G}_1 \rangle \langle \tilde{F}_2 | \tilde{G}_2 \rangle.$$

If $\mathfrak{D}$ is a linear subset of $\mathfrak{H}$, we write $\mathcal{A}^+(\mathfrak{D})$ for the subalgebra of $\mathcal{A}^+(\mathfrak{H})$, which is generated by vectors in $\mathfrak{D}$ (i.e., $\mathcal{A}^+(\mathfrak{D})$ is spanned by symmetric tensor products of vectors in $\mathfrak{D}$).

For a more precise estimate of the symmetric tensor product we introduce the following family of Hilbert norms for a tensor $F = \sum_{p=0}^\infty F_p$ with $F_p \in \Gamma_p^+(\mathfrak{H})$

$$\|F\|_{(\alpha)}^2 = \sum_{p=0}^\infty (p!)^\alpha \, \|F_p\|_p^2, \, \alpha \geq 0. \tag{8.54}$$

The completion of $\Gamma_{fin}^+(\mathfrak{H})$ with the norm $\|F\|_{(\alpha)}$ is called the **Fock scale** $\Gamma_{(\alpha)}^+(\mathfrak{H})$. These spaces have the inclusions $\Gamma_{fin}^+(\mathfrak{H}) \subset \Gamma_{(\alpha)}^+(\mathfrak{H}) \subset \Gamma_{(\beta)}^+(\mathfrak{H}) \subset \Gamma_{(0)}^+(\mathfrak{H}) = \Gamma^+(\mathfrak{H})$ if $\alpha \geq \beta \geq 0$. The symmetric tensor product on $\Gamma_{fin}^+(\mathfrak{H})$ can be extended to a continuous mapping from $\Gamma_{(\alpha)}^+(\mathfrak{H}) \times \Gamma_{(\alpha)}^+(\mathfrak{H})$ into $\Gamma_{(\beta)}^+(\mathfrak{H})$ if $\alpha > \beta$; see the appendix A of Kupsch and Smolyanov (1998) or Kupsch and Smolyanov (2000). Hence the linear space $\Gamma_>^+(\mathfrak{H}) = \cup_{\alpha>0} \Gamma_{(\alpha)}^+(\mathfrak{H}) \subset \Gamma^+(\mathfrak{H})$ is an algebra with the symmetric tensor product.

8.3.3 Basis Systems

If e_k, $k \in \mathbb{K} \subset \mathbb{N}_+$ is an ON basis of $\mathfrak{H}$.[4] Let $\mathbf{n} = (n_1, n_2, \ldots) \in (\mathbb{N}_+)^{\mathbb{K}}$ be an occupation number sequence with $N(\mathbf{n}) = \sum_{k \in \mathbb{K}} n_k$. If $N(\mathbf{n}) = n$, then we may define a symmetric tensor of degree n by

$$F_+(\mathbf{n}) \triangleq \underbrace{e_1 \vee \cdots \vee e_1}_{n_1} \vee \underbrace{e_2 \vee \cdots \vee e_2}_{n_2} \vee \cdots$$

$$= e_1^{\vee n_1} \vee e_2^{\vee n_2} \vee \cdots . \tag{8.55}$$

---

[4] If $\dim \mathcal{H} = \infty$, then $\mathbb{K}$ is the set $\mathbb{N}$. If $\dim \mathcal{H} = K < \infty$, the index set is $\{1, 2, \ldots, K\}$.

Only a finite number of occupation numbers, at most n, can be larger than zero. A factor with $n_k = 0$ is, of course, understood as $e_k^{\vee 0} = 1$. From (8.51), we obtain that the tensors $F_+(m_1, m_2 \ldots)$ and $F_+(n_1, n_2 \ldots)$ are orthogonal if the sequences $(m_1, m_2 \ldots)$ and $(n_1, n_2 \ldots)$ differ. The norm is $\|F_+(n_1, n_2 \ldots)\|^2 = \prod_{k \in \mathbb{K}} n_k!$; see (8.53).

Since the linear span of the tensors $e_{k_1} \vee \cdots \vee e_{k_n}$ with $(k_1, \ldots, k_n) \in \mathbb{K}^n$ is dense in $\Gamma_n^+(\mathfrak{H})$ and any tensor $e_{k_1} \vee \cdots \vee e_{k_n}$ coincides with one of the tensors $F(n_1, n_2, \ldots)$, we see that the tensors $\left\{ F_+(n_1, n_2, \ldots) \mid n_k \in \mathbb{N}_+, \sum_{k \in \mathbb{K}} n_k = n \right\}$ form an orthogonal basis of $\Gamma_n^+(\mathfrak{H})$. An ON basis of $\Gamma_n^+(\mathfrak{H})$ is therefore given by

$$e_+(\mathbf{n}) \triangleq \left(\prod_{k \in \mathbb{K}} n_k! \right)^{-\frac{1}{2}} e_1^{\vee n_1} \vee e_2^{\vee n_2} \vee \cdots \tag{8.56}$$

with the occupation numbers satisfying $N(\mathbf{n}) = n$. The whole Fock space $\Gamma^+(\mathfrak{H}) = \bigoplus_{n=0}^{\infty} \Gamma_n^+(\mathfrak{H})$ has the ON basis $e_+(n_1, n_2 \ldots)$ with $n_k \in \mathbb{N}_+$ and $\sum_{k \in \mathbb{K}} n_k < \infty$. That is, the index set of this basis is the set of all *terminating* sequences $\mathbf{n} = (n_1, n_2 \ldots) \in (\mathbb{N}_+)^{\mathbb{K}}$.

The ON basis for the Boson Fock space constructed in this manner is labeled by the (occupation numbers of) bags drawn from $\mathbb{K}$.

If $1 \leq \dim \mathfrak{H} = K < \infty$ also, the tensor spaces $\Gamma_n^+(\mathfrak{H}) = \mathcal{A}_n^+(\mathfrak{H})$ have a finite dimension $\dim \mathcal{A}_n^+(\mathfrak{H}) = \binom{K+n-1}{n}$, $n \geq 0$. But the Fock space has always infinite dimension.

8.4 Antisymmetric Tensors

8.4.1 Fock Space of Antisymmetric Tensors

For a permutation $\sigma \in \mathfrak{S}_n$, we define the number $[\sigma]$ as $[\sigma] = 0$ if σ is an even permutation and $[\sigma] = 1$ for odd permutations. The linear mapping

$$P_-^{(n)} f_1 \otimes \cdots \otimes f_n \triangleq \frac{1}{n!} \sum_{\sigma \in \mathfrak{S}_n} (-1)^{[\sigma]} f_{\sigma(1)} \otimes f_{\sigma(2)} \otimes \cdots \otimes f_{\sigma(n)} \tag{8.57}$$

with $n \in \mathbb{N}$ defines an n-linear mapping on the tensor space $\mathfrak{H}^{\otimes n}$, which is antisymmetric against the exchange of any two of the vectors f_k. More precisely, since $(-1)^{[\tau\sigma]} = (-1)^{[\tau]}(-1)^{[\sigma]}$ for $\tau, \sigma \in \mathfrak{S}_n$, we obtain

$$P_-^{(n)} f_1 \otimes f_2 \otimes \cdots \otimes f_n = (-1)^{[\tau]} P_-^{(n)} f_{\tau(1)} \otimes f_{\tau(2)} \otimes \cdots \otimes f_{\tau(n)} \tag{8.58}$$

for any permutations $\tau \in \mathfrak{S}_n$.

Excercise Prove the identities

$$P_-^{(n)}\left(P_-^{(n)}f_1 \otimes f_2 \otimes \cdots \otimes f_n\right) = P_-^{(n)}f_1 \otimes f_2 \otimes \cdots \otimes f_n \tag{8.59}$$

and

$$\begin{aligned}
&\langle P_-^{(n)}f_1 \otimes f_2 \otimes \cdots \otimes f_n \mid g_1 \otimes g_2 \otimes \cdots \otimes g_n \rangle \\
&= \langle f_1 \otimes f_2 \otimes \cdots \otimes f_n \mid P_-^{(n)}g_1 \otimes g_2 \otimes \cdots \otimes g_n \rangle
\end{aligned} \tag{8.60}$$

for arbitrary $f_k, g_k \in \mathfrak{H}$, $k = 1, \ldots, n$.

By linearity these identities imply the operator identities

$$P_-^{(n)2} = P_-^{(n)} \quad \text{and} \quad P_-^{(n)*} = P_-^{(n)} \tag{8.61}$$

on $\mathfrak{H}^{\underline{\otimes} n}$. Hence $P_-^{(n)}$ is a projection operator on $\mathfrak{H}^{\otimes n}$, which can be extended to a projection operator on $\mathfrak{H}^{\otimes n}$. We can now continue as in the symmetric case. The image

$$\Gamma_n^-(\mathfrak{H}) \triangleq P_-^{(n)}\mathfrak{H}^{\otimes n} \subset \mathfrak{H}^{\otimes n} \tag{8.62}$$

is a closed subspace of $\mathfrak{H}^{\otimes n}$, and the elements $F \in \Gamma_n^-(\mathfrak{H})$ are the **antisymmetric tensors** of degree $n \in \mathbb{N}$.

With $P_-^{(0)}\lambda \triangleq \lambda \in \mathbb{C} = \Gamma_0^-(\mathfrak{H})$ and $P_-^{(1)}f \triangleq f \in \mathfrak{H} = \Gamma_1^-(\mathfrak{H})$, we now define a projection operator P_- on the Fock space $\Gamma(\mathfrak{H})$ by

$$F = \bigoplus_{n=0}^{\infty} F_n \mapsto P_-F \triangleq \bigoplus_{n=0}^{\infty} P_-^{(n)}F_n, \qquad \text{with } F_n \in \mathfrak{H}^{\otimes n} \tag{8.63}$$

The restriction of this operator to $\mathfrak{H}^{\otimes n}$ agrees with $P_-^{(n)}$.

The closed linear subspace

$$\Gamma^-(\mathfrak{H}) \triangleq P_-\Gamma(\mathfrak{H}) = \bigoplus_{n=0}^{\infty} P_-^{(n)}\mathfrak{H}^{\otimes n} \tag{8.64}$$

is the **Fock space of antisymmetric tensors** or the **Fermion Fock space**.

As an example, take the Hilbert space $\mathfrak{H} = L^2(\mathbb{R}, d\mu)$ as in the Fock example. Then the tensors in $\Gamma_n^-(\mathfrak{H})$, $n \in \mathbb{N}$, *are* functions $f_n^-(x_1, \ldots, x_n) \in L^2(\mathbb{R}^n, d^n\mu)$ that are antisymmetric with respect to the exchange of the variables $x_j \leftrightarrow x_k$. An element of the Fock space $\Gamma^+(\mathfrak{H})$ is a sequence $f = \{f_0, f_1, \ldots\}$ of tensors $f_0 \in \mathbb{C}$ and antisymmetric functions $f_n \in L^2(\mathbb{R}^n, d^n\mu)$ with a finite norm (8.27).

8.4.2 The Grassmann Algebra

The algebraic subspace of antisymmetric tensors of degree n is denoted by $\mathcal{A}_n^-(\mathfrak{H})$, i.e. $\mathcal{A}_n^-(\mathfrak{H}) = P_-^{(n)}\mathfrak{H}^{\otimes n} = \underline{\Gamma}(\mathfrak{H}) \cap \Gamma_n^-(\mathfrak{H}) \subset \mathfrak{H}^{\otimes n}$. The tensor product (8.22) can be defined for arbitrary tensors of the space

$$\mathcal{A}^-(\mathfrak{H}) \triangleq P_-\underline{\Gamma}(\mathfrak{H}) = \bigoplus_{n \geq 0} \mathcal{A}_n^-(\mathfrak{H}) = \underline{\Gamma}(\mathfrak{H}) \cap \Gamma^-(\mathfrak{H}), \tag{8.65}$$

and the bilinear mapping

$$F, G \in \mathcal{A}^-(\mathfrak{H}) \mapsto P_-\,(F \otimes G) \in \mathcal{A}^-(\mathfrak{H}) \tag{8.66}$$

is a product within $\mathcal{A}^-(\mathfrak{H})$. The antisymmetrization operator satisfies the identity

$$P_-\,(P_-F \otimes P_-G) = P_-\,(F \otimes G) \tag{8.67}$$

for tensors $F \in \mathfrak{H}^{\otimes m}$ and $G \in \mathfrak{H}^{\otimes n}$ and $m, n \in \mathbb{N}$. For the proof of this statement, see e.g. Dieudonné (1972).

As a consequence of (8.67), the product (8.66) is associative. Similar to the symmetric case, one can define many associative products on $\mathcal{A}^-(\mathfrak{H})$, but there is exactly one – denoted as **exterior product** or **Grassmann product** $F \wedge G$ – that has the desired factorization property. This exterior product is defined by

$$F \in \mathcal{A}_m^-, \; G \in \mathcal{A}_n^- \mapsto F \wedge G = \sqrt{\frac{(m+n)!}{m!\,n!}} P_-\,(F \otimes G) \in \mathcal{A}_{m+n}^- \tag{8.68}$$

for tensors of fixed degree $m, n = 0, 1, \ldots$, and it is extended to a product on $\mathcal{A}^-(\mathfrak{H})$ by linearity. The proof of the associativity follows as in Proposition 8.3.1 for the symmetric tensor algebra. Moreover, the product (8.68) satisfies the following decomposition given in equation (8.69).

Lemma 8.4.1 *If $\mathfrak{H}_1$ and $\mathfrak{H}_2$ are two orthogonal subspaces of $\mathfrak{H}$, then the inner product of the tensors $F_1 \wedge F_2$ and $G_1 \wedge G_2$ with $F_k, G_k \in \mathcal{A}^-(\mathfrak{H}_k)$, $k = 1, 2$, factorizes into*

$$\langle F_1 \wedge F_2 \mid G_1 \wedge G_2 \rangle = \langle F_1 \mid G_1 \rangle \langle F_2 \mid G_2 \rangle. \tag{8.69}$$

This identity can be derived with essentially the same arguments as given in the proof of Lemma 8.3.2. But one can also use identities for determinants; see the appendix in Dieudonné (1972). The space $\mathcal{A}^-(\mathfrak{H})$ with the product (8.68) is called the Grassmann algebra. Its unit is the vector $1 \in \mathbb{C} = \mathcal{A}_0^-(\mathfrak{H})$.

As a consequence of (8.58), the exterior product satisfies the rule

$$F \wedge G = (-1)^{mn} G \wedge F \tag{8.70}$$

if $F \in \mathcal{A}_m^-$ and $G \in \mathcal{A}_n^-$. For vectors in $\mathfrak{H}$, this product is antisymmetric $f \wedge g = -g \wedge f$ and all powers $f \wedge f$, $f \wedge f \wedge f$, and so on of a vector $f \in \mathfrak{H}$ vanish.

From the definition (8.68) and the norm identity (8.25), we obtain

$$\|F_m \wedge G_n\| \leq \sqrt{\frac{(m+n)!}{m!\,n!}} \, \|F_m\| \, \|G_n\| \tag{8.71}$$

for $F \in \mathcal{A}_m^-(\mathfrak{H})$ and $G \in \mathcal{A}_n^-(\mathfrak{H})$. Hence the tensor product of $\mathcal{A}_m^-(\mathfrak{H})$ and $\mathcal{A}_n^-(\mathfrak{H})$ is continuous, and it can be extended to the Hilbert spaces $\Gamma_m^-(\mathfrak{H})$ and $\Gamma_n^-(\mathfrak{H})$ by continuity. For arbitrary tensors $F_m \in \Gamma_m^-(\mathfrak{H})$ and $G_n \in \Gamma_n^-(\mathfrak{H})$, we then obtain $F_m \wedge G_n \in \Gamma_{m+n}^-(\mathfrak{H})$ with the norm estimate (8.71). But the exterior product is not continuous on $\mathcal{A}^-(\mathfrak{H})$, and it has no continuous extension onto the whole Fock space $\Gamma^-(\mathfrak{H})$.

As in the case of symmetric tensors, we derive some useful results for decomposable antisymmetric tensors. The tensors $f_1 \wedge \cdots \wedge f_m = \sqrt{m!}P_-f_1 \otimes \cdots \otimes f_m$ and $g_1 \wedge \cdots \wedge g_n = \sqrt{n!}P_-g_1 \otimes \cdots \otimes g_n$ with $f_j, g_k \in \mathfrak{H}$ have the inner product $\langle f_1 \wedge \cdots \wedge f_m \mid g_1 \wedge \cdots \wedge g_n \rangle = 0$ if $m \neq n$ and

$$\langle f_1 \wedge \cdots \wedge f_n \mid g_1 \wedge \cdots \wedge g_n \rangle = n! \, \langle f_1 \otimes \cdots \otimes f_n \mid P_A g_1 \otimes \cdots \otimes g_n \rangle$$

$$= \sum_{\sigma \in \mathfrak{S}_n} (-1)^{[\sigma]} \langle f_1 \otimes \cdots \otimes f_n \mid g_{\sigma(1)} \otimes \cdots \otimes g_{\sigma(n)} \rangle$$

$$= \sum_{\sigma \in \mathfrak{S}_n} (-1)^{[\sigma]} \prod_{j=1}^{n} \langle f_j \mid g_{\sigma(j)} \rangle.$$

Hence the general rule is

$$\langle f_1 \wedge \cdots \wedge f_m \mid g_1 \wedge \cdots \wedge g_n \rangle = \begin{cases} 0 & \text{if } m \neq n \\ \det\big((f_j \mid g_k)\big) & \text{if } m = n. \end{cases} \tag{8.72}$$

As in the case of the symmetric tensor algebra, there are other normalizations of the inner product and of the exterior product. But these definitions should finally lead to (8.72).

8.4.3 Basis Systems

If e_k, $k \in \mathbb{K} \subset \mathbb{N}$ is an ON basis of $\mathfrak{H}$, the nonvanishing exterior products of n basis vectors are the tensors $\pm e_{k_1} \wedge e_{k_2} \wedge \cdots \wedge e_{k_n}$, where the indices $k_1, k_2, \ldots, k_n$ are n different numbers. If $\mathbb{J} \subset \mathbb{K}$ is a subset of $|\mathbb{J}| = n$ elements $\{k_1, k_2, \ldots, k_n\}$, one can uniquely order these numbers $k_1 < \cdots < k_n$. Then we use the notation $e_{\mathbb{J}} \triangleq e_{k_1} \wedge e_{k_2} \wedge \cdots \wedge e_{k_n}$. As a consequence of (8.72), these tensors are normalized, $\|e_{\mathbb{J}}\| = 1$, and $\langle e_{\mathbb{J}} \mid e_{\mathbb{K}} \rangle = 0$ if $\mathbb{J} \neq \mathbb{K}$. Hence the tensors $e_{\mathbb{J}}$ with $\mathbb{J} \subset \mathbb{K}$, $|\mathbb{J}| = n$ form an ON basis of $\Gamma_n^-(\mathfrak{H})$. With the notation

$e_\emptyset = 1 \in \mathbb{C}$ for the basis vector of $\Gamma_0^-(\mathfrak{H})$, the Fock space of antisymmetric tensors $\Gamma^-(\mathfrak{H}) = \bigoplus_{n=0}^\infty \Gamma_n^-(\mathfrak{H})$ has the basis $\{e_\mathbb{J} \mid \mathbb{J} \subset \mathbb{K}, \ |\mathbb{J}| < \infty\}$. The index set of this basis is the powerset Power($\mathbb{K}$), i.e. the set of all finite subsets of $\mathbb{K}$.

If $\dim \mathfrak{H} = d < \infty$, the set $\mathbb{K} = \{1, 2, \ldots, K\}$ has $K = d$ elements, and the set of all subsets $\mathbb{J} \subset \mathbb{M}$ with $|\mathbb{J}| = n$ has $\binom{d}{n}$ elements. The space of antisymmetric tensors of degree p has therefore the dimension $\dim \Gamma_n^-(\mathfrak{H}) = \binom{d}{n}$ if $0 \le n \le d$, and $\dim \Gamma_n^-(\mathfrak{H}) = 0$ if $n > d$. Hence the Fock space has the finite dimension $\dim \Gamma^-(\mathfrak{H}) = 2^d$.

For many calculations, it is convenient to use a notation based on occupation numbers, as in the case of symmetric tensors. With the definitions $e_k^{\wedge 0} = 1$, $e_k^{\wedge 1} = e_k$, we can write the basis $e_\mathbb{J}$ in the form

$$e_-(n_1, n_2 \ldots) \triangleq e_1^{\wedge n_1} \wedge e_2^{\wedge n_2} \wedge \cdots, \tag{8.73}$$

where the antisymmetry restricts the occupation numbers as

$$n_k \in \mathbb{N}_- \triangleq \{0, 1\}. \tag{8.74}$$

In this notation, the basis of $\Gamma_n^-(\mathfrak{H})$ is

$$e_-(n_1, n_2 \ldots) \quad \text{with} \ (n_1, n_2 \ldots) \in \mathbb{N}_-^\infty, \ \sum_{k \in \mathbb{M}} n_k = n. \tag{8.75}$$

The basis of the Fock space $\Gamma^-(\mathfrak{H})$ is given by (8.73) with the set of all terminating sequences $(n_1, n_2 \ldots) \in (\mathbb{N}_-)^\mathbb{K}$ as the index set.

The norm estimate (8.71) implies that the exterior product is well defined on the linear span of the Hilbert spaces $\Gamma_n^-(\mathfrak{H})$, $n \ge 0$. Hence the space $\Gamma_{\text{fin}}^-(\mathfrak{H}) = \Gamma^-(\mathfrak{H}) \cap \Gamma_{\text{fin}}(\mathfrak{H})$ is an algebra with the exterior product. If $\dim \mathfrak{H}$ is infinite, the algebra $\Gamma_{\text{fin}}^-(\mathfrak{H})$ is strictly larger than $\mathcal{A}^-(\mathfrak{H})$.

The ON basis for Fermion Fock space may be alternatively labeled by the finite subsets of $\mathbb{K}$, as each state e_k may occur at most once.

The Fermion Fock space is finite dimensional whenever its one-particle space is finite dimensional. However, when the one-particle space $\mathfrak{H}$ is infinite dimensional, both the Boson and Fermion Fock space over $\mathfrak{H}$ are infinite dimensional too, and so are isomorphic as Hilbert spaces. Indeed, one may construct unitary maps between the Boson and Fermion Fock spaces in this case. As an example, we take the Fock spaces $\Gamma^+(\mathfrak{H})$ and $\Gamma^-(\mathfrak{H})$ over $\mathfrak{H} = L^2(\mathbb{R}, d\mu)$, where $d\mu$ is a nonatomic measure on $\mathbb{R}$. The tensors in $\Gamma_n^\pm(\mathfrak{H})$, $n \in \mathbb{N}$, are functions $f_n^\pm(x_1, \ldots, x_n) \in L^2(\mathbb{R}^n, d^n\mu)$, which are symmetric/antisymmetric in the variables $(x_1, \ldots, x_n) \in \mathbb{R}^n$; see the corresponding examples for symmetric and antisymmetric Fock spaces. Then $\Gamma_0^+(\mathfrak{H}) \simeq \Gamma_0^-(\mathfrak{H}) \simeq \mathbb{C}$ and the the tensor spaces $\Gamma_n^\pm(\mathfrak{H})$, $n \in \mathbb{N}$, are isomorphic with the simple identification $f_n^+ \simeq f_n^-$ if $f_n^+(x_1, \ldots, x_n) = f_n^-(x_1, \ldots, x_n)$ for $x_1 < x_2 < \cdots < x_n$. The diagonals $x_j = x_k$ do not count, since the measure is nonatomic. These isometric isomorphisms imply an isometric isomorphism between the Fock spaces $\Gamma^+(\mathfrak{H})$ and $\Gamma^-(\mathfrak{H})$. In both cases of symmetric and antisymmetric tensors, the norm (8.27) can be written as

$$\|f\|^2 = |f_0|^2 + n! \sum_{n=1}^{\infty} \int_{x_1 < \ldots < x_n} |f_n(x_1, \ldots, x_n)|^2 \, d^n\mu. \qquad (8.76)$$

9

Operators and Fields on the Boson Fock Space

In this chapter, we introduce more advanced structures defined on Fock spaces, especially on the Boson Fock space. These objects include the algebras of creation and annihilation operators, of Weyl operators, and of Segal field operators. Moreover, some important distributions of quantum fields, as Gaussian fields or thermal fields, are presented. The chapter is completed with a short insight into q-deformed structures.

9.1 Operators on Fock Spaces

9.1.1 The Tensor Product of Operators

In this section, we give some general constructions on tensor spaces and define operators for both the Boson and the Fermion Fock space.

If $S_1 \in \mathcal{L}(\mathfrak{E}_1)$ is a linear operator on the inner product space $\mathfrak{E}_1$ and $S_2 \in \mathcal{L}(\mathfrak{E}_2)$ a linear operator on the inner product space $\mathfrak{E}_2$, then we define the **tensor product** $S_1 \otimes S_2$ of theses operators as linear operators on the algebraic tensor space $\mathfrak{E}_1 \underline{\otimes} \mathfrak{E}_2$ first on the set of decomposable tensors $f_1 \otimes f_2$, $f_1 \in \mathfrak{E}_1$, $f_2 \in \mathfrak{E}_2$, according to the rule

$$(S_1 \otimes S_2)(f_1 \otimes f_2) \triangleq (Sf_1) \otimes (Sf_2). \tag{9.1}$$

This operator can then be defined on the algebraic tensor product $\mathfrak{E}_1 \underline{\otimes} \mathfrak{E}_2$ by linear extension. If $T_k \in \mathcal{L}(\mathfrak{E}_k)$, $k = 1, 2$ are two other operators on these spaces, then the definition (9.1) implies

$$(S_1 \otimes S_2)(T_1 \otimes T_2) = S_1 T_1 \otimes S_2 T_2. \tag{9.2}$$

Proposition 9.1.1 *If S_1 and S_2 are bounded operators, then also $S_1 \otimes S_2$ is a bounded operator with the norm*

$$\|S_1 \otimes S_2\| = \|S_1\| \|S_2\| . \tag{9.3}$$

Proof On the set of decomposable tensors, we have

$$\langle g_1 \otimes g_2 \mid (S_1 \otimes S_2)(f_1 \otimes f_2)\rangle = \langle g_1 \mid S_1 f_1\rangle \langle g_2 \mid S_2 f_2\rangle . \tag{9.4}$$

From the general formula (8.4) for the norm of an operator on an inner product space follows that the norm of $S_1 \otimes S_2$ has the lower bound

$$\|S_1 \otimes S_2\| \geq \sup \mathrm{Re}\, \langle g_1 \otimes g_2 \mid (S_1 \otimes S_2)(f_1 \otimes f_2)\rangle , \tag{9.5}$$

where the supremum is calculated for all normalized vectors $f_k, g_k \in \mathfrak{E}_k$, $k = 1, 2$. Then (9.4) implies

$$\|S_1 \otimes S_2\| \geq \sup \mathrm{Re}\, \langle g_1 \mid S_1 f_1\rangle \cdot \sup \mathrm{Re}\, \langle g_2 \mid S_2 f_2\rangle = \|S_1\| \|S_2\| . \tag{9.6}$$

Let $\{e_\mu^k\}$ be orthonormal basis systems for $\mathfrak{E}_k$, with $k = 1, 2$ respectively. For $F = \sum_{\mu\nu} c_{\mu\nu} e_\mu^1 \otimes e_\nu^2 \in \mathfrak{E}_1 \otimes \mathfrak{E}_2$, we calculate the norm of $(S_1 \otimes I_2) F = \sum_{\mu\nu} c_{\mu\nu} (S_1 e_\mu^1) \otimes e_\nu^2$

$$\|(S_1 \otimes I_2) F\|^2 = \sum_\nu \left\| \sum_\mu c_{\mu\nu} S_1 e_\mu^1 \right\|^2 \leq \sum_\nu \|S_1\|^2 \sum_\mu |c_{\mu\nu}|^2$$
$$\leq \|S_1\|^2 \|F\|^2 .$$

Hence $\|S_1 \otimes I_2\| \leq \|S_1\|$ is true. By the same arguments, we derive $\|I_1 \otimes S_2\| \leq \|S_2\|$. From (9.2), we obtain $S_1 \otimes S_2 = (S_1 \otimes I_2)(I_1 \otimes S_2)$. The norm of this product is therefore bounded by

$$\|S_1 \otimes S_2\| = \|(S_1 \otimes I_2)\| \|(I_1 \otimes S_2)\| \leq \|S_1\| \|S_2\| . \tag{9.7}$$

The inequalities (9.6) and (9.7) imply (9.3).   □

If the spaces $\mathfrak{H}_1$ and $\mathfrak{H}_2$ are Hilbert spaces, the relation (9.3) implies that the tensor product $S_1 \otimes S_2$ of operators can be extended from the algebraic tensor space $\mathfrak{H}_1 \underline{\otimes} \mathfrak{H}_2$ onto the completed space $\mathfrak{H}_1 \otimes \mathfrak{H}_2$ by continuity. The norm identity (9.3) remains valid.

From (9.1) and the definition of the inner product (8.11), it immediately follows that the adjoint operator of $S_1 \otimes S_2$ is given by

$$(S_1 \otimes S_2)^* = S_1^* \otimes S_2^* . \tag{9.8}$$

The tensor product of self-adjoint operators on Hilbert spaces is therefore again self-adjoint.[1] The relation (9.8) together with (9.2) yields that the tensor product of unitary operators is again unitary.

If S is a bounded linear operator on the Hilbert space $\mathfrak{H}$, then we define the operators $S^{\otimes n} \triangleq \underbrace{S \otimes \cdots \otimes S}_{n}$ on the set of decomposable tensors by

$$S^{\otimes n} (f_1 \otimes \cdots \otimes f_n) = Sf_1 \otimes \cdots \otimes Sf_n \qquad (9.9)$$

and extend it to the tensor space $\mathfrak{H}^{\otimes n}$ by linearity and closure. Using (9.3), we see that the norm is

$$\left\| S^{\otimes n} \right\| = \|S\|^n . \qquad (9.10)$$

A simple calculation yields

$$P_{\pm}^{(n)} (Sf_1 \otimes \cdots \otimes Sf_n) = S^{\otimes n} \left(P_{\pm}^{(n)} (f_1 \otimes \cdots \otimes f_n) \right).$$

Hence $S^{\otimes n}$ maps the subspaces $\Gamma_n^{\pm}(\mathfrak{H})$ of symmetric/antisymmetric tensors into themselves. In fact, on these subspaces, the identities

$$S^{\otimes n} (f_1 \vee \cdots \vee f_n) = Sf_1 \vee \cdots \vee Sf_n \qquad (9.11)$$

or

$$S^{\otimes n} (f_1 \wedge \cdots \wedge f_n) = Sf_1 \wedge \cdots \wedge Sf_n \qquad (9.12)$$

hold.

If S is a bounded linear operator on the Hilbert space $\mathfrak{H}$, then we define the linear operator $\Gamma(S)$ on the Fock space $\Gamma(\mathfrak{H})$ by

$$\Gamma(S)|_{\mathfrak{H}^{\otimes n}} = S^{\otimes n}, \qquad (9.13)$$

with $\Gamma(S)|_{\mathfrak{H}^{\otimes 0}}$ understood as multiplication by 1.

This operator maps the subspaces $\mathfrak{H}^{\otimes n}$ into itself with the norm estimate $\|\Gamma(S)F\| \leq \|S\|^n \|F\|$, $F \in \mathfrak{H}^{\otimes n}$; see (9.10). If S is either a unitary or more generally a contraction, that is, if $\|S\| \leq 1$, then the operator $\Gamma(S)$ can be extended by linearity and continuity to a continuous operator on $\Gamma(\mathfrak{H})$.

Since $P_{\pm}\Gamma(S) = \Gamma(S)P_{\pm}$, the operator $\Gamma(S)$ has well-defined restrictions on the Boson and on the Fermion Fock spaces $\Gamma^{\pm}(\mathfrak{H})$. These restrictions can be defined using the identities (9.11) or (9.12) without reference to the full Fock space $\Gamma(\mathfrak{H})$. To make the distinction of these operators more explicit, we also

---

[1] If these operators are unbounded and essentially self-adjoint on domains $\mathcal{D}_1$ and $\mathcal{D}_2$, then $A_1 \otimes A_2$ is essentially self-adjoint on $\mathcal{D}_1 \otimes \mathcal{D}_2$.

write $\Gamma^+(S)$ for the operator on $\Gamma^+(\mathfrak{H})$, and $\Gamma^-(S)$ for the operator on $\Gamma^-(\mathfrak{H})$. The operators $\Gamma^\pm(S)$ are often called the **second quantization** of S.

We now derive some identities for the operator $\Gamma(S)$ on the Boson or the Fermion Fock space $\Gamma^\pm(\mathfrak{H})$. For $S = I$, we obviously obtain

$$\Gamma(I_\mathfrak{H}) = I_{\Gamma^\pm(\mathfrak{H})}. \tag{9.14}$$

Let S and T be two linear operators on $\mathfrak{H}$; then the rule

$$\Gamma(ST) = \Gamma(S)\Gamma(T) \tag{9.15}$$

follows from (9.2). If S has an inverse operator S^{-1}, then

$$(\Gamma(S))^{-1} = \Gamma(S^{-1}) \tag{9.16}$$

is true. The adjoint operator of $\Gamma(S)$ is

$$(\Gamma(S))^* = \Gamma(S^*). \tag{9.17}$$

If U is a unitary operator on $\mathfrak{H}$, then

$$\Gamma(U)\Gamma(U)^* = \Gamma(U)\Gamma(U^*) = \Gamma(UU^*) = \Gamma(I)$$

implies that $\Gamma(U)$ is a unitary operator on $\Gamma^\pm(\mathfrak{H})$. Another useful identity is[2]

$$\Gamma(S)\,(F \circ G) = (\Gamma(S)F) \circ (\Gamma(S)G)\,, \tag{9.18}$$

which is valid for all operators $S \in \mathcal{L}(\mathfrak{H})$ and for all tensors $F, G \in \Gamma^\pm(\mathfrak{H})$ for which the products are defined. This identity is an immediate consequence of (9.11) and (9.12).

If the (unbounded) operator S is defined on the domain $\mathcal{D} \subset \mathfrak{H}$, we can construct the linear operator $\Gamma^\pm(S)$ on the algebra $\mathcal{A}^\pm(\mathcal{D}) \subset \Gamma^\pm(\mathfrak{H})$ of symmetric/antisymmetric tensors with the help of (9.11) or (9.12). Then the domains of definition can be further extended by continuity arguments. If S is essentially self-adjoint, the following result applies; see Reed and Simon (1972, Sect. VIII.10).

Proposition 9.1.2 *If S is essentially self-adjoint on the domain $\mathcal{D} \subset \mathfrak{H}$, the operator $\Gamma^\pm(S)$ is essentially self-adjoint on $\mathcal{A}^\pm(\mathcal{D}) \subset \Gamma^\pm(\mathfrak{H})$.*

Let M be a linear operator on $\mathfrak{H}$. Then we define a linear operator $d\Gamma(M)$ on $\Gamma(\mathfrak{H})$ by $d\Gamma(M)|_{\Gamma^0(\mathfrak{H})=\mathbb{C}} \triangleq 0$, $d\Gamma(M)|_{\Gamma^1(\mathfrak{H})=\mathfrak{H}} \triangleq M$, and

$$d\Gamma(M)|_{\Gamma^n(\mathfrak{H})} \triangleq \underbrace{M \otimes I \otimes \cdots \otimes I}_{n} + \cdots + \underbrace{I \otimes \cdots \otimes I \otimes M}_{n}\,. \tag{9.19}$$

---

[2] Here the symbol $\circ$ stands for either the symmetric tensor product $\vee$ or for the antisymmetric tensor product $\wedge$, depending on the algebra in which it operates.

The restriction $P_{\pm} d\Gamma(M) = d\Gamma(M) P_{\pm}$ maps $\Gamma^{\pm}(\mathfrak{H})$ into itself and it is also denoted as $d\Gamma(M)$. The operator $d\Gamma(M)$ is often called the **differential second quantization** of M.

As with Proposition 9.1.2, the following result can be obtained.

Proposition 9.1.3 *If M is essentially self-adjoint on the domain $\mathcal{D} \subset \mathfrak{H}$, then the operator $d\Gamma(M)$ is essentially self-adjoint on $\mathcal{A}^{\pm}(\mathcal{D}) \subset \Gamma^{\pm}(\mathfrak{H})$.*

To understand the relation between $\Gamma(.)$ and $d\Gamma(.)$, we consider a one-parameter group of unitary operators with a self-adjoint generator M on $\mathfrak{H}$:

$$t \in \mathbb{R} \mapsto U_1(t) \triangleq \exp(-iMt).$$

Then $t \in \mathbb{R} \mapsto U(t) \triangleq \Gamma(U_1(t))$ is a one-parameter group of unitary operators on $\Gamma(\mathfrak{H})$. The Stone generator of this group is the self-adjoint operator $i\frac{d}{dt}\Gamma(U_1(t))|_{t=0}$. To calculate this generator, we take $F = f_1 \otimes f_2 \otimes \cdots \otimes f_n \in \mathfrak{H}^{\otimes n}$. Then the Leibniz rule for differentiation yields

$$
\begin{aligned}
\frac{d}{dt} \quad & \Gamma(U_1)F|_{t=0} \\
&= (\dot{U}_1 f_1)|_{t=0} \otimes f_2 \otimes \cdots \otimes f_n + \cdots + f_1 \otimes f_2 \otimes \cdots \otimes (\dot{U}_1 f_n)|_{t=0} \\
&= -i(Mf_1) \otimes f_2 \otimes \cdots \otimes f_n - \cdots - if_1 \otimes f_2 \otimes \cdots \otimes (Mf_n),
\end{aligned}
$$

and we have derived the operator identity

$$\frac{d}{dt}\Gamma(U_1(t))|_{t=0} = -i\, d\Gamma(M)$$

or

$$U(t) = \Gamma(\exp(-iMt)) = \exp(-it\, d\Gamma(M)) \qquad (9.20)$$

on $\Gamma(\mathfrak{H})$. The same result can be obtained for the operators on the symmetrized Fock spaces $\Gamma^{\pm}(\mathfrak{H})$.

The relation between $d\Gamma(.)$ and the derivative can be formulated more abstractly.

A linear operator D on the algebra $\mathcal{A}^{\pm}(\mathfrak{H})$ is called a **derivation**, if the Leibniz rule

$$D(F \circ G) = (DF) \circ G + F \circ (DG) \qquad (9.21)$$

is valid for all $F, G \in \mathcal{A}^{\pm}(\mathfrak{H})$.

If we choose $F = G = \Omega \in \mathcal{A}^{\pm}(\mathfrak{H})$ as the unit of the algebra, then (9.21) implies $D\Omega = D(\Omega \circ \Omega) = (D\Omega) \circ \Omega + \Omega \circ (D\Omega) = 2D\Omega$. Hence a derivation always satisfies

$$D\,\Omega = 0. \qquad (9.22)$$

We can characterize derivations by the properties listed in the following proposition.

Proposition 9.1.4 *A linear operator D on the algebra $\mathcal{A}^{\pm}(\mathfrak{H})$ with the properties*

(i) $D\,\Omega = 0$,
(ii) *and*

$$D\,(f \circ G) = (Df) \circ G + f \circ (DG), \tag{9.23}$$

if $f \in \mathfrak{H}$, $G \in \mathcal{A}_n^+(\mathfrak{H})$, $n = 1, 2, \ldots$.

is a derivation.

Proof From (9.23), we obtain

$$
\begin{aligned}
D(f_1 \circ \cdots \circ f_n) &= (Df_1) \circ \cdots \circ f_n + f_1 \circ D(f_2 \cdots \circ f_n) \\
&= (Df_1) \circ \cdots \circ f_n + f_1 \circ (Df_2) \circ f_3 \cdots \circ f_n \\
&\quad + f_1 \circ f_2 \circ D(f_3 \cdots \circ f_n) \\
&= \sum_{j=1}^{n} f_1 \circ \cdots f_{j-1} \circ (Df_j) \circ f_{j+1} \cdots \circ f_n
\end{aligned}
\tag{9.24}
$$

for $n = 2, 3, \ldots$. This identity implies (9.21) for decomposable tensors $F = f_1 \circ \cdots \circ f_m \in \mathcal{A}_m^{\pm}(\mathfrak{H})$ and $G = f_{m+1} \circ \cdots \circ f_{m+n} \in \mathcal{A}_n^{\pm}(\mathfrak{H})$. But then linearity yields (9.21) for all tensors $F, G \in \mathcal{A}^{\pm}(\mathfrak{H})$. $\square$

Corollary 9.1.5 *A derivation D on $\mathcal{A}^{\pm}(\mathfrak{H})$ is uniquely determined by its action on $\mathfrak{H}$, i.e. by the values Df, $f \in \mathfrak{H}$.*

Proposition 9.1.4 implies the following corollary for the differential second quantization.

Corollary 9.1.6 *The operator $d\Gamma(M)$ is a derivation D on the algebra $\mathcal{A}^{\pm}(\mathfrak{H})$ with $Df = Mf$, $f \in \mathfrak{H}$.*

If the Hilbert space of a single particle is denoted by $\mathfrak{H}$, the Hilbert space for an n-particle system of this species is the space $\Gamma_n^+(\mathfrak{H})$ for Bosons, and it is $\Gamma_n^-(\mathfrak{H})$ for Fermions. The quantum field theory with an arbitrary number of particles can then be formulated on the Fock space $\Gamma^{\pm}(\mathfrak{H})$. Thereby, the operators $\Gamma(S)$ and $d\Gamma(M)$ play an important role.

For $M = I$, the operator $d\Gamma(I)$ has the property

$$d\Gamma(I)\,(f_1 \circ \cdots \circ f_n) = nf_1 \circ \cdots \circ f_n, \qquad n \geq 1.$$

By linearity and closure, we obtain $d\Gamma(I)F = nF$ if $F \in \Gamma_n^\pm(\mathfrak{H}), n \geq 0$. Hence $d\Gamma(I)$ is the particle number operator and will often be denoted as N.

Let H_1 be the one-particle Hamilton operator on the Hilbert space $\mathfrak{H}$. If there are no interactions between the particles, the Hamilton operator for systems of an arbitrary number of such particles is

$$H = d\Gamma(H_1). \tag{9.25}$$

As a consequence of (9.20), the unitary groups $U_1(t) = \exp(-iH_1 t)$ on $\mathfrak{H}$ and $U(t) = \exp(-iHt)$ on $\Gamma^\pm(\mathfrak{H})$ are related by

$$U(t) = \Gamma(U_1(t)). \tag{9.26}$$

9.1.2 Creation and Annihilation Operators

If $f \in \mathfrak{H}$, then the **creation operator** $A^*(f)$ and of the vector f is defined as linear operators on the Fock space $\Gamma^\pm(\mathfrak{H})$ by

$$F \in \Gamma_n^\pm(\mathfrak{H}) \mapsto A^*(f)F \triangleq f \circ F \in \Gamma_{n+1}^\pm(\mathfrak{H}), \ n \geq 0, \tag{9.27}$$

and the **annihilation operator** $A(f)$ is the adjoint operator

$$A(f) = \left(A^*(f)\right)^*, \tag{9.28}$$

which maps $\Gamma_n^\pm(\mathfrak{H})$ into $\Gamma_{n-1}^\pm(\mathfrak{H}), \ n \geq 1$, and $\Gamma_0^\pm(\mathfrak{H})$ onto 0. The operator $A^*(f)$ depends linearly on $f \in \mathfrak{H}$, and $A(f)$ depends antilinearly on f. The norm estimates (8.50) and (8.71) imply that the operators $A^*(f)$ and $A(f)$ are well defined on the algebra $\Gamma_{\text{fin}}^\pm(\mathfrak{H}) = \oplus_{n \geq 0}\Gamma_n^\pm(\mathfrak{H})$, the linear span of the Hilbert spaces $\Gamma_n^\pm(\mathfrak{H})$.

If the annihilation operator is applied to a decomposable tensor we obtain with $G \in \Gamma_{\text{fin}}^\pm(\mathfrak{H})$

$$\langle G \mid A(f)(f_1 \circ \cdots \circ f_n)\rangle = \langle f \circ G \mid f_1 \circ \cdots \circ f_n\rangle$$
$$= \sum_{j=1}^n (\pm 1)^{j+1} \langle f \mid f_j\rangle\langle G \mid f_1 \circ \cdots \circ f_{j-1} \circ f_{j+1} \circ \cdots \circ f_n\rangle. \tag{9.29}$$

Hence the action of $A(f)$ on decomposable tensors is

$$A(f)f_1 \circ \cdots \circ f_n = \sum_{j=1}^n (\pm 1)^{j+1} \langle f \mid f_j\rangle f_1 \circ \cdots \circ f_{j-1} \circ f_{j+1} \circ \cdots \circ f_n. \tag{9.30}$$

If $A(f), f \in \mathfrak{H}$, is applied to the vacuum Ω, we obtain

$$\langle G \mid A(f)\,\Omega \rangle = \langle f \circ G \mid \Omega \rangle = 0,$$

since $f \circ G$ is orthogonal to the vacuum for any $G \in \Gamma^{\pm}(\mathfrak{H})$. Hence

$$A(f)\,\Omega = 0 \tag{9.31}$$

follows.

The annihilation operators have an interesting algebraic property: they are derivations in the Boson case and antiderivations in the Fermion case.

A linear operator D on the Grassmann algebra $\mathcal{A}^-(\mathfrak{H})$ is called an **antiderivation**, if

$$D(F \wedge G) = (DF) \wedge G + (-1)^n F \wedge (DG) \tag{9.32}$$

is valid for $F \in \mathcal{A}_n(\mathfrak{H})$, $n = 0, 1, 2, \ldots$, and $G \in \mathcal{A}(\mathfrak{H})$.

Take $F = G = \Omega$ in (9.32), then $D\,\Omega = (D\,\Omega) \wedge \Omega + \Omega \wedge (D\,\Omega) = 2D\,\Omega$ follows. Hence an antiderivation satisfies $D\,\Omega = 0$.

Proposition 9.1.7 *The annihilation operator $A(f)$ is a derivation on the symmetric tensor algebra $\mathcal{A}^+(\mathfrak{H})$, and it is an antiderivation on the Grassmann algebra $\mathcal{A}^-(\mathfrak{H})$.*

Proof   It is sufficient to prove the identity

$$A(f)(F \circ G) = (A(f)F) \circ G + (\pm 1)^m F \circ (A(f)G) \tag{9.33}$$

for $F \in \Gamma_m^{\pm}(\mathfrak{H})$ and $G \in \Gamma_n^{\pm}(\mathfrak{H})$, $m, n \geq 0$. The general case follows by linearity. If $m = 0$ or $n = 0$, this identity is obviously true. If $m, n \geq 1$ the proof can be restricted to decomposable tensors $F = f_1 \circ \cdots \circ f_m \in \Gamma_m^{\pm}(\mathfrak{H})$ and $G = g_1 \circ \cdots \circ g_n \in \Gamma_n^{\pm}(\mathfrak{H})$. Then $A(f)(F \circ G)$ is given by (9.30) and (9.31). The sum for $A(f)(f_1 \circ \cdots \circ f_m \circ g_1 \circ \cdots \circ g_n)$ extends over $m + n$ terms. The terms with $j = 1, \ldots, m$ correspond to $(A(f)F) \circ G$ and the terms with $j = m + 1, \ldots, m + n$ give $(\pm 1)^m F \circ (A(f)G)$. $\square$

The creation and annihilation operators satisfy the **canonical (anti) commutation relations**

$$\left[A^*(f), A^*(g)\right]_{\mp} = \left[A(f), A(g)\right]_{\mp} = 0,$$
$$\left[A(f), A^*(g)\right]_{\mp} = \langle f \mid g \rangle\, I, \tag{9.34}$$

for all $f, g \in \mathfrak{H}$, where $[.,.]_{\mp}$ is the commutator $[S, T]_- \triangleq [S, T] = ST - TS$ on the Boson Fock space, and it is the anticommutator $[S, T]_+ \triangleq \{S, T\} = ST + TS$ on the Fermion Fock space.

The proof of the first line of (9.34) follows from $A^*(f)A^*(g)F = f \circ g \circ F = \pm g \circ f \circ F = \pm A^*(g)A^*(f)F$, $F \in \Gamma^\pm_{\text{fin}}(\mathfrak{H})$ and the relation $A(f)A(g) = (A^*(g)A^*(f))^*$. To derive the second line, we take a tensor $F \in \Gamma^\pm_{\text{fin}}(\mathfrak{H})$ and calculate

$$A(f)A^*(g)F = A(f)(g \circ F) \stackrel{(9.33)}{=} (A(f)g) \circ F \pm g \circ A(f)F$$
$$A^*(g)A(f)F = g \circ A(f)F.$$

Since $A(f)g = \langle f \mid g \rangle$, the last identity in (9.34) follows.

Let f_μ, $\mu = 1, 2, \ldots$, be an orthonormal basis of $\mathfrak{H}$. Then $\langle f_\mu \mid f_\nu \rangle = \delta_{\mu\nu}$ and the operators $a^+_\mu = A^*(f_\mu)$ and $a^-_\nu = A(f_\nu)$ satisfy the (anti) commutation relations

$$\left[a^+_\mu, a^+_\nu\right]_\mp = \left[a^-_\mu, a^-_\nu\right]_\mp = 0,$$
$$\left[a^-_\mu, a^+_\nu\right]_\mp = \delta_{\mu\nu} I. \tag{9.35}$$

The Boson creation/annihilation operators $A^*(f)$ and $A(f)$ are well defined on the algebra $\Gamma^+_{\text{fin}}(\mathfrak{H})$; but these operators are unbounded and they cannot be defined on the whole Fock space. To prove this statement, we take a normalized vector $f \in \mathfrak{H}$. Then $F_n = \frac{1}{\sqrt{n!}} f^{\vee n} \in \mathcal{A}^+_n(\mathfrak{H})$ are normalized tensors for $n = 1, 2, \ldots$. The tensor $A^*(f)F_n = \frac{1}{\sqrt{n!}} f^{\vee n+1}$ has the norm $\|A^*(f)F_n\| = \sqrt{\frac{(n+1)!}{n!}} = \sqrt{n+1}$. Hence $\sup_n \|A^*(f)F_n\|$ is infinite and $A^*(f)$ is unbounded. Since $A^*(\lambda f) = \lambda A^*(f)$ for $\lambda \in \mathbb{C}$, the creation operators $A^*(f)$ are unbounded for all $f \neq 0$. As a consequence of (9.28), also $A(f)$ is unbounded for all $f \neq 0$.

In contrast to this result, the Fermion creation/annihilation operators are bounded operators on the Grassmann algebra $\mathcal{A}^-(\mathfrak{H})$. That statement follows from the canonical anticommutation relations (9.34). As a consequence of (9.28), the products $A(f)A^*(f)$ and $A^*(f)A(f)$ are positive operators for all $f \in \mathfrak{H}$. Hence we get the bounds

$$0 \leq A(f)A^*(f) + A^*(f)A(f) = \left\{A(f), A^*(f)\right\} \stackrel{(9.34)}{=} \|f\|^2 I,$$

and the operators $A(f)A^*(f)$ and $A^*(f)A(f)$ are bounded with an operator norm $\|A(f)A^*(f)\| \leq \|f\|^2$ and $\|A^*(f)A(f)\| \leq \|f\|^2$. Therefore,

$$\left\|A^*(f)F\right\|^2 = \langle A^*(f)F \mid A^*(f)F \rangle = \langle F \mid A(f)A^*(f)F \rangle \leq \|f\|^2 \|F\|^2$$

is valid for all $f \in \mathfrak{H}$ and $F \in \mathcal{A}^-(\mathfrak{H})$. Hence $A^*(f)$ is bounded with operator norm $\|A^*(f)\| \leq \|f\|$. Since $A^*(f) 1 = f$, we even have the stronger statement $\|A^*(f)\| = \|f\|$. The adjoint operator (9.28) has the same operator norm $\|A(f)\| = \|A^*(f)\| = \|f\|$. The Fermion creation and annihilation operators have therefore continuous extensions on the whole Fock space $\Gamma^-(\mathfrak{H})$.

For many calculations in quantum field theory, it is convenient to introduce the **Segal field operator**:

$$\hat{\Phi}(f) \triangleq A^*(f) + A(f), \qquad f \in \mathfrak{H}. \tag{9.36}$$

The operator $\hat{\Phi}(f), f \in \mathfrak{H}$ is essentially self-adjoint on the algebra $\Gamma_{\text{fin}}^{\pm}(\mathfrak{H})$ and it satisfies the relations

$$\hat{\Phi}(f + g) = \hat{\Phi}(f) + \hat{\Phi}(g),$$
$$\hat{\Phi}(\alpha f) = \alpha \,\hat{\Phi}(f) \,\text{if} \,\alpha \in \mathbb{R}.$$

Using the fact that $f \mapsto A^*(f)$ is linear and $f \mapsto A(f)$ is antilinear, the creation and annihilation operators can be recovered from the Segal field operator as

$$A^*(f) = 2^{-1} \left(\hat{\Phi}(f) - i \,\hat{\Phi}(if) \right),$$
$$A(f) = 2^{-1} \left(\hat{\Phi}(f) + i \,\hat{\Phi}(if) \right), \tag{9.37}$$

and the canonical (anti)commutation relations (9.34) turn out to be equivalent to the identity

$$\left[\hat{\Phi}(f), \hat{\Phi}(g) \right]_{\varepsilon} = \begin{cases} 2i \,\text{Im} \,\langle f \mid g \rangle, & \varepsilon = - \\ 2 \,\text{Re} \,\langle f \mid g \rangle, & \varepsilon = + \end{cases}. \tag{9.38}$$

To present still another version of the canonical (anti)commutation relations, we introduce an antilinear involution $f \mapsto f^*$ on the Hilbert space $\mathfrak{H}$. In Section 11.1, such an involution is given for the Hilbert space of relativistic spin zero bosons. The vectors with $f = f^*$ may be called the "real vectors," and those with $f = -f^*$ are the "imaginary vectors." Any vector $f \in \mathfrak{H}$ can be decomposed into $f = f_1 + if_2$ with vectors $f_1 = 1/2 \,(f + f^*)$ and $f_2 = 1/(2i) \,(f - f^*)$ in the real Hilbert space $\mathfrak{H}_{\mathbb{R}} = \{f \in \mathfrak{H} : f = f^*\}$. The operator (9.36) is then decomposed into

$$\hat{\Phi}(f) = \hat{Q}(f_1) + \hat{P}(f_2). \tag{9.39}$$

The self-adjoint operators $\hat{Q}(f)$ and $\hat{P}(f)$ are defined for $f \in \mathfrak{H}_{\mathbb{R}}$ as $\hat{Q}(f) \triangleq \hat{\Phi}(f)$ and $\hat{P}(f) \triangleq \hat{\Phi}(if)$. They satisfy the commutation relations

$$\left[\hat{Q}(f), \hat{Q}(g) \right]_{-} = \left[\hat{P}(f), \hat{P}(g) \right]_{-} = 0,$$
$$\left[\hat{Q}(f), \hat{P}(g) \right]_{-} = i \,\langle f \mid g \rangle \,I, \tag{9.40}$$

for $f, g \in \mathfrak{H}_{\mathbb{R}}$ on $\Gamma_{\text{fin}}^{+}(\mathfrak{H})$, and the anticommutation relations

$$\left[\hat{Q}(f), \hat{Q}(g) \right]_{+} = \left[\hat{P}(f), \hat{P}(g) \right]_{+} = \langle f \mid g \rangle \,I$$
$$\left[\hat{Q}(f), \hat{P}(g) \right]_{+} = 0 \tag{9.41}$$

for $f, g \in \mathfrak{H}_{\mathbb{R}}$ on $\Gamma^-(\mathfrak{H})$. In the case of the symmetric tensor algebra, the operators (9.39) are the generalizations of the position and momentum operators of Section 3.2 to infinite dimensions.

By definition of the creation operators, the linear span of the tensors

$$A^*(f_1)A^*(f_2)\ldots A^*(f_n)\Omega = f_1 \circ \cdots \circ f_n \in \mathcal{A}_n^\pm(\mathfrak{H}) \qquad (9.42)$$

with arbitrary vectors $f_j \in \mathfrak{H}$ and $n = 0, 1, \ldots$ is the full algebra $\mathcal{A}^\pm(\mathfrak{H})$. As a consequence of this fact, the representation of the canonical (anti) commutation relations (9.34) is irreducible.

Proposition 9.1.8 *If S is a bounded operator on $\Gamma^\pm(\mathfrak{H})$ with the property that the operator S commutes on $\mathcal{A}^\pm(\mathfrak{H})$ with $A(f)$ and $A^*(f)$ for all $f \in \mathfrak{H}$, then S is a multiple of the identity.*

Proof   Assume $\Phi \in \mathcal{A}^\pm(\mathfrak{H})$ is a vector with

$$A(f)\Phi = 0$$

for all $f \in \mathfrak{H}$. Then $\langle F \mid A(f)\Phi \rangle = \langle f \circ F \mid \Phi \rangle = 0$ for all $f \in \mathfrak{H}$ and $F \in \mathcal{A}^\pm(\mathfrak{H})$. Since the linear span of $f \circ F$ is dense in $\oplus_{n\geq 1}\mathcal{A}_n^\pm(\mathfrak{H})$, the vector Φ is an element of $\mathcal{A}_0^\pm(\mathfrak{H})$:

$$\Phi = \lambda\,\Omega \text{ with } \lambda \in \mathbb{C}.$$

Then the assumptions on S imply $A(f)S\,\Omega = SA(f)\,\Omega = 0$, and $S\,\Omega = \lambda\,\Omega$ follows. Moreover, we obtain the relations $SA^*(f)\,\Omega = A^*(f)S\,\Omega = \lambda A^*(f)\,\Omega$ and $SA^*(f_1)\ldots A^*(f_n)\,1 = A^*(f_1)\ldots A^*(f_n)S\,1 = \lambda A^*(f_1)\ldots A^*(f_n)\,1$ for arbitrary $f_j \in \mathfrak{H}, j = 1,\ldots,n$ with $n \in \mathbb{N}$. Hence $SF = \lambda F$ follows for all decomposable tensors $F = f_1 \circ \cdots \circ f_n$, and the identity

$$S = \lambda I$$

is derived on $\mathcal{A}^\pm(\mathfrak{H})$. Since S is bounded, this identity is true on the Fock space $\Gamma^\pm(\mathfrak{H})$. $\square$

The creation and annihilation operators are building blocks of more complicated operators. An important example is the representation of the derivations (9.19). Let us first take a finite rank operator S on $\mathfrak{H}$. Then there exist vectors $f_m, g_m \in \mathfrak{H}$ such that

$$Sh = \sum_{m=1}^{M} f_m \langle g_m \mid h \rangle \qquad (9.43)$$

for all $h \in \mathfrak{H}$.

Proposition 9.1.9 *The derivation* $d\Gamma(S)$ *generated by the operator (9.43) has the representation*

$$d\Gamma(S) = \sum_m A^*(f_m)A(g_m). \tag{9.44}$$

Proof We first prove that $A^*(f_m)A(g_m)$ is a derivation on $\Gamma_{\mathrm{fin}}^{\pm}(\mathfrak{H})$. The necessary condition $A^*(f_m)A(g_m) 1 = 0$ is satisfied. For $f \in \mathfrak{H}$ and $G \in \Gamma_{\mathrm{fin}}^{\pm}(\mathfrak{H})$, we calculate

$$A^*(f_m)A(g_m) (f \circ G) \overset{(9.33)}{=} A^*(f_m) ((A(g_m)f) \circ G \pm f \circ A(g_m)G)$$
$$= \langle g_m \mid f \rangle f_m \circ G \pm f_m \circ f \circ A(g_m)G$$
$$= (A^*(f_m)A(g_m)f) \circ G + f \circ (A^*(f_m)A(g_m)G),$$

and $A^*(f_m)A(g_m)$ satisfies the conditions of Proposition 9.1.4. Hence we see that $A^*(f_m)A(g_m)$ and the sum $\sum_m A^*(f_m)A(g_m)$ are derivations.

The relation (9.30) implies $A^*(f_m)A(g_m)h = \langle g_m \mid h \rangle f_m$, and

$$\sum_m A^*(f_m)A(g_m)h = Ah$$

follows from (9.43). Hence the derivations $d\Gamma(A)$ and $\sum_m A^*(f_m)A(g_m)$ coincide on the one-particle space $\mathfrak{H}$. Then Corollary 9.1.5 yields (9.44). $\square$

The identity (9.44) can be generalized to operators S which can be approximated by finite rank operators in the strong operator topology,[3] i.e. to all closed operators. An explicit construction can be given with help of an orthonormal basis $\{f_\mu\}$ of $\mathfrak{H}$. The operator S has the representation

$$Sf = \sum_{\mu,\nu} \alpha_{\mu\nu} f_\mu (f_\nu \mid f) \quad \text{with } \alpha_{\mu\nu} = (f_\mu \mid Af_\nu), \tag{9.45}$$

and $d\Gamma(A)$ follows from (9.44) as

$$d\Gamma(S) = \sum_{\mu,\nu} \alpha_{\mu\nu} a_\mu^+ a_\nu^-, \tag{9.46}$$

with the operators $a_\mu^+ = A^*(f_\mu)$ and $a_\nu^- = A(f_\nu)$.

9.1.3 Generalized Creation and Annihilation Operators

Creation and annihilation operators of singular vectors are often used in the physics literature. Here we present the case related to generalized eigenvectors of the momentum operator in some detail. The calculations are restricted to

---

[3] The (bounded) operators S_N converge in the strong operator topology to a closed operator S, if $\lim_{N \to \infty} \|S_N f - Sf\| = 0$ for all $f \in \mathcal{D}_S$.

the case of bosons. A rather extensive presentation of generalized vectors and operator-valued generalized functions can be found in Bogoliubov et al. (1990, chapter 2). A different point of view is given in Reed and Simon (1972, sect. VIII.10).

The one-particle Hilbert space is taken as $\mathfrak{H} = L^2(\mathbb{R}^3, d\sigma(\mathbf{k}))$. The measure on $\mathbb{R}^3$ is $d\sigma(\mathbf{k}) = \gamma^{-1}(\mathbf{k}) d^3k$ with a continuous positive, increasing and polynomially bounded weight function $\gamma(\mathbf{k})$ such that $\gamma^{-1}(\mathbf{k}) = (\gamma(\mathbf{k}))^{-1}$ is also continuous and polynomially bounded. Examples are the momentum representation of nonrelativistic quantum mechanics with $\gamma \equiv 1$, or the momentum representation of a relativistic particle with $\gamma(\mathbf{k}) = 2\sqrt{m^2 + \mathbf{k}^2}$, $m > 0$; see Section 11.1.3. The elements of $\mathfrak{H}$ are functions $f(\mathbf{k})$ of the variable $\mathbf{k} \in \mathbb{R}^3$. The inner product is

$$\langle f \mid g \rangle = \int f(\mathbf{k})^* g(\mathbf{k}) d\sigma(\mathbf{k}),$$

and $\langle f^* \mid g \rangle = \int f(\mathbf{k}) g(\mathbf{k}) d\sigma(\mathbf{k})$ is a bilinear symmetric form. The space $S(\mathbb{R}^3)$ of smooth strongly decreasing functions is dense in $L^2(\mathbb{R}^3, d\sigma(\mathbf{k}))$.

The generalized vector

$$\chi_{\mathbf{p}}(\mathbf{k}) = \gamma(\mathbf{p})\delta^3(\mathbf{k} - \mathbf{p}), \ \mathbf{p} \in \mathbb{R}^3 \tag{9.47}$$

is an element of $S'(\mathbb{R}_k^3)$, and the pairing $\langle \chi_{\mathbf{p}} \mid \varphi \rangle = \varphi(\mathbf{p})$ is a continuous bounded function of $\mathbf{p}$ for $\varphi \in S(\mathbb{R}_k^3)$. But there is another point of view on this distribution: given $u \in S(\mathbb{R}_p^3)$, the integral

$$\int u(\mathbf{p})\chi_{\mathbf{p}}(\mathbf{k}) d\sigma(p) = u(\mathbf{k})$$

is an element of the Hilbert space $\mathfrak{H}$, i.e. χ is a distribution in $\mathbf{p}$ with values in $\mathfrak{H}$. The formal inner product between two of these vectors

$$\langle \chi_{\mathbf{p}_1} \mid \chi_{\mathbf{p}_2} \rangle = \gamma(\mathbf{p}_2)\delta^3(\mathbf{p}_1 - \mathbf{p}_2) \tag{9.48}$$

is the singular kernel function of the positive identity operator on $\mathfrak{H}$. The generalized functions (9.47) are the eigenvectors of the momentum operator.

The creation/annihilation operators of these singular vectors $a_{\mathbf{p}}^+ = A^*(\chi_{\mathbf{p}})$ and $a_{\mathbf{p}}^- = A(\chi_{\mathbf{p}})$ are operator-valued generalized functions. Integrating with test functions $u(p) \in S(\mathbb{R}_p^3)$, we obtain

$$\int u(\mathbf{p})a_{\mathbf{p}}^+ \, d\sigma(\mathbf{p}) = A^*(u), \qquad \int u(\mathbf{p})^* a_{\mathbf{p}}^- \, d\sigma(\mathbf{p}) = A(u). \tag{9.49}$$

The class of test functions can obviously be extended to functions in $\mathfrak{H}$. Using this interpretation, the canonical commutation relations (9.34) are equivalent to the commutation rules

$$\left[a_{\mathbf{k}}^-, a_{\mathbf{p}}^-\right] = \left[a_{\mathbf{k}}^+, a_{\mathbf{p}}^+\right] = 0$$
$$\left[a_{\mathbf{k}}^-, a_{\mathbf{p}}^+\right] = \gamma(\mathbf{p})\delta^3(\mathbf{k} - \mathbf{p})\,I. \tag{9.50}$$

The singular creation/annihilation operators are often used to represent operators. If M is an integral operator on $\mathfrak{H}$ with the kernel $M(\mathbf{k}, \mathbf{k}')$ such that

$$(Mf)(\mathbf{k}) = \int M(\mathbf{k}, \mathbf{k}') f(\mathbf{k}')d\sigma(\mathbf{k}') \tag{9.51}$$

is true, the representation (9.44) can be generalized to

$$d\Gamma(M) = \iint a_{\mathbf{k}}^+ M(\mathbf{k}, \mathbf{k}') a_{\mathbf{k}'}^- \, d\sigma(\mathbf{k})d\sigma(\mathbf{k}'). \tag{9.52}$$

To establish (9.52) we take a kernel $M(\mathbf{k}, \mathbf{k}') = \sum_m f_m(\mathbf{k})g_m^*(\mathbf{k}')$ with functions $f_m, g_m \in \mathfrak{H}$. Then (9.49) implies $d\Gamma(M) = \sum_m A^*(f_m)A(g_m)$ in agreement with (9.44). If M is a multiplication operator

$$(Mf)(\mathbf{k}) = \omega(\mathbf{k})f(\mathbf{k}) \tag{9.53}$$

with a numerical function ω, the kernel is $M(\mathbf{k}, \mathbf{k}') = \omega(\mathbf{k})\delta^3(\mathbf{k} - \mathbf{k}')$ and (9.52) leads to

$$d\Gamma(M) = \int \omega(\mathbf{k})a_{\mathbf{k}}^+ a_{\mathbf{k}}^- \, d\sigma(\mathbf{k}). \tag{9.54}$$

Representations of operators on the Fock space such as (9.52) or (9.54) are often used in the physics literature and helpful, at least for formal arguments. A simple example is the number operator

$$d\Gamma(I) = \int a_{\mathbf{k}}^+ a_{\mathbf{k}}^- \, d\sigma(\mathbf{k}). \tag{9.55}$$

If $\mathfrak{H} = L^2(\mathbb{R}^3, d\sigma(\mathbf{k}))$ is the momentum representation of the one-particle Hilbert space with the Hamilton operator (9.53), then (9.54) is the Hamilton operator of the free dynamics.

Another class of generalized vectors is related to the position in configuration space. Take two copies of $\mathbb{R}^3$ in duality by the bilinear pairing $\mathbf{x}\mathbf{k} = \sum_{j=1}^3 x_j k_j$, $\mathbf{x} \in \mathbb{R}_x^3$, $\mathbf{k} \in \mathbb{R}_k^3$. The functions

$$\phi_{\mathbf{x}}(\mathbf{k}) = (2\pi)^{-\frac{3}{2}} e^{-i\mathbf{k}\mathbf{x}} \tag{9.56}$$

are generalized vectors in $\mathfrak{H} = L^2(\mathbb{R}_k^3, d\sigma(\mathbf{k}))$ depending on the parameter $\mathbf{x} \in \mathbb{R}_x^3$. The integration $\int \varphi(\mathbf{x})\phi_{\mathbf{x}}(\mathbf{k})d^3x = \tilde{\varphi}(\mathbf{k})$ with a test function $\varphi(\mathbf{x}) \in S(\mathbb{R}_x^3)$

gives an element of $\mathfrak{H}$. The functions (9.56) are therefore generalized functions with values in $\mathfrak{H}$. The formal inner product

$$\langle \phi_{\mathbf{x}} \mid \phi_{\mathbf{y}} \rangle = (2\pi)^{-3} \int e^{i\mathbf{k}(\mathbf{x}-\mathbf{y})} d\sigma(\mathbf{k}) = R(\mathbf{x} - \mathbf{y}) \tag{9.57}$$

yields a positive definite kernel function R on $S(\mathbb{R}^3_x)$. In nonrelativistic quantum mechanics, the momentum representation is normalized with $\gamma \equiv 1$, and the functions (9.56) are the eigenvectors of the position operator, normalized to $R(\mathbf{x} - \mathbf{y}) = \delta^3(\mathbf{x} - \mathbf{y})$. The relativistic case is discussed in Section 11.1.3.

More details about generalized creation and annihilation operators are given in Section 13.1.

9.1.4 The Dynamics of Free Particles

Let H_1 be the Hamilton operator on the one-particle Hilbert space $\mathfrak{H}$. If the particles have no interaction the Hamilton operator, for many particle systems on the Fock space $\Gamma^\pm(\mathfrak{H})$ is $H = d\Gamma(H_1)$. The unitary time evolution on $\Gamma^\pm(\mathfrak{H})$ is then given by

$$U(t) = \exp(-iHt) = \Gamma(U_1(t)) \text{ with } U_1(t) = \exp(-iH_1 t). \tag{9.58}$$

The vacuum is invariant under this evolution

$$U(t)\,\Omega = \Omega, \tag{9.59}$$

and a decomposable tensor $f_1 \circ \cdots \circ f_n$ with $f_j \in \mathfrak{H}, j = 1, \ldots, n$, remains decomposable for all times

$$U(t)\,(f_1 \circ \cdots \circ f_n) = (U_1(t)f_1) \circ \cdots \circ (U_1(t)f_n). \tag{9.60}$$

Let $A^*(f)$ be the creation operator of the vector $f \in \mathfrak{H}$, then the Heisenberg evolution $U^*(t)A^*(f)U(t)$ is derived from the identity

$$U^*(t)A^*(f)U(t)\,F = U^*(t)\,(f \circ U(t)\,F)$$
$$= \left(U_1^*(t)f\right) \circ F = A^*(U_1^*(t)f)\,F,$$

which is true for all $F \in \Gamma^\pm(\mathfrak{H})$. Hence the dynamics of the creation operator in the Heisenberg picture is $U^*(t)A^*(f)U(t) = A^*(U_1^*(t)f)$. The dynamics of the annihilation operator follows from the adjoint identity. The general result is

$$A^\#(f;t) \triangleq \Gamma(U_1^*(t))A^\#(f)\Gamma(U_1(t)) = A^\#(U_1^*(t)f), \tag{9.61}$$

where $A^{\#}$ stands for A^* or A. This rule implies that the equal time canonical (anti)commutation relations (9.34)

$$\left[A^*(f;t), A^*(g;t)\right]_{\mp} = \left[A(f;t), A(g;t)\right]_{\mp} = 0,$$
$$\left[A(f;t), A^*(g;t)\right]_{\mp} = \langle f \mid g \rangle \, I, \; f, g \in \mathfrak{H}, \tag{9.62}$$

are valid for all times $t \in \mathbb{R}$. The operators (9.61) satisfy the differential equation

$$\frac{d}{dt}A^{\#}(f;t) = i\left[H, A^{\#}(f;t)\right] = \pm i A^{\#}(H_1 f;t). \tag{9.63}$$

In the case of a time-independent interaction, the evolution is determined by a group of unitary operators $U(t) = \exp(-iHt)$, where H is the (semibounded) Hamiltonian. The time evolution of the creation/annihilation operators is again $A^{\#}(f;t) = U^*(t)A^{\#}(f)U(t)$, and the canonical (anti)commutation relations (9.62) remain valid at equal times.

9.2 Exponential Vectors and Weyl Operators

In this section, we present the generalization of the Bargmann vectors introduced in Section 3.2.3 to the Boson Fock space $\Gamma^+(\mathfrak{H})$ of infinite dimensional spaces $\mathfrak{H}$.

9.2.1 Exponential Vectors and Coherent States

The exponential series

$$f \in \mathfrak{H} \mapsto \exp(f) \triangleq \sum_{n=0}^{\infty} \frac{1}{n!} f^{\vee n} = \sum_{n=0}^{\infty} \frac{1}{\sqrt{n!}} f^{\otimes n}$$

converges uniformly within $\Gamma^+(\mathfrak{H})$ with the norm

$$\|\exp(f)\|^2 = \sum_{n=0}^{\infty} \left(\frac{1}{n!}\right)^2 \|f^{\vee n}\|^2 = \sum_{n=0}^{\infty} \frac{1}{n!} \|f\|^{2n} = e^{\|f\|^2}. \tag{9.64}$$

The second identity follows from $\|f^{\vee n}\|^2 \overset{(8.53)}{=} n! \, \|f\|^{2n}$. The Bargmann vector of Section 3.2.3 therefore has the following generalization.

The **exponential vector** with test function $f \in \mathfrak{H}$ is the series

$$\exp(f) = \sum_{n=0}^{\infty} \frac{1}{n!} f^{\vee n} \in \Gamma^+(\mathfrak{H}). \tag{9.65}$$

In Section 8.3.2 we have introduced the Fock scales $\Gamma_{(\alpha)}(\mathfrak{H})$ and the algebra $\Gamma_>^+(\mathfrak{H}) = \cup_{\alpha>0}\Gamma_{(\alpha)}^+(\mathfrak{H}) \subset \Gamma^+(\mathfrak{H})$. The exponential vectors are obviously elements of the spaces $\Gamma_{(\alpha)}(\mathfrak{H})$, $0 \leq \alpha < 1$, and of the algebra $\Gamma_>^+(\mathfrak{H})$. The inner product of two exponential vectors follows from (9.64) by polarization:

$$\langle \exp(f) \mid \exp(g) \rangle = e^{\langle f \mid g \rangle}. \tag{9.66}$$

The normalized exponential vectors are usually denoted as **coherent states**:

$$\text{coh}(f) = e^{-\frac{1}{2}\|f\|^2} \exp(f). \tag{9.67}$$

The series (9.65) satisfies the functional relation of an exponential:

$$\exp(f) \vee \exp(g) = \exp(f + g). \tag{9.68}$$

Hence the linear span of the exponential vectors

$$\mathcal{A}_{coh}^+(\mathfrak{H}) \triangleq \textit{Lin}\,\{\exp(f) \mid f \in \mathfrak{H}\} \tag{9.69}$$

is an algebra with the product $\vee$. The following inclusions are obvious:

$$\mathcal{A}_{coh}^+(\mathfrak{H}) \subset \Gamma_{(\alpha)}^+(\mathfrak{H}) \subset \Gamma_>^+(\mathfrak{H}) \subset \Gamma^+(\mathfrak{H}), \qquad 0 < \alpha < 1,$$

where the spaces $\Gamma_{(\alpha)}^+(\mathfrak{H})$ are Fock scales as introduced in Section 8.3.2.

Another direct consequence of the series expansion (9.65) is a simple relation for the operators (9.13) and (9.19). For operators S, M defined on the set $\mathfrak{D} \subset \mathfrak{H}$, we have the identities

$$\Gamma(S)\exp(f) = \exp(Sf), \quad \text{and} \tag{9.70}$$

$$d\Gamma(M)\exp(f) = (Mf) \vee \exp(f) \tag{9.71}$$

are true for $f \in \mathfrak{D}$.

The function $h \in \mathfrak{H} \mapsto \exp(h)$ has an absolutely convergent power series expansion at $h = 0$. It is therefore infinitely differentiable at $h = 0$. As a consequence of (9.68), the mapping $h \in \mathfrak{H} \mapsto \exp(f + h)$ has an absolutely convergent power series expansion at any point $f \in \mathfrak{H}$. The function (9.65) is therefore an analytic function on the Hilbert space $\mathfrak{H}$; see Hille and Phillips (1957, Secs. 3.16–3.18). The estimate

$$\exp(f + h) - \exp(f) = \exp(f) \vee (\exp(h) - 1) = h \vee \exp(f) + O(\|h\|^2)$$

implies that the derivative of $\exp(f)$ at f in direction $h \in \mathfrak{H}$ is $\exp(f) \vee h = h \vee \exp(f)$.

Lemma 9.2.1 *Let $\mathfrak{D}$ be a dense linear subset of $\mathfrak{H}$, then the linear span of the exponential vectors $\{\exp(f) : f \in \mathfrak{D}\}$ is a dense subset of $\Gamma^+(\mathfrak{H})$.*

Proof   We first assume that $\mathfrak{D} = \mathfrak{H}$ and prove the theorem for this case. Let $\mathfrak{T}$ be the completion of the linear span of $\{\exp(f) : f \in \mathfrak{H}\}$. The function $\mathbb{C}^n \ni (\alpha_1, \ldots, \alpha_n) \mapsto \Phi_n(\alpha_1, \ldots, \alpha_n) \triangleq \exp(\alpha_1 f_1 + \cdots + \alpha_n f_n) \in \mathfrak{T} \subset \Gamma^+(\mathfrak{H})$ with vectors $f_j \in \mathfrak{H}, j \in \{1, \ldots, n\}, n \in \mathbb{N}$, is analytic, and all its derivatives are elements of $\mathfrak{T}$. The derivative $\frac{\partial^n}{\partial \alpha^n} \Phi(\alpha_1, \ldots, \alpha_n)$ at $\alpha_1 = \cdots = \alpha_n = 0$ is

$$\frac{\partial^n}{\partial \alpha^n} \Phi_n(\alpha_1, \ldots, \alpha_n) = f_1 \vee \cdots \vee f_n \in \mathcal{A}_n^+(\mathfrak{H}). \tag{9.72}$$

Thereby the symbol $\frac{\partial^n}{\partial \alpha^n}$ means $\frac{\partial^n}{\partial \alpha_1 \partial \alpha_2 \ldots \partial \alpha_k}$. The tensor $\exp(0) = 1$ and the derivatives (9.72) with $n = 1, 2, \ldots$, span the algebra $\mathcal{A}^+(\mathfrak{H})$. Hence $\mathfrak{T}$ coincides with $\Gamma^+(\mathfrak{H})$.

Now assume that $\mathfrak{D}$ is a dense subset of $\mathfrak{H}$. The function $\exp(f)$ is continuous in the norm topologies of the spaces $\mathfrak{H}$ and $\Gamma^+(\mathfrak{H})$. Therefore, the completion of the linear span of $\{\exp(f) : f \in \mathfrak{D}\}$ coincides with the completion of the linear span of $\{\exp(f) : f \in \mathfrak{H}\}$. Hence the Lemma is true. $\qquad \square$

Assume the Hilbert space $\mathfrak{H}$ has an involution (8.5), then $\mathfrak{H}_{\mathbb{R}} = \{f = f^* : f \in \mathfrak{H}\}$ is the Hilbert space of real elements. In the proof of Lemma 9.2.1, we can choose the vectors f_j as elements of $\mathfrak{H}_{\mathbb{R}}$. The $\mathbb{C}$-linear span of all tensors $f_1 \vee \cdots \vee f_n$ with $f_j \in \mathfrak{H}_{\mathbb{R}}$ spans $\mathcal{A}_n^+(\mathfrak{H})$. Hence we obtain the following corollary.

Corollary 9.2.2 *The linear span of the exponential vectors with real test functions, $\{\exp(f) : f \in \mathfrak{H}_{\mathbb{R}}\}$, is a dense subset of $\Gamma^+(\mathfrak{H})$.*

As a consequence of Lemma 9.2.1, a tensor $F \in \Gamma^+(\mathfrak{H})$ is uniquely determined by the function $\mathfrak{H} \ni h \mapsto \langle \exp(h) \mid F \rangle \in \mathbb{C}$.

Corollary 9.2.3 *A linear operator K on $\Gamma^+(\mathfrak{H})$, which is defined on a linear domain including the exponential vectors, is uniquely determined by the images on the set of exponential vectors $K \exp(f)$, $f \in \mathfrak{H}$, or by the matrix elements $\langle \exp(g) \mid K \exp(f) \rangle \in \mathbb{C}$ for $f, g \in \mathfrak{H}$.*

The domain of definition of the creation and annihilation operators (9.27) and (9.28) includes the exponential vectors. We obtain

$$A^*(h) \exp(f) = h \vee \exp(f) = \frac{\partial}{\partial \lambda} \exp(f + \lambda h) \mid_{\lambda=0}, \tag{9.73}$$

where the last identity is valid for real and for complex λ. The relation

$$\langle \exp(g) \mid A(h) \exp(f) \rangle = \frac{\partial}{\partial \lambda} \langle \exp(g + \lambda h) \mid \exp(f) \rangle \mid_{\lambda=0} = e^{\langle g \mid f \rangle} \langle h \mid f \rangle$$

implies the identity

$$A(h) \exp(f) = \langle h \mid f \rangle \exp(f). \tag{9.74}$$

By little algebra, one derives that the canonical commutation relations (9.34) are valid on the linear span of the exponential vectors. If S is a unitary operator on $\mathfrak{H}$, then the relations (9.70) and (9.73) imply the identities

$$\Gamma(S^{-1})A^*(h)\Gamma(S) = A^*(S^{-1}h),$$
$$\Gamma(S^{-1})A(h)\Gamma(S) = A(S^{-1}h). \tag{9.75}$$

9.2.2 Weyl Displacement Operators

We now extend the notion of Weyl unitaries to Fock space.

The **Weyl displacement operators** $D(f), f \in \mathfrak{H}$, are first defined on the linear span of the exponential vectors by

$$D(f)\exp(h) \triangleq e^{-\langle f|h\rangle - \frac{1}{2}\|f\|^2} \exp(f + h). \tag{9.76}$$

For $f = 0$, we obtain $D(0) = I$. The calculation of the product $D(f)D(g)$

$$D(f)D(g)\exp(h)$$

$$= \exp\left(-\langle f+g \mid h\rangle - \frac{1}{2}\|f+g\|^2 + \frac{1}{2}\langle g\mid f\rangle - \frac{1}{2}\langle f\mid g\rangle\right)\exp(f+g+h)$$

$$= \exp\frac{1}{2}\left(\langle g\mid f\rangle - \langle f\mid g\rangle\right)D(f+g)\exp(h)$$

yields first that $D(f)D(-f) = D(-f)D(f) = I$. Hence the operators (9.76) are invertible with $D^{-1}(f) = D(-f)$. Other consequences are the **Weyl relations**

$$D(f)D(g) = e^{\omega(f,g)}D(f+g), \tag{9.77}$$

with the $\mathbb{R}$-bilinear antisymmetric form $\omega(f,g) = -\frac{1}{2}\langle f\mid g\rangle + \frac{1}{2}\langle g\mid f\rangle = -i\,\mathrm{Im}\,\langle f\mid g\rangle$. The matrix element between exponential vectors follows from (9.66) as

$$\langle \exp(g)\mid D(f)\exp(h)\rangle = \exp\left(\langle g\mid f+h\rangle - \langle f\mid h\rangle - \frac{1}{2}\|f\|^2\right). \tag{9.78}$$

A special case of this identity is the vacuum expectation

$$\langle \Omega\mid D(f)\ \Omega\rangle = \exp\left(-\frac{1}{2}\|f\|^2\right).$$

Since the calculation of $(D(-f)\exp(g)\mid \exp(h))$ yields again the result (9.78), we obtain $(D(h))^* = D(-h) = D^{-1}(h)$ on the linear span of the exponential vectors, which is dense in $\Gamma^+(\mathfrak{H})$. The Weyl operators can therefore be extended to unitary operators on the Fock space $\Gamma^+(\mathfrak{H})$.

The Weyl relations imply that for fixed $h \in \mathfrak{H}$ the operators $\mathbb{R} \ni \lambda \mapsto D(\lambda h)$ form a one-parameter unitary group. The generator of this group follows from (9.76) as $\frac{d}{d\lambda}D(\lambda f)\exp(h)\,|_{\lambda=0} = -\langle f \mid h \rangle \exp(h) + f \vee \exp(h) = (-A(f) + A^*(f))\exp(h)$. Hence the Weyl operator $D(h)$ is given by

$$D(f) = \exp\left(A^*(f) - A(f)\right) \tag{9.79}$$

on the entire Fock space $\Gamma^+(\mathfrak{H})$. The operator (9.79) is the extension of the displacement operator defined in Section 3.2.1 to arbitrary finite and infinite dimensions. The exponential series $\exp A^*(h)$ and $\exp A(h)$, $h \in \mathfrak{H}$, are continuous mappings from the spaces $\Gamma^+_{(\alpha)}(\mathfrak{H})$ into $\Gamma^+_{(\beta)}(\mathfrak{H})$ if $1 > \alpha > \beta \geq 0$. The identity

$$D(h) = \exp\left(A^*(h)\right)\exp\left(-A(h)\right)\,e^{-\frac{1}{2}\|h\|^2}$$

can be easily verified on the linear span of the exponential vectors. It is therefore also true on $\Gamma^+_>(\mathfrak{H})$.

If U is a unitary operator on $\mathfrak{H}$, then (9.75) and (9.79) imply the identity

$$\Gamma(U^*)D(h)\Gamma(U) = D(U^*h). \tag{9.80}$$

The relations (9.73) and (9.76) imply

$$D(h)A^*(f)D(-h)\exp(g) = e^{\langle h|g\rangle - \frac{1}{2}\|h\|^2}\frac{d}{d\lambda}D(h)\exp(g - h + \lambda f) \tag{9.81}$$

$$= e^{\langle h|g\rangle - \frac{1}{2}\|h\|^2}\frac{d}{d\lambda}e^{-\langle h|g\rangle - \frac{1}{2}\|h\|^2}\exp(g + \lambda f)\,|_{\lambda=0} \tag{9.82}$$

$$= -\langle h \mid f \rangle \exp(g) + f \vee \exp(g). \tag{9.83}$$

Hence $D(h)A^*(f)D^{-1}(h) = A^*(f) - \langle h \mid f \rangle\,I$ follows for all $h \in \mathfrak{H}$ and $f \in \mathfrak{H}$, and we have derived the identities

$$\begin{aligned} D(h)A^*(f)D^{-1}(h) &= A^*(f) - \langle h \mid f \rangle\,I, \\ D(h)A(f)D^{-1}(h) &= A(f) - \langle f \mid h \rangle\,I. \end{aligned} \tag{9.84}$$

There is an alternative version of the Weyl operator that uses the self-adjoint Segal field operator (9.36) $\hat{\Phi}(h) = A^*(h) + A(h)$:

$$W(h) \triangleq \exp\left(i\,\hat{\Phi}(h)\right) = D\,(ih)\,. \tag{9.85}$$

For subsequent use, we note the Weyl relations (9.77) for the operators $W(h)$:

$$W(f)W(g) = e^{-i\mathrm{Im}\,\langle f|g\rangle}\,W(f + g). \tag{9.86}$$

The relations (9.84) are equivalent to

$$W(h)\hat{\Phi}(f)W^{-1}(h) = \hat{\Phi}(f) - 2\,\mathrm{Im}\,\langle h \mid f \rangle\,I. \tag{9.87}$$

If $W(h)$, $h \in \mathfrak{H}$, are unitary operators on $\Gamma^+(\mathfrak{H})$, which satisfy the Weyl relations (9.86), then $\mathbb{R} \ni \lambda \mapsto W(\lambda h)$ is a one-parameter group $W(\lambda h) = \exp\left(i\lambda\hat{\Phi}(h)\right)$. The self-adjoint generators $\hat{\Phi}(h)$ are $\mathbb{R}$-linear in h and fulfill the canonical commutation relations (9.38). This statement follows from

$$W(\lambda h)W(\gamma f)W(-\lambda h) \overset{(9.86)}{=} \exp\left(-\lambda\gamma\left(\langle h \mid f \rangle - \langle f \mid h \rangle\right)\right) W(\gamma f)$$

by differentiation with respect to the real parameters γ and λ. A canonical structure is therefore uniquely defined by the Weyl operators.

9.2.3 Free Dynamics of Weyl Operators

Let H_1 be the positive one-particle Hamiltonian on $\mathfrak{H}$. The Hamiltonian on the Fock space $\Gamma^+(\mathfrak{H})$ is $H = d\Gamma(H_1)$ and the dynamics on $\Gamma^+(\mathfrak{H})$ is induced by the unitary group $U(t) = \exp(-iHt) = \Gamma(U_1(t))$ with $U_1(t) = \exp(-iH_1t)$. As a consequence of (9.70), the time evolution of a coherent state is

$$U(t)\exp(f) = \exp(U_1(t)f). \tag{9.88}$$

The time evolution of a Weyl operator follows from (9.80):

$$U^*(t)D(h)U(t) = D(U_1^*(t)h). \tag{9.89}$$

The skew-symmetric form, $\mathrm{Im}\,\langle f \mid g \rangle = \mathrm{Im}\left\langle U_1^*(t)f \mid U_1^*(t)g\right\rangle$, of the Weyl relations (9.77) does not depend on the time, and the Weyl relations are invariant against the free dynamics. This result agrees with the invariance of the canonical commutation relations derived in Section 9.1.4.

9.3 Distributions of Boson Fields

9.3.1 Gaussian Fields

We recall the definition of the Segal operator field $\hat{\Phi}(f) = A^*(f) + A(f)$ from (9.36). In the vacuum state, we have the matrix elements

$$\langle\Omega| \hat{\Phi}(f_n)\dots\hat{\Phi}(f_1)\, \Omega\rangle = \sum_{\varepsilon}\langle\Omega| A^{\varepsilon(n)}(f_n)\dots A^{\varepsilon(1)}(f_1)\,\Omega\rangle,$$

where $f_n,\dots,f_1$ are given one-particle vectors. Here we have that $\varepsilon(k)$ denotes the presence or absence of $*$ on $A(f_k)$, and we have a sum over the 2^n possibilities $\varepsilon = (\varepsilon(n),\dots,\varepsilon(1))$.

Note that not all sequences ε contribute. If we set $s(k) = -1$ when the kth operator is an annihilator, and $s(k) = +1$ when it is a creator, then the partial sums $\sum_{k=1}^{m} s(k)$ must never go negative ($1 \leq m \leq n$), and $\sum_{k=1}^{n} s(k)$ must be

annihilator vertex at p

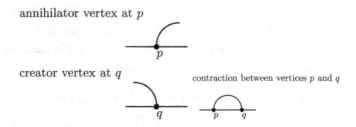

creator vertex at q

contraction between vertices p and q

Figure 9.1 Creation and annihilation vertices.

zero. Such sequences are called Catalan sequences, and these are the only ones that are not zero.

We give a simple argument to compute $\langle \Omega | A^{\varepsilon(k)}(f_n) \ldots A^{\varepsilon(1)}(f_1) \Omega \rangle$: every time we encounter an expression $\ldots A(f_i)A^*(f_j)\ldots$, we replace it with $\ldots [\langle f_i | f_j \rangle + A^*(f_j) A(f_i)]\ldots$; the term $\langle f_i | f_k \rangle$ is a scalar and can be brought outside the expectation, leaving a product with two fewer fields to average. Ultimately, we must pair up every creator with an annihilator; otherwise, we get an expression that averages zero due to (9.31). Therefore, only the even moments are nonzero and we obtain the identity

$$\sum_\varepsilon \langle \Omega | A^{\varepsilon(2n)}(f_{2n}) \ldots A^{\varepsilon(1)}(f_1) \Omega \rangle = \sum_{\text{Pair}(2n)} \prod_{k=1}^n \langle f_{p_k} | f_{q_k} \rangle. \qquad (9.90)$$

Here $(p_k, q_k)_{j=1}^n \in \text{Pair}(2n)$ is a pair partition: the p_k correspond to annihilators and the q_k to creators, so we must have $p_k > q_k$ for each j; the ordering of the pairs is unimportant, so for definiteness we take $q_n > \cdots > q_2 > q_1$. We may picture this as follows: for each $i \in \{1, 2, \ldots, 2n\}$, we have a vertex; with $A^*(f_i)$ we associate a creator vertex with weight f_i and with $A(f_i)$ we associate an annihilator vertex with weight f_i.

A matched creation/annihilation pair (p_k, q_k) is called a *contraction* over the creator vertex q_k and annihilator vertex p_k, it corresponds to a multiplicative factor $\langle f_{p_k} | f_{q_k} \rangle$, and is shown pictorially as a single line (Figure 9.1).

We then consider a sum over all possible diagrams. As an illustrative example, the fourth-order expression is

$$\langle \Omega | \hat{\Phi}(f_4)\hat{\Phi}(f_3) \hat{\Phi}(f_2) \hat{\Phi}(f_1) \Omega \rangle$$
$$= \langle f_4 | f_3 \rangle \langle f_2 | f_1 \rangle + \langle f_4 | f_2 \rangle \langle f_3 | f_1 \rangle + \langle f_4 | f_1 \rangle \langle f_3 | f_2 \rangle. \qquad (9.91)$$

Here we have $|\text{Pair}(4)| = \frac{4!}{2^2 2!} = 3$, and the three pair partitions, which are $\{\{4, 3\}, \{2, 1\}\}$, $\{\{4, 2\}, \{3, 1\}\}$, and $\{\{4, 1\}, \{3, 2\}\}$, may be sketched as in Figure 9.2.

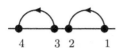 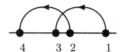 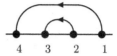

Figure 9.2 The diagrams corresponding to fourth-order moments.

Setting all the f_k equal to a fixed test function f, we obtain

$$\langle \Omega| \hat{\Phi}(f)^k \; \Omega\rangle = \langle \Omega| [A^*(f) + A(f)]^k \; \Omega\rangle = \|f\|^k |\text{Pair}(k)|.$$

We then see that the observable $Q(f)$ has a mean-zero Gaussian distribution of variance $\|f\|^2$ for the Fock vacuum state. We could have obtained this more expressly by noting that

$$\langle \Omega| e^{it\hat{\Phi}(f)} \; \Omega\rangle \overset{(9.85)}{=} \langle \Omega| D(tf) \; \Omega\rangle = e^{-\frac{1}{2}\|f\|^2 t^2}.$$

However, we have also obtained a more general moment formula by our current calculation.

The result of the calculation can be summarized in the formulas

$$\langle \Omega| \hat{\Phi}(f_n)\dots\hat{\Phi}(f_1) \; \Omega\rangle = 0, \; n \text{ odd},$$

$$\langle \Omega| \hat{\Phi}(f_{2m})\dots\hat{\Phi}(f_1) \; \Omega\rangle = \sum_{\text{Pair}(2m)} \prod_{j=1}^{n} \langle\Omega|\hat{\Phi}(f_{p_k})\hat{\Phi}(f_{q_k}) \Omega\rangle, \; n \text{ even}. \quad (9.92)$$

The summation extends over pair partitions $(p_k, q_k)_{j=1}^m \in \text{Pair}(2m)$ with $p_k > q_k$ and $q_m > \cdots > q_2 > q_1$. This structure of the vacuum expectation values characterizes a Gaussian field. It will appear again in Section 11.1, where the free relativistic Boson field is investigated.

9.3.2 Poissonian Fields

Now let us introduce the field observables

$$N(f,g) = (A(f) + 1)^* (A(g) + 1)$$
$$= A^*(f) A(g) + A(g) + A^*(f) + 1$$
$$\equiv \sum_{\alpha,\beta\in\{0,1\}} [A^*(f)]^\alpha [A(g)]^\beta$$

where we employ a convention that $[x]^0 = 1$ and $[x]^1 = 1$ for any algebraic variable x.

We now wish to study matrix elements of the form

$$\langle\Omega| N(f_n,g_n)\dots N(f_1,g_1) \; \Omega\rangle.$$

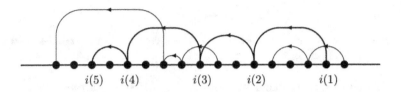

$$i(5)\qquad i(4)\qquad\qquad i(3)\qquad\quad i(2)\qquad\qquad i(1)$$

Figure 9.3 Typical Poissonian diagram; see the text.

By expanding each term, we may write this as a sum over all expressions of the type

$$\langle\Omega| A^*(f_n)^{\alpha(n)} A(g_n)^{\beta(n)} \ldots A^*(f_1)^{\alpha(1)} A(g_1)^{\beta(1)} \Omega\rangle$$

taken over all $\alpha, \beta \in \{0, 1\}^n$. This time, in the diagrammatic description, we have n vertices, with each vertex being one of four possible types $A^*A, A^*, A, 1$:

scatterer   emitter   absorber   constant

A typical situation is depicted in Figure 9.3.

Evidently we must again join up all creation and annihilation operators into pairs. However, get creation, multiple scattering, and annihilation as the rule; otherwise, we have a stand-alone constant term of unity at a vertex. In Figure 9.3, we can think of a particle being created at vertex $i(1)$, then scattered at $i(2), i(3), i(4)$ successively before being annihilated at $i(5)$. (This component has been highlighted using thick lines.) Each such component corresponds to a unique part, here $\{i(5), i(4), i(3), i(2), i(1)\}$, having two or more elements; singletons may also occur, and these are just the constant term vertices. Therefore, every such diagram corresponds uniquely to a partition of $\{1, \ldots, n\}$. Once this observation is made, it is easy to see that

$$\sum_{\alpha,\beta\in\{0,1\}^n} \langle\Omega| A^*(f_n)^{\alpha(n)} A(g_n)^{\beta(n)} \ldots A^*(f_1)^{\alpha(1)} A(g_1)^{\beta(1)} \Omega\rangle$$

$$= \sum_{\pi\in\mathrm{Part}(n)} \prod_{\{i(k)>\cdots>i(2)>i(1)\}\in\pi} \langle g_{i(k)}|f_{i(k-1)}\rangle \cdots \langle g_{i(3)}|f_{i(2)}\rangle\langle g_{i(2)}|f_{i(1)}\rangle. \quad (9.93)$$

If we now take all the f_k and g_k equal to a fixed f, then we arrive at[4]

$$\langle\Omega| N(f,f)^n \Omega\rangle = \sum_{m=0}^{n} \left\{ \begin{matrix} n \\ m \end{matrix} \right\} \|f\|^{2(n-m)}.$$

---

[4] Note that a part of size k contributes $\|f\|^{2(k-1)}$, so a partition of n vertices into m blocks of sizes $k_1, \ldots k_m$ contributes $\|f\|^{2(k_1+\cdots+k_m-m)} = \|f\|^{2(n-m)}$.

It therefore follows that, for each normalized vector $\phi \in \mathfrak{h}$ and $\lambda > 0$, the observable

$$M(\phi, \lambda) = \lambda N\left(\frac{1}{\sqrt{\lambda}}\phi, \frac{1}{\sqrt{\lambda}}\phi\right) = (A^*(\phi) + \sqrt{\lambda})(A(\phi) + \sqrt{\lambda}) \quad (9.94)$$

has a Poisson distribution of intensity λ in the Fock vacuum state.

Again, we can arrive at this from an alternative point of view. We see that

$$D(f)^* A^*(\phi) A(\psi) D(f) = [D(f)^* A(\phi) D(f)]^* D(f)^* A(\psi) D(f)$$

$$= A^*(\phi) A(\psi) + \langle \psi | f \rangle A^*(\phi) + \langle f | \phi \rangle A(\psi) + \langle f | \phi \rangle \langle \psi | f \rangle.$$

So we set $\psi = \phi$ with $\|\phi\| = 1$ and $f = \sqrt{\lambda}\phi$ to get

$$D(f)^* A^*(\phi) A(\phi) D(f) = A^*(\phi) A(\phi) + \sqrt{\lambda} A^*(\phi) + \sqrt{\lambda} A(\phi) + \lambda$$

$$\equiv M(\phi, \lambda),$$

and so

$$\langle \Omega | e^{itN(\phi, \lambda)} \Omega \rangle = \langle D(f)\Omega | e^{itA^*(\phi)A(\phi)} D(f)\Omega \rangle$$

$$= e^{-\|f\|^2} \langle \exp(f) | \Gamma(e^{it|\phi\rangle\langle\phi|}) \exp(f) \rangle$$

$$= e^{-\|f\|^2} \langle \exp(f) | \exp(f + \langle \phi | f \rangle (e^{it} - 1)\phi) \rangle$$

$$= \exp\{\lambda(e^{it} - 1)\}.$$

9.3.3 Exponentially Distributed Fields

Is there a similar interpretation for Stirling numbers of the first kind as well? Here we should be dealing with cycles within permutations rather than parts in a partition. Consider the following representation of a cycle $(i(1), i(2), \ldots, i(6))$.

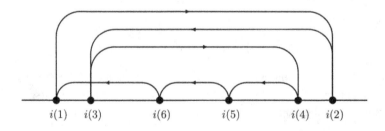

To make the sense of the cycle clear, we are forced to use arrows and therefore we have two types of lines. In any such diagrammatic representation of a permutation, we may encounter any of the following five possible of vertices:

An uncontracted (constant) vertex indicates a fixed point for the permutation.

Let us consider the one-dimensional case first. The preceding suggests that we should use two independent (that is, commuting) Bose variables, say b_1, b_2. Let us set

$$b \triangleq b_1 \otimes 1_2 + 1_1 \otimes b_2^*.$$

Then we see that b^* will actually commute with b, and Wick ordering b^*b gives

$$b^*b = b_1^*b_1 \otimes 1_2 + 1_2 \otimes b_2^*b_2 + b_1 \otimes b_2 + b_1^* \otimes b_2^* + 1_1 \otimes 1_2,$$

and here we see the five vertex terms we need.

Let Ω_1 and Ω_2 be the vacuum state for b_1 and b_2 respectively, then let $\Omega = \Omega_1 \otimes \Omega_2$ be the joint vacuum state. We wish to show that $N = b^*b$ has a Gamma distribution of unit power (an exponential distribution!) in this state.

First of all, let $q = b^* + b$ so that $q = q_1 \otimes 1_2 + 1_1 \otimes q_2$, where $q_k = b_k + b_k^*$. Now each q_k has a standard Gaussian distribution in the corresponding vacuum state ($\langle \Omega_k | e^{tq_k} \Omega_k \rangle = e^{t^2/2}$), and so

$$\langle \Omega | e^{tQq} \Omega \rangle = \langle \Omega_1 | e^{tq_1} \Omega_1 \rangle \langle \Omega_2 | e^{tq_2} \Omega_2 \rangle = e^{t^2}.$$

Therefore, $\langle \Omega | q^{2n} \Omega \rangle = 2^n |\mathrm{Pair}(2n)| = \frac{(2n)!}{n!}$.

However, $q^{2n} = \sum_m \binom{2n}{m} (b^*)^m (b)^{2n-m}$ (remember that b^* and b commute!), and so $\langle \Omega | q^{2n} \Omega \rangle \equiv \binom{2n}{n} \langle \Omega | (b^*)^n (b)^n \Omega \rangle = \binom{2n}{n} \langle \Omega | (b^*b)^n \Omega \rangle$. Therefore,

$$\langle \Omega | (b^*b)^n \Omega \rangle = n!$$

and so, for $t < 1$,

$$\langle \Omega | \exp\{tb^*b\} \Omega \rangle = \frac{1}{1-t}.$$

Therefore, $N = b^*b$ has an exponential distribution in the joint vacuum state.

The generalization of this result to Bosonic fields over a Hilbert space $\mathfrak{h}$ is straightforward enough.

Theorem 9.2.4 *Let $\mathfrak{h}$ be a separable Hilbert space and let J be an conjugation on $\mathfrak{h}$. Define operator fields $B(.)$ on the double Fock space $\Gamma(\mathfrak{h}) \otimes \Gamma(\mathfrak{h})$ by*

$$B(f) = B_1(f) \otimes 1_2 + 1_1 \otimes B_2^*(Jf), \tag{9.95}$$

where the $B_i(.)$ are usual Bosonic fields on the factors $\Gamma(\mathfrak{h})$, $i(= 1, 2)$. Then $[B(f), B^*(g)] = 0$ for all $f, g \in \mathfrak{h}$ and if $N(f, g) \triangleq B^*(f)B(g)$, then we have the following expectations in the joint Fock vacuum state $\Omega = \Omega_1 \otimes \Omega_2$:

$$\langle \Omega | N(f_n, g_n) \dots N(f_1, g_1) \Omega \rangle$$
$$= \langle B(f_n) \dots B(f_1)\Omega | B(g_n) \dots B(g_1)\Omega \rangle$$
$$= \text{perm} \left(\langle f_j | g_k \rangle \right) = \text{perm} \left(\langle g_j | f_k \rangle \right), \tag{9.96}$$

where perm *is the permanent (8.52).*

The proof should be obvious at this point, so we omit it. The sum is over all permutations $\sigma \in \mathfrak{S}_n$, and each permutation is decomposed into its cycles; the product is then over all cycles $(i(1), i(2), \dots, i(k))$ making up a particular permutation. We note that the representation corresponds to a type of infinite dimensional limit to the double-Fock representation for thermal states.

9.4 Thermal Fields

We now come to a well-known trick for representing thermal states of Bose systems. For definiteness, let $\mathfrak{h} = L^2(\mathbb{R}^3)$ be the space of wave functions in the momentum representation, then the thermal state $\langle \cdot \rangle_{\beta,\mu}$ is characterized as the Gaussian (sometimes referred to as quasifree) mean zero state with

$$\langle A^*(f)A(f) \rangle_{\beta,\mu} = \int |f(k)|^2 n_{\beta,\mu}(k) d^3k,$$

where

$$n_{\beta,\mu}(k) = (e^{\beta[E(k)-\mu]} - 1)^{-1}.$$

(We have the physical interpretation of β as inverse temperature, μ as chemical potential, and $E(k)$ the energy spectrum function.) Let us denote by $n_{\beta,\mu}$ the operation of pointwise multiplication by the function $n_{\beta,\mu}(\cdot)$ on $\mathfrak{h}$, that is, $(n_{\beta,\mu}f)(k) = n_{\beta,\mu}(k)f(k)$. We check that $\langle [A^*(f) + A(f)]^2 \rangle_{\beta,\mu} = \langle f | C_{\beta,\mu}f \rangle$ with $C_{\beta,\mu} = 2n_{\beta,\mu} + 1 \equiv \coth \dfrac{\beta[E - \mu]}{2}$.

Theorem 9.4.1 (Araki–Woods) *Let $\mathfrak{h}$ be a separable Hilbert space and $\jmath$ an conjugation on $\mathfrak{h}$. Define operator fields $B(.)$ on $\Gamma(\mathfrak{h}) \otimes \Gamma(\mathfrak{h})$ by*

$$B(f) = B_1(\sqrt{n + 1}f) \otimes 1_2 + 1_1 \otimes B_2^*(\jmath\sqrt{n}f), \tag{9.97}$$

where the $B_i(.)$ are usual Bosonic fields on the factors $\Gamma(\mathfrak{h})$, $(i = 1, 2)$, and n is a positive operator on $\mathfrak{h}$. Then the fields satisfy the canonical commutation

relations (9.34) and their moments in the joint Fock vacuum state $\Omega = \Omega_1 \otimes \Omega_2$
are precisely the same as for the thermal state when we take $n = n_{\beta,\mu}$.

To see this, note that

$$\langle \Omega | e^{i[A^*(f)+A(f)]} \Omega \rangle = \langle \Omega_1 | e^{i[A_1^*(f)+A_1(f)]} \Omega_1 \rangle \langle \Omega_2 | e^{i[A_2^*(f)+A_2(f)]} \Omega_2 \rangle$$

$$= \exp \left\{ -\frac{1}{2} \langle f | (n+1) f \rangle \right\} \exp \left\{ -\frac{1}{2} \langle f | (n) f \rangle \right\}$$

$$= \exp \left\{ -\frac{1}{2} \langle f | C f \rangle \right\},$$

where $C = 2n+1$ is the covariance of the state. The first factor describes the so-called spontaneous emissions and absorptions while the second factor describes the so-called stimulated emissions and absorptions. In the zero temperature limit, $\beta \to \infty$, we find that $n \to 0$, and so the second-factor fields (stimulated emission and absorption) become negligible.

We remark that for this state $\langle A^*(f)A(f) \rangle = \langle f | nf \rangle \geq 0$, and so n must be a positive operator. In particular, we must therefore have the constraint

$$C \geq 1. \tag{9.98}$$

The representation (9.97) has been introduced by Araki and Woods (1963). It has become an important ingredient of quantum field theory at finite temperature, see e.g. Chap. 2 of the review article Landsman and Weert (1987).

9.5 q-deformed Commutation Relations

So far, we have presented the two staple examples from quantum field theory – Bose and Fermi fields. However, there has been much interest in other possibilities. Let us postulate **canonical q-deformed commutation relations** of the form

$$A(f)A(g)^* = q A(g)^*A(f) + \langle f | g \rangle. \tag{9.99}$$

The cases $q = \pm 1$ are the Bose and Fermi cases. The case $q = 0$ is of particular importance and leads to what is now known as **Free statistics**. While there are no fundamental particles obeying free statistics, it turns out that these models capture noncommutative central limit results that are of enormous importance to random matrix theory. In particular, $\Gamma^0(\mathfrak{H})$ will be the Fock space introduced in Section 8.2.3. We refer the reader to the lecture notes of Voiculescu et al. (1975) and of Nica and Speicher (2006).

Roughly speaking, for $-1 < q < 1$ it is possible to construct a Hilbert space $\Gamma^q(\mathfrak{H})$ of decomposable tensors on which we may define concrete operators satisfying (9.99) for $-1 \leq q \leq 1$; see Bozejko et al. (1997) and Saitoh and Yoshida (2000b). We define a sesquilinear form

$$\langle f_1 \circ \cdots \circ f_n | g_1 \circ \cdots g_m \rangle_q \triangleq \delta_{n,m} \sum_{\sigma \in \text{Perm}(n)} q^{i(\sigma)} \langle f_1 | g_{\sigma(1)} \rangle \cdots \langle f_n | g_{\sigma(n)} \rangle,$$

where $i(\sigma) \triangleq \#\{(j,k) : 1 \leq j < k \leq n, \sigma(j) > \sigma(k)\}$ is the number of inversions of the permutation σ, and take $\Gamma^q(\mathfrak{H})$ to be the completion of the finite number vectors with respect to the corresponding norm.

The creator is defined by

$$A(g)^* f_1 \circ \cdots \circ f_n = g \circ f_1 \circ \cdots \circ f_n, \tag{9.100}$$

while the annihilator $A(g)$ is the following generalization of (9.30):

$$A(g) f_1 \circ \cdots \circ f_n = \sum_{j=1}^{n} (q)^{j-1} \langle g \mid f_j \rangle f_1 \circ \cdots \circ f_{j-1} \circ f_{j+1} \circ \cdots \circ f_n. \tag{9.101}$$

In addition, we have the vacuum vector Ω, and here we have

$$A(g)^* \Omega = g; \qquad A(g) \Omega = 0. \tag{9.102}$$

The operators $A(g)$ and $A(g)^*$ are then bounded for $-1 \leq q < 1$ with norm $\|g\| / \max\{1, \sqrt{1-q}\}$.

9.5.1 q-deformed Gaussian Distribution

One may then use (9.99) repeatedly to get a q-deformed Wick ordering rule. We will consider $\langle \Omega | A^{\varepsilon(2n)}(f_{2n}) \ldots A^{\varepsilon(1)}(f_1) \Omega \rangle$ for a Catalan sequence ε.

To this end, let us express $\pi \in \text{Pair}(2n)$ as a collection of pairs $(p_k, q_k)_{j=1}^n$ with the p_k correspond to annihilators and the q_k to creators, so we must have $p_k > q_k$ for each j and fix the ordering of the pairs as $q_n > \cdots > q_2 > q_1$. The number of crossings associated with the pair partition is defined to be

$$N_c(\pi) \triangleq \left\{ (j,k) : p_j > p_k > q_k > q_j \right\}. \tag{9.103}$$

The reason for referring to these as crossings is immediate from the situation in Figure 9.4. Note that we have focused on just the jth and kth pair for clarity, and in general there will be other pairs (omitted).

One finds a remarkable generalization of equation (9.90).

$$\sum_{\varepsilon} \langle \Omega | A^{\varepsilon(2n)}(f_{2n}) \ldots A^{\varepsilon(1)}(f_1) \Omega \rangle = \sum_{\text{Pair}(2n)} q^{N_c(\pi)} \prod_{k=1}^{n} \langle f_{p_k} | f_{q_k} \rangle.$$

$$\tag{9.104}$$

$$p_j \quad p_k \quad q_j \quad q_k$$

Figure 9.4 A crossing pair of partitions.

From this, we obtain the generating function for the distribution of the q-deformed Segal operator field $\hat{\Phi}(f) = A^*(f) + A(f)$:

$$\langle \Omega | \, e^{it\hat{\Phi}(f)} \, \Omega \rangle = \sum_{n=0}^{\infty} \frac{\|f\|^{2n}}{(2n)!} \sum_{\pi \in \text{Pair}(2n)} q^{N_c(\pi)}. \tag{9.105}$$

Let the distribution be $\mathbb{P}$, so that $\langle \Omega | \, e^{it\hat{\Phi}(f)} \, \Omega \rangle = \int_{-\infty}^{\infty} e^{itx} \mathbb{P}[dx]$, then one may obtain a continued fraction expression; see Bozejko and Yoshida (2006).

$$\int_{-\infty}^{\infty} \frac{1}{z-x} \mathbb{P}[dx] = \cfrac{[0]_q}{z - \cfrac{[1]_q}{z - \cfrac{[2]_q}{z - \cfrac{[3]_q}{\ddots}}}} \tag{9.106}$$

where $[n]_q = \frac{1-q^n}{1-q} = 1 + q + \cdots + q^{n-1}$, $[0]_q = 1$.

In the case $q = 0$, we have the simple identity for $H(z) = \int_{-\infty}^{\infty} \frac{1}{z-x} \mathbb{P}[dx]$ that $H(z) = \frac{1}{z-H(z)}$, which leads to $H(z) = \frac{1}{2}(z \pm \sqrt{z^2 - 4})$, and one may calculate the Hilbert transform at the boundary to get $\mathbb{P}[dx] \equiv \varrho(x)\,dx$, where

$$\varrho(x) \equiv \frac{1}{\pi} \text{Im}\, H(x - i0^+)\,dx$$
$$= \begin{cases} \frac{1}{2\pi}\sqrt{4 - x^2}, & |x| \le 2; \\ 0, & |x| > 2. \end{cases} \tag{9.107}$$

This is the well-known Wigner semicircle law.

It turns out that for $-1 < q < 1$, the distributions are always absolutely continuous and compactly supported. For $q = 1$, we have, of course, the standard Gaussian.

For the Fermi case, $q = -1$, we have that $[n]_{-1}$ is one, for n odd, a zero, for $n > 0$ even, leading to the terminating continued fraction $H(z) = \frac{1}{z - 1/z}$ which is easily seen to correspond to the discrete distribution $\frac{1}{2}\delta_1[dx] + \frac{1}{2}\delta_{-1}[dx]$. (This is, of course, a fair coin – but also corresponds to what we termed the standard Fermionic Gaussian distribution, for reasons that ought to be by now apparent.)

In general, for $-1 < q < 1$, the density has been shown by Bozejko et al. (1997) to take the form

$$\varrho(x) = \begin{cases} \frac{1}{\pi}\sqrt{1-q}\,\sin\theta\prod_{n=1}^{\infty}\left\{(1-q^n)|1-e^{2i\theta}q^n|^2\right\}, & |x| \leq \frac{2}{\sqrt{1-q}}; \\ 0, & \text{otherwise}, \end{cases}$$

where $\theta \in [0, \pi]$ is defined by $x = \frac{2}{\sqrt{1-q}}\cos\theta$.

9.5.2 q-deformed Poisson Distribution

We may define a number operator by

$$N = \left(A\left(f\right)^* + \sqrt{\lambda}\right)\left(A\left(f\right) + \sqrt{\lambda}\right)$$

and we similarly find that

$$\langle\Omega|\,e^{itN}\,\Omega\rangle = \sum_{n=0}^{\infty}\frac{\|f\|^n}{n!}\sum_{\pi\in\text{Part}(n)}q^{N_c(\pi)}\,(it)^{N(\pi)},$$

where again we have a modification of the Bose formula involving the deformation parameter q raised to the power of the crossing number for partitions.

Remarkably, there is a closed formula for the probability distribution due to Saitoh and Yoshida (2000a). Taking $\|f\| = 1$, this will be

$$\mathbb{P}[dx] = \varrho(x)\,dx + \sum_{n=0}^{N_{\max}}p_n\delta_{\nu_n}[dx],$$

where the density is

$$\varrho(x) = \frac{(1-q)}{2\pi x}\sqrt{\frac{4\lambda}{1-q} - \left(x - \lambda - \frac{1}{1-q}\right)^2}$$

$$\times \prod_{n=1}^{\infty}(1-q^n)\frac{\lambda(1+q^n)^2 - (1-q)q^n\left(x - \lambda - \frac{1}{1-q}\right)^2}{q^n\left(x - \lambda - \frac{1}{1-q}\right) + \lambda + \frac{q^{2n}}{1-q}},$$

for $\left(\sqrt{\lambda} - \frac{1}{\sqrt{1-q}}\right)^2 < x < \left(\sqrt{\lambda} + \frac{1}{\sqrt{1-q}}\right)^2$, and $\varrho(x) = 0$ otherwise;

$$N_{\max} = \sup\left\{n : q^{2n} \geq \lambda(1-q)\right\},$$

$$p_n = \left(1 - \frac{\lambda(1-q)}{q^{2k}}\right)\frac{\left(\lambda q^{-n+1}\right)^n}{[1]_q[2]_q\ldots[n]_q}\frac{1}{\exp_q\left(\lambda q^{-n+1}\right)},$$

$$\nu_n = [n]_q + \lambda\left(1 - \frac{1}{q^n}\right),$$

and $\exp_q(\lambda) \triangleq \sum_{n=0}^{\infty} \frac{1}{[0]_q[1]_q\cdots[n]_q}\lambda^n$ defines the q-deformed exponential. Note that $\exp_q(\lambda) \equiv \prod_{n=0}^{\infty}(1-(1-q)q^n x)^{-1}$.

For the "Bose" case $q \to 1^-$, we find that the absolutely continuous part vanishes, and the remainder is the usual Poisson distribution of intensity λ: $v_n \equiv n$ and $p_n \equiv \frac{\lambda^n}{n!}e^{-\lambda}$.

For the Free case $q \to 0$, we find that density is

$$\varrho_{\text{free}}(x) = \begin{cases} \frac{1}{2\pi x}\sqrt{4\lambda - (x - \lambda - 1)^2}, & \left(\sqrt{\lambda} - 1\right)^2 < x < \left(\sqrt{\lambda} + 1\right)^2; \\ 0, & \text{otherwise}; \end{cases}$$

but, for the discrete part, only the $n = 0$ term may survive with $v_0 = 0$ and $p_0 = \max\{1 - \lambda, 0\}$.

The Fermi case, $q = -1$, is just the Fermionic Poisson distribution, which we found as the law of the 2×2-matrix N_λ; recall (3.2), in the ground state.

10

L^2-Representations of the Boson Fock Space

As we have already seen in Section 4.2, the Boson Fock space has representations as L^2-space that are fequently used in quantum field theory and in stochastic analysis. In this chapter, we recapitulate the most important cases: the Bargmann–Fock or complex wave representation and the Wiener–Itō–Segal representation. The latter representation is used in a more abstrct version, called the Wiener–Segal or real wave representation, and in a stochastic setting first introduced by Itō.

10.1 The Bargmann–Fock Representation

From the discussion of Lemma 9.2.1, we know that any tensor $F \in \Gamma^+(\mathfrak{H})$ is uniquely determined by the function

$$\mathfrak{H} \ni z \mapsto \varphi_F(z) \triangleq \langle \exp(z) \mid F \rangle \in \mathbb{C}. \tag{10.1}$$

Since the exponential series is a holomorphic function, the function $\varphi_F(z)$ is antiholomorphic. The unit $1 \in \mathcal{A}_0^+(\mathfrak{H})$ of the tensor algebra is mapped onto the constant function $\varphi_1(z) \equiv 1$, and tensors in $\mathcal{A}^+(\mathfrak{H})$ are represented by antiholomorphic polynomials.

Proposition 10.1.1 *For tensors F and G in the algebra $\Gamma_>^+(\mathfrak{H})$, the following product formula is true:*

$$\varphi_{F \vee G}(z) = \varphi_F(z) \cdot \varphi_G(z). \tag{10.2}$$

Proof The product $F \vee G$ is defined for tensors in the algebra $\Gamma_>^+(\mathfrak{H})$, which includes the exponential vectors. We choose $F = \exp(f)$ and $G = \exp(g)$. Then

$$\langle \exp(h) \mid \exp(f) \vee \exp(g) \rangle = \langle \exp(h) \mid \exp(f + g) \rangle = e^{\langle h \mid f + g \rangle}$$
$$= e^{\langle h \mid f \rangle} e^{\langle h \mid g \rangle} = \langle \exp(h) \mid \exp(f) \rangle \langle \exp(h) \mid \exp(g) \rangle$$

189

follows, and (10.2) is true. By linearity, this result can be extended to tensors F and G in the linear span of the exponential vectors. Since this set is dense in $\Gamma^+(\mathfrak{H})$ and the inner product is continuous, the result is true for all tensors, for which the product $F \vee G$ is defined. □

Hence the mapping $F \mapsto \varphi_F(z)$ is an isomorphism between the tensor algebra $\mathcal{A}^+(\mathfrak{H})$ and the multiplicative algebra of antiholomorphic polynomials $\varphi_F(z)$, $F \in \mathcal{A}^+(\mathfrak{H})$, with unit $\varphi(z) \equiv 1$.

We now define an L^2-inner product for these antiholomorphic functions. Let dv be the canonical Gaussian premeasure on (the underlying real space of) $\mathfrak{H}$. This measure has the Fourier–Laplace transform

$$\int_{\mathfrak{H}} e^{\overline{\langle v|z \rangle} + \langle u|z \rangle} dv = e^{\langle u|v \rangle} \text{ with } u, v \in \mathfrak{H}. \tag{10.3}$$

Proposition 10.1.2 *If F and G are tensors in $\mathcal{A}^+_{coh}(\mathfrak{H})$, then the following identity is true:*

$$\int \varphi_F(z)^* \varphi_G(z) dv = \langle F \mid G \rangle. \tag{10.4}$$

Proof The functions $\varphi_F(z)$ with $F \in \mathcal{A}^+_{coh}(\mathfrak{H})$ are cylinder functions and they are measurable with respect to dv. If F and G are exponential vectors, the identity (10.4) is a consequence of (9.66) and (10.3). The extension to $F, G \in \mathcal{A}^+_{coh}(\mathfrak{H})$ follows by (anti)linear continuation. □

A special case of (10.4) is $\int |\varphi_F(z)|^2 dv = \|F\|^2$. Hence (10.1) is an isometric mapping from $\mathcal{A}^+_{coh}(\mathfrak{H})$ into the pre-Hilbert space of the linear span of antiholomorphic exponential functions, which are square integrable with respect to dv. The completion of this pre-Hilbert space is denoted by $L^2_a(\mathfrak{H}, dv)$. By continuity, the mapping (10.1) can be extended to an isometric isomorphism between $\Gamma^+(\mathfrak{H})$ and $L^2_a(\mathfrak{H}, dv)$. This representation of the Fock space by antiholomorphic functions is called the **Bargmann–Fock** or **complex wave representation** (Bargmann, 1961; Segal, 1962). In the literature, one often assumes that $\mathfrak{H}$ has an conjugation $z \mapsto z^*$. Then one can use the holomorphic functions $\varphi_F(z^*)$ for the representation.

If $\mathfrak{H}$ is finite dimensional, $\mathfrak{H} \cong \mathbb{C}^n$, the promeasure is the Gaussian measure $dv = \frac{1}{\pi^n} e^{-\|z\|^2} d^n x d^n y$ on $\mathbb{C}^n$, where $x \in \mathbb{R}^n$ and $y \in \mathbb{R}^n$ are the real and the imaginary part of $z = x + iy \in \mathbb{C}^n$.

If $\mathfrak{H}$ is infinite dimensional, one can extend the promeasure dv to a σ-additive measure on a space $\mathfrak{H}_>$, which is strictly larger than $\mathfrak{H}$. The functions $\varphi_F(z)$, $z \in \mathfrak{H}_>$, are then square integrable with respect to this measure for all $F \in \Gamma^+(\mathfrak{H})$, but only a dense subset of these functions is continuous and antiholomorphic in the variable $z \in \mathfrak{H}_>$.

10.2 Wiener Product and Wiener–Segal Representation

Another important L^2-representation of the Boson Fock space is the **Wiener–Segal representation** or **real wave representation** that has been introduced by Wiener (1930) and Segal (1956). A stochastic version of this isomorphism – first derived by Itô (1951) – is presented in the next section.

Let $\mathfrak{H}$ be a Hilbert space with conjugation $\mathfrak{H} \ni f \mapsto f^* \in \mathfrak{H}$. Then

$$B(f,g) \triangleq \langle f^*, g \rangle \in \mathbb{C} \tag{10.5}$$

is a bilinear nondegenerate symmetric form with modulus $|\beta(f,g)| \leq \|f\| \|g\|$. On the linear span of the exponential vectors $\mathcal{A}_{coh}^+(\mathfrak{H})$, we define a linear invertible mapping Υ by

$$\Upsilon \exp(f) \triangleq \exp(f - 2^{-1} B(f,f)) = e^{-\frac{1}{2}B(f,f)} \exp(f). \tag{10.6}$$

Since $\exp(f)$ and $\exp(f - 2^{-1}B(f,f))$ are analytic in f, we can derive the action of Υ on the algebra $\mathcal{A}^+(\mathfrak{H})$ by power counting. This calculation begins with

$$\Upsilon 1 = 1, \qquad \Upsilon f = f, \qquad \Upsilon (f \vee f) = f \vee f - B(f,f) 1.$$

A closer investigation of this mapping is given in Kupsch and Smolyanov (1998), where it is called **Wick ordering** or **normal ordering** (of tensors). The mapping (10.6) can be extended to a continuous linear mapping from the Hilbert space $\Gamma_{(\alpha)}^+(\mathfrak{H})$ into the space $\Gamma_{(\beta)}^+(\mathfrak{H})$ if $\alpha > \beta \geq 0$, see Appendix A of Kupsch and Smolyanov (1998).

Starting from (10.6), it is possible to define a product with symbol ∇ on $\mathcal{A}_{coh}^+(\mathfrak{H})$ by

$$(\exp f) \nabla (\exp g) \triangleq e^{-B(f,g)} \exp(f + g). \tag{10.7}$$

This product is symmetric and associative on $\mathcal{A}_{coh}^+(\mathfrak{H})$. Both sides of (10.7) are analytic functions of f and g. Calculating the linear parts in f and g, we obtain the bilinear mapping

$$\mathfrak{H} \times \mathfrak{H} \ni (f,g) \mapsto f \nabla g \triangleq f \vee g - B(f,g)\Omega \in \mathcal{A}^+(\mathfrak{H}), \tag{10.8}$$

which can be extended to an associative and commutative product $F \nabla G$ on $\mathcal{A}^+(\mathfrak{H})$ called the **Wiener product**. With $\Omega \nabla F = F \nabla \Omega = F$, this algebra has the unit element $\Omega = 1 \in \mathcal{A}_0^+(\mathfrak{H})$. Using the mapping (10.6), the formula (10.7) can be written as

$$F \triangledown G \triangleq \Upsilon^{-1}\left(\Upsilon F \vee \Upsilon G\right), \tag{10.9}$$

with $F = \exp(f)$ and $G = \exp(g)$. By linear extension to $F, G \in \mathcal{A}^+_{coh}(\mathfrak{H})$, this formula defines a bilinear symmetric product on $\mathcal{A}^+_{coh}(\mathfrak{H})$. With continuity arguments, this product can be extended to a product on the spaces $\Gamma^+_{(\alpha)}(\mathfrak{H})$, $\alpha > 0$.

From Corollary 9.2.2, we know that a tensor $F \in \Gamma^+(\mathfrak{H})$ is uniquely determined by the function $\mathfrak{H}_\mathbb{R} \ni x \mapsto \langle \exp(x) \mid F \rangle \in \mathbb{C}$. Given $F \in \mathcal{A}^+_{coh}(\mathfrak{H}) \subset \Gamma^+(\mathfrak{H})$, we define the function

$$\psi_F(x) \triangleq \langle \exp(x) \mid \Upsilon F \rangle. \tag{10.10}$$

Thereby the tensor $F = 1$ is represented by the function $\psi_F(x) \equiv 1$.

Proposition 10.2.1 *For tensors F and G in $\mathcal{A}^+_{coh}(\mathfrak{H})$, the following product formula is true:*

$$\psi_{F \triangledown G}(x) = \psi_F(x) \cdot \psi_G(x). \tag{10.11}$$

Proof The formula (10.11) follows from (10.2) and (10.9). □

The functions (10.10) with $F \in \mathcal{A}^+_{coh}(\mathfrak{H})$ are cylinder functions on $\mathfrak{H}_\mathbb{R}$, and they are measurable with respect to the canonical Gaussian pro-measure $d\mu(x)$ on $\mathfrak{H}_\mathbb{R}$. This pro-measure has the Fourier–Laplace transform

$$\int_{\mathfrak{H}_\mathbb{R}} e^{\langle x \mid f \rangle} d\mu(x) = e^{B(f,f)/2}, f \in \mathfrak{H}. \tag{10.12}$$

Lemma 10.2.2 *If F and G are tensors in $\mathcal{A}^+_{coh}(\mathfrak{H})$, then the following identity is true*

$$\int \psi_F(x)^* \psi_G(x) d\mu(x) = \langle F \mid G \rangle. \tag{10.13}$$

Proof It is sufficient to prove the identity (10.13) for $F = \exp(f)$ and $G = \exp(g)$ with $\psi_F(z) = \langle \exp(x) \mid \Upsilon \exp(f) \rangle = e^{-B(f,f)/2} e^{\langle x \mid f \rangle}$ and $\psi_G(z) = e^{-B(g,g)/2} e^{\langle x \mid g \rangle}$. Then the integral in (10.13) is calculated as

$$\exp\left(-\tfrac{1}{2} B(f^\star, f^\star) - \tfrac{1}{2} B(g, g)\right) \int e^{\langle x \mid f^\star + g \rangle} d\mu(x)$$
$$\overset{(10.13)}{=} \exp B(f^\star, g) \overset{(10.5)}{=} \exp \langle f \mid g \rangle,$$

and the identity (10.13) is true for exponential vectors. The extension to tensors in $\mathcal{A}^+_{coh}(\mathfrak{H})$ follows by (anti)linearity. □

A special case of (10.13) is $\int |\psi_F(x)|^2 d\mu(x) = \|F\|^2$. Hence

$$F \mapsto \psi_F(x) = \langle \exp(x) \mid \Upsilon F \rangle \tag{10.14}$$

is an isometric mapping from $\mathcal{A}^+_{coh}(\mathfrak{H}) \subset \Gamma^+(\mathfrak{H})$ into a pre-Hilbert space of functions ψ_F on $\mathfrak{H}_\mathbb{R}$, which are square integrable with respect to $d\mu$. The completion of this pre-Hilbert space is denoted by $L^2(\mathfrak{H}_\mathbb{R}, d\mu)$. By continuity, the mapping (10.14) can be extended to an isometric isomorphism between $\Gamma^+(\mathfrak{H})$ and $L^2(\mathfrak{H}_\mathbb{R}, d\mu)$. The representation of the Fock space is called the **Wiener–Segal representation** or **real wave representation**. If $\mathfrak{H} = \mathbb{C}^n$ is finite dimensional, the pro-measure $d\mu(x)$ is the Gaussian measure $d\mu(x) = (2\pi)^{-\frac{n}{2}} \exp\left(-\frac{1}{2}\|x\|^2\right) d^n x$ on $\mathfrak{H}_\mathbb{R} = \mathbb{R}^n$. In this case, one can absorb the weight function $(2\pi)^{-\frac{n}{2}} \exp\left(-\frac{1}{2}\|x\|^2\right)$ into the wave functions ψ, and one obtains the standard Schrödinger representation on $L^2(\mathbb{R}^n, d^n x)$.

10.3 Itō–Fock Isomorphism

If the one-particle Hilbert space is an L^2-space with measure given by the covariance of a martingale, the Wiener–Segal isomorphism can be constructed by stochastic integrals, a method first given by Itō in the case of the canonical Wiener process (Itô, 1951).

Let X_t be a martingale with canonical probability space $(\Omega_X, \mathcal{F}_X, \mathbb{P})$ and let $\mathfrak{h}_X = L^2(\Omega_X, \mathcal{F}_X, \mathbb{P})$. We consider the function $F(t) = \mathbb{E}\left[X_t^2\right]$ which then defines a monotone increasing function. We shall understand $dF(t)$ to be the Stieltjes integrator in the following.

It turns out that our considerations so far allow us to construct a natural isomorphism between $\mathfrak{h}_X$ and the Fock space $\Gamma_+\left(L^2\left(\mathbb{R}^+, dF\right)\right)$ (see, e.g., Parthasarathy, 1992).

For $f \in L^2\left(\mathbb{R}^+, dF\right)$, we define the random variable

$$X(f) = \int_{[0,\infty)} f(s)\, dX_s$$

and a process $X_t(f) = X\left(1_{[0,t]}f\right)$.

Lemma 10.3.1 *Let* $X_t^{(n)}(f) = \int_{\Delta^n(t)} dX_{t_n}(f) \ldots dX_{t_1}(f)$, *then*

$$\mathbb{E}\left[X_t^{(n)}(f) X_s^{(m)}(g)\right] = \frac{1}{n!}\left[\int_0^{t\wedge s} f(u) g(u)\, dF(u)\right]^n \delta_{n,m}.$$

Proof For simplicity, we ignore the intensities. Let

$$X_t^{(n)} = \int_{\Delta^n(t)} dX_{t_n} \ldots dX_{t_1}.$$

Then we have $\mathbb{E}\left[X_t^{(n)}X_s^{(0)}\right] = \mathbb{E}\left[X_t^{(n)}\right] = 0$ whenever $n > 0$. Next suppose that n and m are positive integers, then

$$\mathbb{E}\left[X_t^{(n)}X_s^{(m)}\right] = \mathbb{E}\left[\int_0^t dX_u\, X_{u^-}^{(n-1)} \int_0^s dX_v\, X_{v^-}^{(m-1)}\right]$$
$$= \int_0^{t\wedge s} dF\,(u)\,\mathbb{E}\left[X_u^{(n-1)}X_u^{(m-1)}\right]$$

and we may reiterate until we reduce at least one of the orders to zero. We then have

$$\mathbb{E}\left[X_t^{(n)}X_s^{(m)}\right] = \delta_{n,m}\int_0^{t\wedge s} dF\,(u_n)\int_0^{u_n} dF\,(u_{n-1})\cdots\int_0^{u_2} dF\,(u_1)$$
$$= \frac{1}{n!}F\,(t\wedge s)^n\,\delta_{n,m}.$$

The proof with the intensities from $L^2\left(\mathbb{R}^+,dF\right)$ included is a straightforward generalization. $\qquad\square$

Theorem 10.3.2 *The Hilbert space $\mathfrak{h}_X = L^2\left(\Omega_X,\mathcal{F}_X,\mathbb{P}\right)$ and the Fock space $\Gamma_+\left(L^2\left(\mathbb{R}^+,dF\right)\right)$ are naturally isomorphic.*

Proof Consider the map into the exponential vectors given by

$$e^{X_t(f)} \equiv \sum_{n\geq 0}\frac{1}{n!}X_t^{(n)}\,(f) \mapsto \exp\,(f)$$

for each $f \in L^2\left(\mathbb{R}^+,dF\right)$, and denote the $t \to \infty$ limit as $e^{X(f)}$. We know that the exponential vectors are dense in Fock space and in a similar way the exponential processes $e^{X_t(f)}$ are dense in $\mathfrak{h}_X$. The map may then be extended to one between the two Hilbert spaces.

Unitarity follows from the observation that

$$\mathbb{E}\left[e^{X(f^*)}e^{X(g)}\right] = e^{\int_{[0,\infty)}f^*g\,dF},$$

which is an immediate consequence of the previous lemma. $\qquad\square$

Now for $F \in \mathfrak{h}_X = L^2\left(\Omega_X,\mathcal{F}_X,\mathbb{P}\right)$, we will have the so-called **chaotic expansion**

$$F = \sum_n\int_{\Delta_n}\widehat{F}(t_n,\ldots,t_1)\,dX_{t_n}\ldots dX_{t_1}. \tag{10.15}$$

Following our ideas from Chapter 4, we introduce the shorthand notation

$$F = \oint_{\mathrm{Power}(\mathbb{R}_+)}\widehat{F}(T)\,dX_T. \tag{10.16}$$

For instance, the diagonal-free exponentials are

$$\phi^{\int f(t) dX_t} = \fint_{\text{Power}(\mathbb{R}_+)} f^T \, dX_T.$$

10.3.1 Stochastic Convolutions

The **stochastic Wick convolution** $\diamond_X$ is defined by

$$FG \equiv \fint_{\text{Power}(\mathbb{R}_+)} [\widehat{F} \diamond_X \widehat{G}](T) \, dX_T, \qquad (10.17)$$

whenever $F = \fint \widehat{F}(T) \, dX_T$ and $G = \fint \widehat{G}(T) \, dX_T$. Under favorable circumstances, we may deduce a closed form for the convolution.

The Wiener Wick Convolution

Starting from the exponential $\phi^{\int f(t) dW_t} = e^{\int f(t) dW_t - \frac{1}{2} \int f(t)^2 dt}$, we have

$$
\begin{aligned}
\phi^{\int f(t) dW_t} \phi^{\int g(t) dW_t} &= e^{\int f(t) g(t) dt} \phi^{\int [f(t) + g(t)] dW_t} \\
&= \fint_{\text{Power}(\mathbb{R}_+)} f^S g^S \, dS \fint_{\text{Power}(\mathbb{R}_+)} [f + g]_T dW_T \\
&= \fint_{\text{Power}(\mathbb{R}_+)} dW_T \fint_{\text{Power}(\mathbb{R}_+)} dS \sum_{T_1 + T_2 = T} f^{T_1 + S} g^{T_2 + S}.
\end{aligned}
$$

As the exponential vectors are dense, we generalize to

$$[\widehat{F} \diamond_W \widehat{G}](T) = \fint_{\text{Power}(\mathbb{R}_+)} dS \sum_{T_1 + T_2 = T} \widehat{F}(T_1 + S) \widehat{G}(T_2 + S). \qquad (10.18)$$

The Poisson Wick Convolution

This time we have $\phi^{\int f(t) dN_t} = \exp \int \ln(1 + f) dN$, and so

$$
\begin{aligned}
\phi^{\int f(t) dN_t} \phi^{\int g(t) dN_t} &= \exp \int \ln(1 + f + g + fg) dN \\
&= \phi^{\int [f(t) + g(t) + f(t) g(t)] dN_t} \\
&= \fint_{\text{Power}(\mathbb{R}_+)} (f + g + fg)^T dN_T,
\end{aligned}
$$

but we may write

$$
\begin{aligned}
(f + g + fg)^T &= \sum_{T_1 + T_2 + T_3 = T} f^{T_1} g^{T_3} (fg)^{T_2} \\
&\equiv \sum_{T_1 + T_2 + T_3 = T} f^{T_1 + T_2} g^{T_3 + T_2}.
\end{aligned}
$$

Again we generalize to get

$$[\widehat{F} \diamond_N \widehat{G}](T) = \sum_{T_1+T_2+T_3=T} \widehat{F}(T_1 + T_2)\widehat{G}(T_3 + T_2). \qquad (10.19)$$

10.3.2 Wiener–Itō–Segal Isomorphism

Let us specify the case where $(\Omega_W, \mathcal{F}_W, \mathbb{P}_W)$ is the probability space of the canonical Wiener process with $\Omega_W = C^0[0, \infty)$. Here we obtain the Wiener–Itō–Segal isomorphism

$$L^2 (\Omega_W, \mathcal{F}_W, \mathbb{P}_W) \cong \Gamma \left(L^2 (\mathbb{R}^+, dt) \right). \qquad (10.20)$$

Let $f \in L^2 (\mathbb{R}^+, dt)$ with real and imaginary parts f' and f'', then we have $W(f) = W(f') + iW(f'')$, that is, $W(\cdot)$ is complex linear in its argument. We recall that the Wiener exponential is given by

$$\mathcal{e}^{W(f)} = e^{W(f) - \frac{1}{2} \int_0^\infty f^2}.$$

Note that this is analytic in f, and in particular the integral term is $\int_0^\infty f^2$ and not $\int_0^\infty |f|^2$.

We may introduce some notation at this stage. For $x \in C^0[0, \infty)$, a Wiener trajectory, that is, $x = x(t)$, is a continuous path in $t \geq 0$ with $x(0) = 0$. We set

$$\langle x | \exp (f) \rangle \triangleq \mathcal{e}^{W(f)} (x), \qquad (10.21)$$

which of course corresponds to the identification $\mathcal{e}^{W(f)}$ with the exponential vector $\exp (f)$ behind the isomorphism.

The set of operators $\{Q(f) : f \in S\}$ on the Fock space will be commutative if and only if $\text{Im}\langle f | g \rangle = 0$, for all $f, g \in S$. We fix $S = L^2_{\mathbb{R}} (\mathbb{R}^+, dt)$, the real-valued L^2-functions.

Lemma 10.3.3 *In the preceding setting of the Wiener–Itō–Segal representation for the Fock space $\Gamma \left(L^2 (\mathbb{R}^+, dt) \right)$ by $L^2 (\Omega_W, \mathcal{F}_W, \mathbb{P}_W)$, the operators $Q(g)$, with test function $g \in L^2_{\mathbb{R}} (\mathbb{R}^+, dt)$, correspond to multiplication by the random variable $W(g) = \int_0^\infty g(t) \, dW_t$.*

Proof Starting with the identity $e^{iQ(g)} = D(ig)$, we note that for $g \in L^2_{\mathbb{R}} (\mathbb{R}^+, dt)$ we have

$$\left\langle x|e^{iQ(g)}\exp\left(f\right)\right\rangle = e^{-\frac{1}{2}\int g^2 - i\int gf}\left\langle x|\exp\left(f+ig\right)\right\rangle$$

$$= e^{-\frac{1}{2}\int g^2 - i\int gf}\phi^{W(f+ig)}\left(x\right)$$

$$= e^{-\frac{1}{2}\int g^2 - i\int gf}e^{W(f+ig)}\left(x\right)e^{-\frac{1}{2}\int(f+ig)^2}$$

$$= e^{iW(g)}\left(x\right)\left\langle x|\exp\left(f\right)\right\rangle.$$

□

The identification (10.21) can be rephrased by the following resolution of identity:

$$I = \int_{C^0[0,\infty)} |x\rangle\,\langle x|\,\mathbb{P}_W\,[dx]. \tag{10.22}$$

Technically, we should give the rigged Hilbert space underlying this resolution. The particular Gelfand triple is similar in construction to the Schwartz spaces and is due to Hida (2008) and forms the basis of the mathematical theory known as white noise analysis.

Theorem 10.3.4 (Cameron–Martin–Girsanov) *Let $F[x]$ denote a functional of a trajectories $x \in C^0[0,\infty)$. For $h \in L^2_{\mathbb{R}}\left(\mathbb{R}^+, dt\right)$, we define the function $H(t) = \int_0^t h(s)\,ds$, then*

$$\mathbb{E}\left[F[x+H]\right] = \mathbb{E}\left[F[x]\phi^{W(h)}(x)\right],$$

where $\phi^{W(h)}(x) = e^{\int_0^\infty h\,dx - \frac{1}{2}\int_0^\infty h^2}$.

Proof This is actually easy to prove on the Fock space, where the shift by the function H is then implemented by the displacement operator. We note that

$$\mathbb{E}\left[F[x+H]\right] = \langle\Omega|F[Q+H]\Omega\rangle,$$

where now we consider a functional of the operator-valued process $Q_t + H_t = Q_t + \langle h|1_{[0,t]}\rangle \equiv D\left(\frac{h}{2}\right)^* Q_t D\left(\frac{h}{2}\right)$. Therefore,

$$\langle\Omega|F[Q+H]\Omega\rangle = \left\langle\Omega|D(\frac{h}{2})^* F[Q] D(\frac{h}{2})\Omega\right\rangle$$

$$= e^{-\frac{1}{4}\|h\|^2}\left\langle\exp(\frac{h}{2})|F[Q]\exp(\frac{h}{2})\right\rangle$$

$$= e^{-\frac{1}{4}\int h^2}\int_{\Omega_W}|\langle x|\exp\left(f\right)\rangle|^2\,F[Q(x)]\,\mathbb{P}[dx]$$

$$= e^{-\frac{1}{2}\int h^2}\int_{\Omega_W}e^{W(h)}(x)\,F[Q(x)]\,\mathbb{P}[dx].$$

□

11

Local Fields on the Boson Fock Space: Free Fields

The theory of general quantum fields was first presented by Heisenberg and Pauli (1929, 1930) on the basis of the canonical formalism of fields. Free quantum fields are obtained in this way, and a successful perturbation theory has been developed for interacting quantum fields using the canonical formalism. But as shown by Haag – see Haag's Theorem in Streater and Wightman (1964) – the interaction picture does not exist for quantum field theories, which are invariant against space-time translations and rotations. Therefore, relativistic quantum field theories with interaction need a foundation beyond canonical formalism.

It is the aim of this chapter to emphasize an alternative approach to relativistic quantum field theory that is based on the representation theory of the inhomogeneous Lorentz group (Poincaré group) and on locality without reference to canonical equal time commutators. In Section 11.1, the Hilbert space of a scalar (spin zero) particle is given by an irreducible unitary (ray) representation of the Poincaré group. The Fock space of this Hilbert space carries the n-particle states. The free field $\Phi(x)$ with x being an element of the Minkowski space is constructed as linear combinations of the corresponding creation and annihilation operators in such a way that it transforms covariantly under Poincaré transformations. The additional essential demand is locality, which means that two fields $\Phi(x)$ and $\Phi(y)$ commute if x and y are spacelike separated. The free field is then determined up to unessential phase factors. Such an approach is presented for any spin in Weinberg (1964, 1995). As additional literature, we refer to Itzykson and Zuber (1980), to Kastler (1961, chap. V), and to Reed and Simon (1975, sec.X.7).

Since the canonical formalism is often used in the literature, we present canonical fields and their equal time commutators in Section 11.2. This approach is applied in Chapter 12.

11.1 The Free Scalar Field

11.1.1 Lorentz and Poincaré Group

The four-dimensional Minkowski spacetime is denoted by $\mathbb{M}$. The elements of $\mathbb{M}$ are the vectors $x = (x^0, \mathbf{x}) = (x^0, x^1, x^2, x^3) \in \mathbb{M}$.

Here we have $x^0 = t \in \mathbb{R}$, $\mathbf{x} \in \mathbb{R}^3$, and use units with $\hbar = c = 1$.

The structure of $\mathbb{M}$ is given by the indefinite bilinear form

$$\mathbb{M} \ni x, y \longrightarrow x \cdot y = x^0 y^0 - \mathbf{x}.\mathbf{y} \in \mathbb{R}. \tag{11.1}$$

The forward light cone is $\mathcal{V}_0^+ = \{x \mid x^2 \equiv x \cdot x = 0, \; x^0 \geq 0\}$, and a light ray starting at $x = 0$ and propagating forward in time can reach points on $\mathcal{V}_0^+$. The convex cone $\mathcal{V}^+ = \{x \mid x^2 > 0, \; x^0 > 0\}$ is called the (open) forward cone, and $\mathcal{V}^- = -\mathcal{V}^+$ is the backward cone; see Figure 11.1. Their closures are $\overline{\mathcal{V}^+}$ and $\overline{\mathcal{V}^-}$.

The spacelike hypersurfaces $\mathcal{V}_{i\lambda} = \{x \mid x^2 = -\lambda^2\}$, $\lambda > 0$, are connected, whereas the timelike hypersurfaces $\mathcal{V}_m = \{x \mid x^2 = m^2\}$, $m > 0$, split into the disconnected hyperboloids $\mathcal{V}_m^+ = \{x \mid x^2 = m^2, \; x^0 > 0\}$ and $\mathcal{V}_m^- = -\mathcal{V}_m^+$; see Figure 11.2.

Lorentz transformations Λ are linear transformation of the Minkowski space that satisfy

$$(\Lambda x) \cdot (\Lambda y) = x \cdot y \text{ for all } x, y \in \mathbb{M}. \tag{11.2}$$

The set of all these transformations form a group – the Lorentz group – which is denoted by $\mathcal{L}$. The determinant of a Lorentz transformation is either $\det \Lambda = +1$ or $\det \Lambda = -1$. A transformation $\Lambda \in \mathcal{L}$ with $\det \Lambda = 1$ is called a proper Lorentz transformation, and the corresponding subset of $\mathcal{L}$ is $\mathcal{L}_+$. A Lorentz transformation is called orthochronous, if

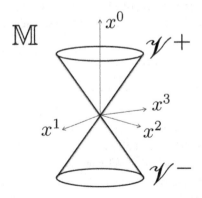

Figure 11.1 The open future and past cones $\mathcal{V}^\pm$ in Minkowski spacetime $\mathbb{M}$.

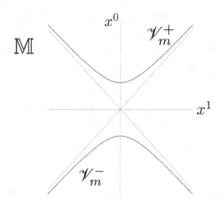

Figure 11.2 The spacelike hyperboloids $\mathscr{V}_m^{\pm}$ (cross-section in the (x^0, x^1) plane).

$$\Lambda \mathscr{V}^+ = \mathscr{V}^+, \tag{11.3}$$

the corresponding subset of $\mathcal{L}$ is $\mathcal{L}^{\uparrow}$. The set $\mathcal{L}$, and its subsets $\mathcal{L}^{\uparrow}$, $\mathcal{L}_+$ and $\mathcal{L}_+^{\uparrow} = \mathcal{L}^{\uparrow} \cap \mathcal{L}_+$ are groups. The proper orthochronous Lorentz group $\mathcal{L}_+^{\uparrow}$ is the connectivity component of the identity. It acts transitively on the hyperboloids $\mathscr{V}_m^+, \mathscr{V}_m^-$ and $\mathscr{V}_{i\lambda}$, $m > 0, \lambda > 0$.

The Minkowski space can be extended to the complex Minkowski space $\mathbb{M}_C \simeq \mathbb{C}^4$. The complex linear transformations Λ, which preserve the $\mathbb{C}$-bilinear form $z \cdot w = z^0 w^0 - \sum_{j=1}^{3} z^j w^j$ and which have the determinant $\det \Lambda = 1$, form the complex Lorentz group $\mathcal{L}_{\mathbb{C}}$.

The symmetry group of relativistic theories is the inhomogeneous Lorentz group or Poincaré group $\mathcal{P}_+^{\uparrow} = \mathbb{M} \times \mathcal{L}_+^{\uparrow}$ of all translations $y \in \mathbb{M}$ and all proper orthochronous Lorentz transformations $\Lambda \in \mathcal{L}_+^{\uparrow}$. This group is the connectivity component of the identity. The full Poincaré group $\mathcal{P} = \mathbb{M} \times \mathcal{L}$ acts on the Minkowski space as $\mathbb{M} \ni x \mapsto \Lambda x + y \in \mathbb{M}$. This definition implies the product rule

$$(\mathbb{M} \times \mathcal{L}) \times (\mathbb{M} \times \mathcal{L}) \mapsto \mathbb{M} \times \mathcal{L} :$$
$$(y, \Lambda), (y', \Lambda') \mapsto (y, \Lambda) \circ (y', \Lambda') = (y + \Lambda y', \Lambda \Lambda'). \tag{11.4}$$

In the following, we will at the slight risk of overkill introduce the notation $\widehat{\mathbb{M}}$ for the set of four momemta $(k^0, \mathbf{k})$, where k^0 is the energy variable and $\mathbf{k}$ is the three-momentum. The set of all three momenta will be denoted as $\widehat{\mathbb{R}}^3$. We do this to distinguish spacetime and energy-momenta coordinates; however, the same Minkowski metric applies to $\widehat{\mathbb{M}}$ along with the associated Lorentz and Poincaré group actions. We will use the same notation for hyperboloids in both cases.

11.1.2 The Klein–Gordon Equation

The Klein–Gordon (KG) equation is the differential equation

$$\left(\Box + m^2\right)\phi(x) = 0 \tag{11.5}$$

with the parameter $m > 0$. The symbol $\Box$ stands for the Lorentz invariant second-order differential operator

$$\Box = \left(\frac{\partial}{\partial x^0}\right)^2 - \sum_{j=1}^{3}\frac{\partial^2}{\partial x_j^2} = \frac{1}{c^2}\frac{\partial^2}{\partial t^2} - \Delta.$$

Going over to the Fourier representation

$$\phi(x) = (2\pi)^{-\frac{3}{2}}\int \tilde{\phi}(k)e^{-ik\cdot x}d^4k,$$

the differential equation (11.5) is replaced by the algebraic equation

$$\left(-k^2 + m^2\right)\tilde{\phi}(k) = 0.$$

The solutions $\tilde{\phi}(k)$ have support in the double hyperboloid $\mathscr{V}_m = \mathscr{V}_m^+ \cup \mathscr{V}_m^-$ in $\widehat{\mathbb{M}}$. Specifically, we have

$$\mathscr{V}_m^+ = \left\{k \in \widehat{\mathbb{M}} : k^2 = m^2, \, k^0 > 0\right\} = \left\{k \in \widehat{\mathbb{M}} : k^0 = \omega(\mathbf{k})\right\}, \tag{11.6}$$

where

$$\omega(\mathbf{k}) = \sqrt{\mathbf{k}^2 + m^2},$$

while $\mathscr{V}_m^- = -\mathscr{V}_m^+$, and they can be factorized as $\tilde{\phi}(k) = f(k)\delta(k^2 - m^2)$ with functions $\mathscr{V}_m \ni k \mapsto f(k) \in \mathbb{C}$. The measure $\delta(k^2 - m^2)d^4k$ is concentrated on $\mathscr{V}_m$ and is invariant under Lorentz transformations. The functions $f(k)$ are taken as elements of the Hilbert space $\mathcal{H}_m$ with the inner product

$$\langle f_1 \mid f_2\rangle_m = \int f_1(k)^*f_2(k)\,\delta(k^2 - m^2)d^4k. \tag{11.7}$$

With these functions, the Fourier representation of the solutions of the KG-equation is given by

$$\phi(x) = (2\pi)^{-\frac{3}{2}}\int f(k)e^{-ik\cdot x}\delta(k^2 - m^2)d^4k. \tag{11.8}$$

The linear space of functions (11.8) with $f \in \mathcal{H}_m$ is denoted as $\mathcal{K}_m$. The Hilbert spaces $\mathcal{H}_m^\pm = \left\{f \in \mathcal{H}_m : \mathrm{supp}f \subset \mathscr{V}_m^\pm\right\}$ are isomorphic orthogonal subspaces of $\mathcal{H}_m$. The restriction of the measure $\delta(k^2 - m^2)d^4k$ to the hyperboloid $\mathscr{V}_m^+$ is

$$d\mu_m(k) = \delta_+\left(k^2 - m^2\right)d^4k = (2\omega(\mathbf{k}))^{-1}d^3k, \tag{11.9}$$

with $\delta_+(k^2 - m^2) \triangleq \Theta(k^0)\delta(k^2 - m^2)$. The Hilbert space $\mathcal{H}_m^+$ is therefore isomorphic to $L^2(\widehat{\mathbb{R}}^3, \frac{d^3k}{2\omega(\mathbf{k})})$. The image of $\mathcal{H}_m^\pm$ under the transformation (11.8) is called $\mathcal{K}_m^\pm$.

We define on $\mathcal{K}_m$ an inner product $(\phi_1 \mid \phi_2)_m$ with the property

$$(\phi_1 \mid \phi_2)_m = \langle f_1 \mid f_2 \rangle_m, \tag{11.10}$$

if the functions $\phi_j, j = 1, 2$, are the Fourier transforms of f_j. Standard Fourier theorems imply the identities

$$(\phi_1 \mid \phi_2)_m = i \int_{x^0 = t} \left(\phi_1(x)^* \frac{\partial}{\partial x^0} \phi_2(x) - \frac{\partial \phi_1(x)^*}{\partial x^0} \phi_2(x) \right) d^3x$$
$$\text{if } f_1 \in \mathcal{H}_m^+, f_2 \in \mathcal{H}_m, \tag{11.11}$$

$$(\phi_1 \mid \phi_2)_m = -i \int_{x^0 = t} \left(\phi_1(x)^* \frac{\partial}{\partial x^0} \phi_2(x) - \frac{\partial \phi_1(x)^*}{\partial x^0} \phi_2(x) \right) d^3x$$
$$\text{if } f_1 \in \mathcal{H}_m^-, f_2 \in \mathcal{H}_m. \tag{11.12}$$

As a consequence of (11.10), the sesquilinear form (11.11) is the positive definite inner product of the space $\mathcal{K}_m^+$. A real solution $\phi(x)$ of the KG-equation has the Fourier representation (11.8) with a function f, which is real with respect to the antilinear involution

$$f(k) \mapsto f^\natural(k) = f(-k)^*. \tag{11.13}$$

This involution is an antiunitary operator within $\mathcal{H}_m$. The inner product (11.8) is invariant under the transformations

$$f(k) \mapsto e^{ik \cdot y} f(\Lambda^{-1} k) \tag{11.14}$$

for all elements (y, Λ) of the full Poincaré group $\mathcal{P}$. The equivalent transformation on $\mathcal{K}_m$ follows from the Fourier transform (11.8) as

$$\phi(x) \mapsto \phi_{y,\Lambda}(x) \triangleq \phi(\Lambda(x - y)). \tag{11.15}$$

If the group $\mathcal{P}$ is restricted to the orthochronous group $\mathcal{P}^\uparrow$, the transformation (11.14) maps $\mathcal{H}_m^+$ onto $\mathcal{H}_m^+$ and $\mathcal{H}_m^-$ onto $\mathcal{H}_m^-$. Consequently, (11.15) maps $\mathcal{K}_m^+$ onto $\mathcal{K}_m^+$ and $\mathcal{K}_m^-$ onto $\mathcal{K}_m^-$.

If $f(k)$ is a sufficiently decreasing function, for example, if $|\mathbf{k}|^3 f(k) \in \mathcal{H}_m$, then $\left(e^{ik \cdot x} \mid f \right)_m$ is defined, and $(2\pi)^{-\frac{3}{2}} \left(e^{ik \cdot x} \mid f \right)_m = \phi(x)$ is the Fourier integral (11.8). The projection of ϕ onto $\mathcal{K}_m^\pm$ is obtained by

$$\phi^\pm(x) = (2\pi)^{-\frac{3}{2}} \left\langle \Theta(\pm k^0) e^{ik \cdot x} \mid f(k) \right\rangle_m \overset{(11.10)}{=} \pm \left(\Delta_m^\pm(y - x) \mid \phi(y) \right) \tag{11.16}$$

with the generalized functions

$$\Delta_m^+(x) = (2\pi)^{-3} \int e^{-ik\cdot x} d\mu_m(k) = (2\pi)^{-3} \int_{\mathbb{R}^3} e^{i(\mathbf{kx} - \omega(\mathbf{k})x^0)} \frac{d^3k}{2\omega(\mathbf{k})} \tag{11.17}$$

and

$$\Delta_m^-(x) = -\Delta_m^+(-x) = -\Delta_m^+(x)^*. \tag{11.18}$$

The $\Delta_m^\pm(x)$ are invariant under orthochronous Lorentz transformations, $\Delta_m^\pm(\Lambda x) = \Delta_m^\pm(x)$, $\Lambda \in \mathcal{L}^\uparrow$. The integral (11.17) is calculated in Section 11.1.5 with the result

$$\Delta_m^+(x) = \frac{m}{(2\pi)^2} \frac{K_1\left(m\sqrt{-x^2 + i0x^0}\right)}{\sqrt{-x^2 + i0x^0}} \tag{11.19}$$

if $m > 0$. The function $K_1(s)$ is the modified Bessel function of the second kind. The sum

$$\Delta_m(x) = i\left(\Delta_m^+(x) + \Delta_m^-(x)\right) = -2\,\mathrm{Im}\,\Delta_m^+(x)$$
$$= i(2\pi)^{-3} \int e^{-ik\cdot x} \varepsilon(k^0) \delta(k^2 - m^2) d^4k \tag{11.20}$$

is a real $\mathcal{L}^\uparrow$ invariant and antisymmetric distribution, $\Delta(-x) = -\Delta(x)$. The generalized function $\Delta(x)$ has therefore a support restricted to the closed backward and forward cones $\overline{\mathscr{V}^+} \cup \overline{\mathscr{V}^-}$.

The integral representations (11.11) and (11.12) for the inner products (11.16) lead to the following formula for $\phi(x) = \phi^+(x) + \phi^-(x)$:

$$\phi(x) = \int_{y^0 = t_0} \left(\Delta(x - y)\frac{\partial}{\partial y^0}\phi(y) - \frac{\partial\Delta(x - y)}{\partial y^0}\phi(y)\right)d^3y. \tag{11.21}$$

This formula solves the initial value problem. The KG-equation is a second-order differential equation in the time variable. The initial condition at time t_0, say $t_0 = 0$, is given by the function $\phi(0, \mathbf{x}) = \varphi(\mathbf{x})$ and its derivative $\frac{\partial}{\partial x^0}\phi(x^0, \mathbf{x})\mid_{x^0=0} = \psi(\mathbf{x})$. The knowledge of these functions is equivalent to the knowledge of $f(k)$ on both branches of $\mathscr{V}_m$. The formula (11.21) determines $\phi(x)$ at any time $x^0 \in \mathbb{R}$. Moreover, it proves that solutions of the KG-equation propagate causally. Starting at $t = 0$ from an initial state localized in a region $G \subset \mathbb{R}^3$, that is, $\mathrm{supp}\,\varphi \subset G$ and $\mathrm{supp}\,\psi \subset G$, the solution $\phi(x)$ propagates for $t > 0$ into the causal shadow $G + \mathscr{V}^+$ of G. But if the class of solutions is restricted to functions with positive frequencies, there is no causal localization: if $0 \neq \phi(x) \in \mathcal{K}^+$, then for arbitrary $t_1 < t_2$ the space region $\{\mathrm{supp}_\mathbf{x}\phi(t_1, \mathbf{x})\} \cup \{\mathrm{supp}_\mathbf{x}\phi(t_2, \mathbf{x})\}$ is unbounded. The proof of this statement follows from the Fourier representation

$$\phi(t, \mathbf{x}) = (2\pi)^{-\frac{3}{2}} \int f(\omega(\mathbf{k}), \mathbf{k}) \, e^{-i\omega(\mathbf{k})t + i\mathbf{k}\mathbf{x}} \frac{d^3k}{2\omega(\mathbf{k})}. \qquad (11.22)$$

If $\phi(t, \mathbf{x})$ has compact support for $t = t_1$ and $t = t_2$, then both the functions $g(\mathbf{k}) \triangleq \omega^{-1} f(\omega, \mathbf{k}) \exp(-it_1\omega)$ and $g(\mathbf{k}) \exp(-i(t_2 - t_1)\omega)$ with $\omega = \sqrt{\mathbf{k}^2 + m^2}$ would then be entire analytic functions in $\mathbf{k} \in \mathbb{C}^3$. But that is not possible.

11.1.3 The One-Particle Hilbert Space

Momentum Representation

There are straightforward methods to derive the theory of relativistic fields from unitary representations of the inhomogeneous Lorentz group – see, e.g., Bogoliubov et al. (1990) – or from the underlying canonical structure – see, e.g., Baez et al. (1992). Here we start in a less fundamental way from the classical theory of a relativistic free particle with mass m and spin zero. The L^2-space over the classical momentum space of this particle is then chosen as the one-particle Hilbert space.

For a relativistic particle of mass, $m > 0$ the four-momentum $p = (E, \mathbf{p}) \in \widehat{\mathbb{M}}$ is concentrated on the hyperboloid $\mathcal{V}_m^+$. The associated Lorentz invariant measure on the Minkowski space is (11.9). As one-particle Hilbert space $\mathcal{H}_m^+$, we choose the space $L^2(\mathcal{V}_m^+, d\mu_m)$ of functions $\mathcal{V}_m^+ \ni k \mapsto f(k) \in \mathbb{C}$ that are square integrable with respect to this measure. In Section 11.1.2, this space was called $\mathcal{H}_m^+$. The inner product of $\mathcal{H}_m^+$ is

$$\langle f \mid g \rangle_m = \int f(k)^* g(k) d\mu_m(k). \qquad (11.23)$$

Identifying the function $f(k)$, $k \in \mathcal{V}_m^+$, with $f(\omega(\mathbf{k}), \mathbf{k})$, $\mathbf{k} \in \widehat{\mathbb{R}}^3$, the space $\mathcal{H}_m^+$ is isomorphic to $L^2(\widehat{\mathbb{R}}^3, \frac{d^3k}{2\omega(\mathbf{k})})$.

Many calculations of this section are true for all masses $m \geq 0$, and we give some results including $m = 0$. But the quantum field theory for $m = 0$ particles needs some additional work related to the fact that $L^2(\widehat{\mathbb{R}}^3, \frac{d^3k}{2\omega(\mathbf{k})}) \subset L^2(\widehat{\mathbb{R}}^3, d^3k)$ is true for $m > 0$ but not for $m = 0$. Therefore, we restrict to the case $m > 0$ in general.

The Hilbert space $\mathcal{H}_m^+$ carries a unitary representation of the proper orthochronous Poincaré group $\mathcal{P}_+^\uparrow$ that can be read off from the invariance transformations (11.14) of the KG-equation

$$(U_1(y, \Lambda)f)(k) = e^{ik \cdot y} f(\Lambda^{-1}k), \quad (y, \Lambda) \in \mathcal{P}_+^\uparrow. \qquad (11.24)$$

It is straightforward to prove that these transformations are unitary and that they satisfy the identities for a representation of the group $\mathcal{P}_+^{\uparrow}$:

$$U(0, I) = id,$$
$$U(y, \Lambda)U(y', \Lambda') = U(y + \Lambda y', \Lambda\Lambda'). \tag{11.25}$$

The 4-momentum operators $P_{(1)} = (H_1, \mathbf{P}_1)$ are the generators of the space and time translations

$$(\mathbf{P}_1 f)(k) = i\frac{\partial}{\partial \mathbf{y}} \left(e^{-i\mathbf{k}\mathbf{y}}f(k)\right)_{\mathbf{y}=0} = \mathbf{k}f(k), \tag{11.26}$$

$$(H_1 f)(k) = -i\frac{\partial}{\partial t} \left(e^{ik^0 t}f\right)_{t=0} = \omega(\mathbf{k})f(k). \tag{11.27}$$

The spectrum of the 4-momentum operator $P_{(1)} = (H_1, \mathbf{P}_1)$ is concentrated on the hyperboloid $\mathcal{V}_m^+$. The time evolution of a vector $f(k)$ is

$$f(k, t) = (U_1(t)f)(k) = f(k)e^{-i\omega(\mathbf{k})t}.$$

There are two possibilities to extend the representation (11.24) to a representation of the orthochronous group $\mathcal{P}^{\uparrow}$, either by the rule (11.24) applied to transformations in $\mathcal{P}^{\uparrow}$, or by

$$(U_1(y, \Lambda)f)(k) = (\det \Lambda) \, e^{ik \cdot y} f(\Lambda^{-1}k).$$

In the first case, the particles are called scalar particles, and in the second case pseudoscalar particles.

Spacetime Functions

If $f(k)$ is a function in $\mathcal{H}_m^+$, then $\psi(x) = (2\pi)^{-\frac{3}{2}} \int f(k)e^{-ik \cdot x}d\mu_m(k)$ is a solution of the KG equation with positive frequencies. The unitary representation of the proper orthochronous Poincaré group (11.24), cf. (11.14), leads to $\psi(x) \mapsto \psi_{y,\Lambda}(x) = \psi(\Lambda(x - y))$. This is the behavior of classical fields on the Minkowski space; and it is a justification to take the variable x as point in the configuration Minkowski space. But, as already stated at the end of Section 11.1.2, there is no localization of the wave function in a finite region of space (or spacetime) consistent with causality. Hence causal propagation has only a vague meaning for one-particle states. These problems have been investigated in great detail by Hegerfeldt (1974) and by Hegerfeldt and Ruijsenaars (1980). We come back to locality and causality in the context of quantum fields in the Sections 11.1.4 and 12.2.

There is another possibility to specify one-particle vectors by functions of the spacetime variables. Let $\varphi(x)$ be a function in $\mathcal{S}(\mathbb{M})$. Then the Fourier transform[1]

$$\hat{\varphi}(k) = \sqrt{2\pi}\,\mathcal{F}_4\,[\varphi]\,(k) = (2\pi)^{-\frac{3}{2}} \int_{\mathbb{M}} \varphi(x)e^{ik\cdot x}d^4x \qquad (11.28)$$

is a function $\hat{\varphi}(k) \in \mathcal{S}(\widehat{\mathbb{M}})$, and the restriction to the mass hyperboloid $\hat{\varphi}(k)\,|_{k\in\mathcal{V}_m^+}$ determines a unique vector in the one-particle space $\mathcal{H}_m^+$. The hermitean form

$$\langle \hat{\varphi}_2 \mid \hat{\varphi}_1 \rangle_{\widehat{\mathbb{M}}} = \int \hat{\varphi}_2(k)^*\hat{\varphi}_1(k)\delta_+(k^2 - m^2)\,d^4k \qquad (11.29)$$

is an extension of the inner product (11.23) to functions in $\mathcal{S}(\widehat{\mathbb{M}})$. The form (11.29) is positive hermitean with the highly nontrivial null-space

$$\mathcal{S}_0(\widehat{\mathbb{M}}) = \left\{ \hat{\varphi} \in \mathcal{S}(\widehat{\mathbb{M}}) : \hat{\varphi}(k) = 0 \text{ if } k \in \mathcal{V}_m^+ \right\}. \qquad (11.30)$$

The quotient space $\mathcal{S}(\widehat{\mathbb{M}})/\mathcal{S}_0(\widehat{\mathbb{M}})$ is a pre-Hilbert space with the inner product induced by (11.29), and the completion of $\mathcal{S}(\widehat{\mathbb{M}})/\mathcal{S}_0(\widehat{\mathbb{M}})$ is isomorphic to the one-particle Hilbert space $\mathcal{H}_m^+$. The mapping

$$J: \varphi \mapsto \hat{\varphi} = \sqrt{2\pi}\,\mathcal{F}_4\,[\varphi] \mapsto f(k) = \hat{\varphi}(k)\,|_{k\in\mathcal{V}_m^+} \in \mathcal{H}_m^+ \qquad (11.31)$$

is a linear operator from $\mathcal{S}(\mathbb{M})$ into $\mathcal{H}_m^+$. This mapping is continuous with respect to the topologies of $\mathcal{S}(\mathbb{M})$ and of $\mathcal{H}_m^+$, and the image space $J\mathcal{S}(\mathbb{M})$ is a dense linear subspace of $\mathcal{H}_m^+$. The Fourier relations imply the identity

$$\langle J\varphi \mid J\psi \rangle_m = \langle \varphi \parallel \psi \rangle_m, \qquad (11.32)$$

where

$$\langle \varphi_2 \parallel \varphi_1 \rangle_m \triangleq \iint_{\mathbb{M}\times\mathbb{M}} \varphi_2(x)^*\Delta_m^+(x - y)\,\varphi_1(y)\,d^4x\,d^4y \qquad (11.33)$$

is a degenerate positive hermitean form on $\mathcal{S}(\mathbb{M})$ with the null-space $\mathcal{S}_0(\mathbb{M}) = \mathcal{F}_4^{-1}\mathcal{S}_0(\widehat{\mathbb{M}}) = \ker J$. The integral kernel $\Delta_m^+(x - y)$ is the generalized function (11.17). The space $\mathcal{S}(\mathbb{M})$ is a semi-Hilbert space with the form (11.33). Since $\{J\varphi : \varphi \in \mathcal{S}(\mathbb{M})\}$ is dense in $\mathcal{H}_m^+$, the functions $\varphi \in \mathcal{S}(\mathbb{M})$ label in a nonunique way a dense set of one-particle vectors.

The functions $\varphi(x) \in \mathcal{S}(\mathbb{M})$ carry the continuous representation

$$\varphi(x) \mapsto \varphi_{y,\Lambda}(x) = \varphi(\Lambda(x - y)), \ (y, \Lambda) \in \mathcal{P}^\uparrow,$$

---

[1] The additional factor $\sqrt{2\pi}$ is convenient for the subsequent calculations.

of the orthochronous Poincaré group. The Fourier transform $\hat{\varphi}(k)$ behaves like (11.14), and the function $f = J\varphi$ transforms according to the unitary representation (11.24) of the one-particle vectors.

So far, the mappings (11.31) have been defined for functions $\varphi \in S(\mathbb{M})$. But it is possible to extend the domain of this mapping to more general spaces.

Lemma 11.1.1 *The mappings (11.31) can be extended to the following spaces:*

(i) *The Hilbert space $L^2(\mathbb{M}, (1 + (x^0)^2) d^4x)$.*
(ii) *The space of generalized functions $\varphi(x) = \delta(x^0 - t)\chi(\mathbf{x})$ localized at a fixed time t with $\chi(\mathbf{x}) \in L^2(\mathbb{R}^3, d^3x)$. The topology of this space is the L^2-topology of the factor $\chi(\mathbf{x})$.*

The mapping (11.31) is a continuous linear operator from either of these spaces into $\mathcal{H}_m^+$. The identity (11.32) remains valid for these extensions.

Proof We first take functions $\varphi(x) \in L^2(\mathbb{M}, (1 + (x^0)^2) d^4x)$, which have the finite norm

$$\|\varphi\|_L^2 = \int_{\widehat{\mathbb{M}}} |\varphi(x)|^2 \left(1 + (x^0)^2\right) d^4x. \tag{11.34}$$

The Fourier transform $\hat{\varphi} = \sqrt{2\pi} \mathcal{F}_4 [\varphi]$ is an element of the Sobolev space with norm $\|\hat{\varphi}\|_S^2 \triangleq \int_{\mathbb{M}} \left(|\hat{\varphi}(k)|^2 + \left|\frac{\partial}{\partial k^0}\hat{\varphi}(k)\right|^2\right) d^4k$. Then the one-particle norm of $J\varphi$ is well defined and can be estimated by

$$\|J\varphi\|^2 = \int_{\mathbb{M}} |\hat{\varphi}(k)|^2 \delta_+(k^2 - m^2) d^4k$$

$$= 2\text{Re} \int_{\mathbb{R}^3} \frac{d^3k}{2\omega(\mathbf{k})} \int_{\omega(\mathbf{k})}^\infty dk^0 \left(\hat{\varphi}(k)^* \frac{\partial}{\partial k^0}\hat{\varphi}(k)\right)$$

$$\leq \frac{1}{2m} \|\hat{\varphi}\|_S^2 = \frac{\pi}{m} \|\varphi\|_L^2.$$

For the second case, we choose singular functions $\varphi(x) = \delta(x^0 - t)\chi(\mathbf{x})$, which are localized at a fixed time t with a function $\chi(\mathbf{x}) \in L^2(\mathbb{R}^3, d^3x)$. The Fourier transform $\sqrt{2\pi} \mathcal{F}_4 [\varphi]$ is $\hat{\varphi}(k) = \exp(ik^0 t) \tilde{\chi}(\mathbf{k})$ with $\tilde{\chi} \in L^2(\mathbb{R}^3, d^3k)$. The one-particle norm of $J\varphi$ is then estimated by the norm of χ

$$\|J\varphi\|^2 = \int_{\mathbb{M}} |\hat{\varphi}(k)|^2 \delta_+(k^2 - m^2) d^4k = \int_{\mathbb{R}^3} |\tilde{\varphi}(\mathbf{k})|^2 \frac{d^3k}{2\omega(\mathbf{k})}$$

$$\leq (2m)^{-1} \int_{\mathbb{R}^3} |\tilde{\chi}(\mathbf{k})|^2 d^3k = (2m)^{-1} \int_{\mathbb{R}^3} |\chi(\mathbf{x})|^2 d^3x.$$

The Fourier identity $\langle J\varphi \mid J\psi \rangle = \langle \varphi \parallel \psi \rangle$ remains valid for functions in the spaces 1 and 2. $\square$

In the case of functions $\varphi(x) = \delta(x^0 - t)\chi(\mathbf{x})$, the integral (11.33) becomes the positive definite hermitean form

$$q_m(\chi_1 \mid \chi_1) = \iint_{\mathbb{R}^3 \times \mathbb{R}^3} \chi_1(\mathbf{x})^* \Delta_m^+(0, \mathbf{x} - \mathbf{y})\chi_2(\mathbf{y})d^3x d^3y,$$

defined on $L^2(\mathbb{R}^3, d^3x)$ with the locally integrable kernel function

$$\Delta_m^+(0, \mathbf{x}) = (2\pi)^{-3} \int_{\mathbb{R}^3} e^{i\mathbf{k}\mathbf{x}} \frac{d^3k}{2\omega(\mathbf{k})} \overset{(11.19)}{=} (2\pi)^{-2} \frac{m}{|\mathbf{x}|} K_1(m|\mathbf{x}|).$$

The closure of $L^2(\mathbb{R}^3, d^3x)$ with respect to the norm $\sqrt{q(\chi\mid\chi)}$ is the space $\mathcal{F}_3^{-1}\left(L^2(\mathbb{R}^3, \frac{d^3k}{2\omega(\mathbf{k})})\right) = (-\Delta + m^2)^{\frac{1}{4}} L^2(\mathbb{R}^3, d^3x)$, which is the Fourier transform of $L^2(\mathbb{R}^3, \frac{d^3k}{2\omega(\mathbf{k})}) \cong \mathcal{H}_m^+$.

Generalized Vectors

A short introduction to generalized vectors was given in Section 9.1.3. The Hilbert space $\mathcal{H}_m^+ = L^2(\mathcal{V}_m^+, d\mu_m) \simeq L^2(\widehat{\mathbb{R}}^3, \frac{d^3k}{2\omega(\mathbf{k})})$ can be extended to generalized functions in $S'(\mathbb{R}^3)$. The simplest examples are vectors with fixed momentum $p \in \widehat{\mathbb{M}}$

$$\psi_p(k) = 2\omega(\mathbf{p})\delta^3(\mathbf{k} - \mathbf{p}), \qquad p \in \mathcal{V}_m^+, \qquad k \in \mathcal{V}_m^+. \tag{11.35}$$

The inner product $\langle \psi_p \mid f \rangle_m = \int \psi_p(k)f(k)d\mu_m(k) = f(p)$ is well defined for all continuous functions $f \in \mathcal{H}_m^+$. The mapping

$$S(\mathcal{V}_m^+) \ni f \mapsto \int f(p)\psi_p(k)\,d\mu_m(k) = f(k) \in \mathcal{H}_m^+$$

can be extended to functions in $\mathcal{H}_m^+$. The vectors (11.35) are (formally) normalized to

$$\int \psi_p(k)\psi_{p'}(k)\,d\mu_m(k) = 2\omega(\mathbf{p})\delta^3(\mathbf{p} - \mathbf{p}'). \tag{11.36}$$

The distribution (11.35) is the kernel of the $\mathcal{L}^\uparrow$-invariant identity operator, and ψ_p has the simple covariance property $\psi_{\Lambda p}(k) = \psi_p(\Lambda^{-1}k)$ for $\Lambda \in \mathcal{L}^\uparrow$.

A candidate for a state at position $\mathbf{x} \in \mathbb{R}^3$ at time $x^0 = t$ is

$$\phi_x(k) = (2\pi)^{-\frac{3}{2}} e^{ik\cdot x}, \qquad k \in \mathcal{V}_m^+, \qquad x \in \mathbb{M}. \tag{11.37}$$

The inner product (11.23) of ϕ_x with sufficiently decreasing function $f \in L^2(\mathcal{V}_m^+, d\mu_m)$ coincides with the Fourier transform

$$\langle \phi_x \mid f \rangle_m = (2\pi)^{-\frac{3}{2}} \int e^{-ik\cdot x} f(k)d\mu_m(k) \overset{(11.29)}{=} \phi(x).$$

This result can be extended to all $f \in \mathcal{H}_m^+$. Given a function φ in $\mathcal{S}(\mathbb{M})$, the integral $\int \varphi(x)\phi_x(k)\, d^4x = (2\pi)^{-\frac{3}{2}} \int e^{ik\cdot x}\varphi(x)d^4x = f(k)$ yields the image $J\varphi$ of the mapping (11.31). The functions (11.37) are formally normalized to

$$\int \phi_x(k)^* \phi_y(k) d\mu_m(k) = \frac{1}{(2\pi)^3} \int e^{-ik\cdot(x-y)} d\mu_m(k) \overset{(11.17)}{=} \Delta_m^+(x-y).$$

Since $\Delta_m^+(0, \mathbf{x})$ is not a localized distribution, this result indicates again that there is no reasonable localization in position for a relativistic one-particle state.

11.1.4 Fock Space and Field Operators

Creation and Annihilation Operators

The Fock space $\Gamma^+(\mathcal{H}_m^+)$ and the corresponding creation and annihilation operators $A^+(f)$ and $A^-(f)$, $f \in \mathcal{H}_m^+$, can be constructed as in Chapter 9. The vacuum vector is denoted as Ω. A concrete realization of $\Gamma_n^+(\mathcal{H}_m^+)$, $n = 1, 2, \ldots$, is given by functions $F(k_1, \ldots, k_n)$ which are symmetric in the variables $k_j \in \mathcal{V}_m^+$, $j = 1, \ldots, n$, and which are square integrable with respect to the product measure $(d\mu_m)^n$. The creation/annihilation operators satisfy the commutation relations (9.34)

$$\left[A^+(f), A^+(g)\right] = \left[A^-(f), A^-(g)\right] = 0,$$
$$\left[A^-(f), A^+(g)\right] = \langle f \mid g \rangle I \overset{(11.23)}{=} \int f(k)^* g(k) d\mu_m(k) I. \qquad (11.38)$$

The representation (11.14) of the Poincaré group can be extended to the representation $U(y, \Lambda) = \Gamma(U_1(y, \Lambda))$ on the Fock space. The generator of the translations is the total 4-momentum P

$$P = d\Gamma\left(P_{(1)}\right). \qquad (11.39)$$

The point spectrum of P consists of the eigenvalue zero with the vacuum as eigenvector, and the continuous spectrum includes the hyperboloid $\mathcal{V}_m^+$ (the spectrum of $P_{(1)}$) and the closed convex set $Conv\, \mathcal{V}_{2m}^+$ generated by the hyperboloid $\mathcal{V}_{2m}^+$. For all values of the mass $m \geq 0$, the spectrum of P is therefore a subset of the closed forward light cone

$$Spec\, P \subset \overline{\mathcal{V}^+}. \qquad (11.40)$$

The energy component of (11.39) is the many-particle Hamilton operator

$$H = d\Gamma(H_1). \qquad (11.41)$$

The time evolution is given by the unitary group $t \mapsto U(t) = \exp(-iHt)$ parametrized by $t \in \mathbb{R}$, and the dynamics of the creation and annihilation operators is, as shown in (9.61),

$$U^*(t)A^\pm (f(k)) U(t) = A^\pm \left(f(k)e^{i\omega(\mathbf{k})t} \right). \tag{11.42}$$

Following Section 9.1.4, the behavior of the creation/annihilation operators under Poincaré transformations $(y, \Lambda) \in \mathbb{M} \times \mathcal{L}^\uparrow$ is seen to be

$$U(y, \Lambda)A^\pm(f)U^*(y, \Lambda) = A^\pm (U_1(y, \Lambda)f) = A^\pm \left(e^{ik \cdot y} f(\Lambda^{-1}k) \right). \tag{11.43}$$

The creation/annihilation operators of the singular states (11.35)

$$A_p^\pm \triangleq A^\pm \left(2\omega(\mathbf{p}) \delta^3 (\mathbf{k} - \mathbf{p}) \right) \tag{11.44}$$

are operator-valued generalized functions on $\Gamma^+(\mathcal{H}_m^+)$. We obtain the well-defined operators $A^\pm(f)$, $f \in \mathcal{H}_m^+$, by the "integration"

$$A^+(f) = \int f(k)A_k^+ \, d\mu_m(k), \qquad A^-(f) = \int f(k)^* A_k^- \, d\mu_m(k). \tag{11.45}$$

The operators (11.44) satisfy the formal commutation relations

$$\left[A_k^-, A_p^- \right] = \left[A_k^+, A_p^+ \right] = 0,$$
$$\left[A_k^-, A_p^+ \right] = 2\omega(\mathbf{p}) \delta^3 (\mathbf{p} - \mathbf{p}') \, I, \tag{11.46}$$

which after the integration (11.45) lead to (11.38). More explanations are given in Section 9.1.3.

Covariant Field Operators

Given a function $\varphi(x) \in \mathcal{S}(\mathbb{M})$, then the mapping (11.31) determines the one-particle vectors $f = J\varphi$ and $f^\natural = J\varphi^*$. We define **field operators** $\Phi^{(\mp)}(\varphi)$ as the creation/annihilation operators

$$\Phi^{(-)}(\varphi) = A^+(f) \text{ and } \Phi^{(+)}(\varphi) = A^-(f^\natural). \tag{11.47}$$

The reason for the interchange of the superscripts $\pm$ will be explained later. The relation (9.28) leads to the identity on $\mathcal{H}_m^+$

$$\Phi^{(+)}(\varphi) = \left(\Phi^{(-)}(\varphi^*) \right)^*. \tag{11.48}$$

If $\varphi(x)$ is a real function, its Fourier transform satisfies the identity $\hat{\varphi}(k) = \hat{\varphi}(-k)^*$, and the functions f and $f^\natural$ coincide, $f^\natural(k) = f(k) \in \mathcal{H}_m^+$.

The rules (11.24) and (11.43) imply that $\Phi^{(\mp)}(\varphi)$ are covariant field operators under Poincaré transformations

$$U(y, \Lambda)\Phi^{(\mp)}(\varphi)U^*(y, \Lambda) = \Phi^{(\mp)}(\varphi_{y,\Lambda}). \qquad (11.49)$$

Both mappings $\varphi \mapsto \Phi^{(-)}(\varphi)$ and $\varphi \mapsto \Phi^{(+)}(\varphi)$ are linear (and continuous in suitable topologies). We can therefore write

$$\Phi^{(-)}(\varphi) = \int \Phi^{(-)}(x)\varphi(x)\,d^4x = \int A_k^+ f(k)d\mu_m(k),$$

$$\Phi^{(+)}(\varphi) = \int \Phi^{(+)}(x)\varphi(x)\,d^4x = \int A_k^- f(-k)d\mu_m(k), \qquad (11.50)$$

with an operator-valued generalized functions $\Phi^{(\mp)}(x)$. The Fourier integral (11.29) indicates that the operators $\Phi^{(\mp)}(x)$ are the creation/annihilation operators of the singular vector (11.37)

$$\Phi^{(\mp)}(x) = A^\pm\left((2\pi)^{-\frac{3}{2}}e^{ik\cdot x}\right) = (2\pi)^{-\frac{3}{2}}\int A_k^\pm e^{\mp ik\cdot x}d\mu_m(k). \quad (11.51)$$

The creation operator $\Phi^{(-)}(x)$ has only negative frequencies, and the annihilation operator $\Phi^{(+)}(x)$ has only positive frequencies. The superscripts $\mp$ refer to these properties. The relations (11.49) imply the transformation rules for these singular field operators

$$U(y, \Lambda)\Phi^{(\mp)}(x)U^*(y, \Lambda) = \Phi^{(\mp)}(\Lambda x + y). \qquad (11.52)$$

Both the operators $\Phi^{(\pm)}(x)$ have the same covariant transformation property under Poincaré transformations. As the 4-dimensional Fourier transforms of the fields $\Phi^{(\mp)}(x)$ are concentrated on $\mathscr{V}_m^+$ and $\mathscr{V}_m^- = -\mathscr{V}_m^+$, these fields are solutions of the Klein–Gordon equation[2]

$$\left(\Box + m^2\right)\Phi^{(\mp)}(x) = 0. \qquad (11.53)$$

The vacuum expectations of the product of two of these fields are

$$\left\langle \Omega \mid \Phi^{(\pm)}(\varphi), \Phi^{(\pm)}(\psi)\Omega \right\rangle = \left\langle \Omega \mid \Phi^{(-)}(\varphi), \Phi^{(+)}(\psi)\Omega \right\rangle = 0,$$

$$\left\langle \Omega \mid \Phi^{(+)}(\varphi), \Phi^{(-)}(\psi)\Omega \right\rangle = \left\langle \Omega \mid A^-(f^\natural)A^+(g))\Omega \right\rangle = \left\langle f^\natural \mid g \right\rangle_m,$$

with the latter expression equal to $= \langle \varphi^* \parallel \psi \rangle_m$, and we have the one-particle vectors $f^\natural = J\varphi^*$ and $g = J\psi$. The commutation relations of the fields follow as

---

[2] The precise meaning of this equation is that $\int \Phi^{(\mp)}(x)\left(\Box + m^2\right)\varphi(x)d^4x = 0$ is true for all test functions $\varphi(x) \in \mathcal{S}(\mathbb{M})$.

$$\left[\Phi^{(-)}(\varphi), \Phi^{(-)}(\psi)\right] = \left[\Phi^{(+)}(\varphi), \Phi^{(+)}(\psi)\right] = 0,$$

$$\left[\Phi^{(+)}(\varphi), \Phi^{(-)}(\psi)\right] = \langle \varphi^* \parallel \psi \rangle_m,$$

$$\left[\Phi^{(-)}(\varphi), \Phi^{(+)}(\psi)\right] = -\langle \psi^* \parallel \varphi \rangle_m \qquad (11.54)$$

with the bilinear form, as shown in (11.33),

$$\langle \varphi^* \parallel \psi \rangle_m = \iint_{M \times M} \varphi(x)\,\Delta_m^+(x - y)\,\psi(y)d^4x d^4y.$$

These identities are consequences of the relations (11.33), (11.38), and (11.47). The commutators $\left[\Phi^{(\pm)}(x), \Phi^{(\pm)}(y)\right]$ follow from (11.54) by extracting the test functions φ and ψ

$$\left[\Phi^{(-)}(x), \Phi^{(-)}(y)\right] = \left[\Phi^{(+)}(x), \Phi^{(+)}(y)\right] = 0,$$

$$\left[\Phi^{(+)}(x), \Phi^{(-)}(y)\right] = \Delta_m^+(x - y),$$

$$\left[\Phi^{(-)}(x), \Phi^{(+)}(y)\right] = \Delta_m^-(x - y) \qquad (11.55)$$

with the distributions (11.17) and (11.18).

Local Fields

Let $\varphi(x)$ be a test function in $\mathcal{S}(\mathbb{M})$. We introduce the field operator

$$\Phi(\varphi) \triangleq \Phi^{(-)}(\varphi) + \Phi^{(+)}(\varphi) = A^+(J\varphi) + A^-(J\varphi^*), \qquad (11.56)$$

where the operator J has been defined in (11.31). The mapping $\varphi \mapsto \Phi(\varphi)$ is linear, and we can write

$$\Phi(\varphi) = \int \Phi(x)\varphi(x)d^4x \qquad (11.57)$$

with the operator-valued generalized function

$$\Phi(x) = \Phi^{(-)}(x) + \Phi^{(+)}(x) = A^+\left((2\pi)^{-\frac{3}{2}}e^{ik\cdot x}\right) + A^-\left((2\pi)^{-\frac{3}{2}}e^{ik\cdot x}\right).$$
$$(11.58)$$

As a consequence of (11.48) and (11.31), the adjoint operator of $\Phi(\varphi)$ is $(\Phi(\varphi))^* = \Phi(\varphi^*)$. Choosing a real test function φ, we have $f = f^\natural$, and the operator $\Phi(\varphi)$ is self-adjoint and agrees with the Segal field operator (9.36) $\hat{\Phi}(f) = A^+(f) + A^-(f)$. The operator-valued generalized function $\Phi(x)$ is usually also denoted as self-adjoint field operator. With a positive weight function φ, the operator $\Phi(\varphi)$ can be interpreted as a field strength measured over a spacetime region with efficiency $\varphi(x)$.

The operator (11.31) J maps $S(\mathbb{M})$ onto a dense linear subsets of $\mathcal{H}_m^+$. Therefore, the linear span of the vectors $\Phi(\varphi)\Omega = \Phi^{(-)}(\varphi)\Omega$ with $\varphi \in S(\mathbb{M})$ is dense in $\mathcal{H}_m^+$. Since any $\varphi \in S(\mathbb{M})$ can be written as $\varphi = \varphi_1 + i\varphi_2$ with real functions in $S(\mathbb{M})$, also the $\mathbb{C}$-linear span of vectors $\Phi(\varphi)\Omega$ with real $\varphi \in S(\mathbb{M})$ is dense in $\mathcal{H}_m^+$.

The commutation rules (11.54) imply

$$[\Phi(\varphi), \Phi(\psi)] = -i \int \varphi(x)\Delta_m(x - y)\psi(y)d^4x d^4y\, I \qquad (11.59)$$

with the generalized function (11.20) $\Delta_m(x)$. The distribution $\Delta_m(x)$ is invariant under orthochronous Lorentz transformations $\Delta_m(\Lambda x) = \Delta_m(x)$, $\Lambda \in \mathcal{L}^\uparrow$, and it is antisymmetric, $\Delta_m(-x) = -\Delta_m(x)$, $x \in \mathbb{M}$. Such a distribution always vanishes at spacelike distances

$$\Delta_m(x) = 0 \qquad \text{if } x^2 = x \cdot x < 0. \qquad (11.60)$$

Hence $\Phi(\varphi)$ and $\Phi(\psi)$ commute if the supports of φ and ψ are spacelike separated.

The most important properties of the singular operator (11.58) are the following:

(i) It is covariant under inhomogeneous Lorentz transformations $(y, \Lambda) \in \mathbb{M} \times \mathcal{L}^\uparrow$, as shown in (11.52):

$$U(y, \Lambda)\Phi(\varphi)U^*(y, \Lambda) = \Phi(\varphi_{y,\Lambda}), \qquad (11.61)$$

$$U(y, \Lambda)\Phi(x)U^*(y, \Lambda) = \Phi(\Lambda x + y). \qquad (11.62)$$

(ii) It is a local field operator:

$$[\Phi(x), \Phi(y)] = 0 \qquad \text{if } (x - y)^2 < 0. \qquad (11.63)$$

(iii) It is a self-adjoint field operator (in the sense defined previously).
(iv) It is a solution of the Klein–Gordon equation:

$$\left(\Box + m^2\right)\Phi(x) = 0. \qquad (11.64)$$

The locality is an essential property of a relativistic field operator. It guarantees that the field strength and other local observables – such as the energy density – at spacelike distances are causally independent.

The time evolution of the field operators $\Phi(\varphi)$ is given by

$$U^*(t)\Phi(\varphi)U(t) = \Gamma(e^{iH_1t})\Phi(\varphi)\Gamma(e^{-iH_1t}) = \Phi(\varphi_t)$$

with $\varphi_t(x) = \varphi(x^0 - t, \mathbf{x})$, or

$$U^*(t)\Phi(x)U(t) = \Phi(x^0 + t, \mathbf{x}).$$

For real $\varphi \in \mathcal{S}(\mathbb{M})$, the operator (11.56) $\Phi(\varphi)$ is self-adjoint, and it agrees with the Segal field operator (9.36) $\hat{\Phi}(f)$, that is,

$$\Phi(\varphi) \equiv \hat{\Phi}(f).$$

Then the exponential $\exp(i\,\Phi(\varphi))$ is well defined, and the Weyl relations introduced in Section 9.2.2 can be transferred to the local fields

$$\exp(i\,\Phi(\varphi))\exp(i\,\Phi(\psi)) = e^{i\delta(\varphi,\psi)}\exp(i\,\Phi(\varphi+\psi)) \qquad (11.65)$$

with the antisymmetric form

$$\delta(\varphi,\psi) = \frac{1}{2}\iint\limits_{\mathbb{M}\times\mathbb{M}} \varphi(x)\Delta_m(x-y)\psi(y)\,d^4x d^4y. \qquad (11.66)$$

To derive the identity (11.65) from (9.86), we use the formula (11.33) for the inner product and the relation (11.20). The displacement relation (9.87) can be written in terms of the local fields as

$$e^{i\,\Phi(\psi)}\,\Phi(\varphi)\,e^{-i\,\Phi(\psi)} = \Phi(\varphi) - \iint\limits_{\mathbb{M}\times\mathbb{M}} \varphi(x)\Delta_m(x-y)\psi(y)\,d^4x d^4y\, I.$$

$$(11.67)$$

The real functions φ and ψ can be taken from the space $\mathcal{S}(\mathbb{M})$, but see the following subsection for singular arguments. Extracting the test function φ, we obtain

$$e^{i\,\Phi(\psi)}\,\Phi(x)\,e^{-i\,\Phi(\psi)} = \Phi(x) - \int_{\mathbb{M}} \Delta_m(x-y)\psi(y)\,d^4y\, I. \qquad (11.68)$$

Singular Distributions

So far, we have assumed that the test functions φ are elements of the space $\mathcal{S}(\mathbb{M})$. But following the arguments of Section 11.1.3, the field operators (11.56) $\Phi(\varphi)$ are still defined, if φ has the form $\varphi(x) = \delta(x^0 - t)\chi(\mathbf{x})$ with $\chi \in L^2(\mathbb{R}^3, d^3x)$. Likewise $\Phi(\varphi)$ is also meaningful for φ an element of the space $L^2(\mathbb{M}, (1+(x^0)^2)\,d^4x)$. The formulas (11.65) and (11.67) have a well-defined meaning for real test functions from these spaces.

The factorization $\varphi(x) = \delta(x^0 - t)\chi(\mathbf{x})$ leads to field operators $\Phi(t,\chi)$ at fixed times. If $\chi \in L^2(\mathbb{R}^3, d^3x)$ the Fourier transform $\tilde{\chi}$ is an element of $L^2(\mathbb{R}^3, d^3k)$, and $\tilde{\chi}(\mathbf{k})$ and $\omega(\mathbf{k})\tilde{\chi}(\mathbf{k})$ are elements of the one-particle space $L^2(\mathbb{R}^3, \frac{d^3k}{2\omega(\mathbf{k})})$. The operator $\Phi(t,\chi)$ has the representation

$$\Phi(t,\chi) = \int \Phi(t,\mathbf{x})\chi(\mathbf{x})\,d^3x$$

$$= A^+\left(e^{i\omega(\mathbf{k})t}\,\tilde{\chi}(\mathbf{k})\right) + A^-\left(e^{i\omega(\mathbf{k})t}\,\tilde{\chi}(-\mathbf{k})^*\right). \qquad (11.69)$$

Moreover, the derivative $\dot{\Phi}(t, \chi) = \frac{\partial}{\partial t} \Phi(t, \chi)$ is also well defined. If $\chi(\mathbf{x})$ is taken as a test function in $\mathcal{S}(\mathbb{R}^3)$, the operator $\Phi(t, \chi)$ is infinitely often differentiable with respect to t. The time evolution of the operators $\Phi(t, \chi)$ is given by $U^*(t)\Phi(s, \chi)U(t) = \Gamma(e^{iH_1 t})\Phi(s, \chi)\Gamma(e^{-iH_1 t}) = \Phi(s + t, \chi)$.

11.1.5 Two-Point Functions and Propagators

Wightman Functions

The vacuum expectation of the local field operator $\Phi(x)$ vanishes, so that we have $(\Omega \mid \Phi(x)\Omega) = 0$. The vacuum expectation of the product of two fields is not trivial, however, and is known as the two-point **Wightman function**,

$$w_2(x, y) \triangleq \langle \Omega \mid \Phi(x)\Phi(y)\,\Omega \rangle. \tag{11.70}$$

Proposition 11.1.2 *The two-point Wightman function is related to the commutator (11.55) by*

$$w_2(x, y) = \Delta_m^+(x - y) \tag{11.71}$$

and is a distribution depending on the difference variable $x - y$. Moreover, the two-point Wightman function has the invariance property

$$w_2(\Lambda x + a, \Lambda y + a) = w_2(x, y), \qquad a \in \mathbb{M}, \qquad \Lambda \in \mathcal{L}^\uparrow. \tag{11.72}$$

Proof The identity (11.71) is readily established as follows:

$$
\begin{aligned}
\langle \Omega \mid \Phi(x)\Phi(y)\,\Omega \rangle &= \left\langle \Phi^{(-)}(x)\Omega \mid \Phi^{(-)}(y)\Omega \right\rangle \\
&= \left\langle \Omega \mid \Phi^{(+)}(x)\Phi^{(-)}(y)\,\Omega \right\rangle \\
&= \left\langle \Omega \mid \left[\Phi^{(+)}(x), \Phi^{(-)}(y) \right] \Omega \right\rangle \\
&= \Delta_m^+(x - y).
\end{aligned}
$$

The invariance property is then inherited from $\Delta_m^+(x)$. □

The integral representation of $\Delta_m^+(x)$ is, as shown in (11.17),

$$\Delta_m^+(x) = (2\pi)^{-3} \int e^{-ik \cdot x} \delta_+(k^2 - m^2)\, d^4 k. \tag{11.73}$$

The distribution $\Delta_m^+(x)$ is a solution of the Klein–Gordon equation with positive frequencies. It is invariant under orthochronous Lorentz transformations, $\Delta_m^+(\Lambda x) = \Delta_m^+(x)$, $\Lambda \in \mathcal{L}^\uparrow$.

Since $k \cdot y < 0$, if $k \in \mathcal{V}_m^+$ and $y \in \mathcal{V}^-$, the function (11.73) is the boundary value of the analytic function

$$w(\xi) = (2\pi)^{-3} \int e^{-ik\cdot\xi} \delta_+(k^2 - m^2)\, d^4k, \qquad (11.74)$$

which is a holomorphic function inside the tubular complex domain

$$\mathcal{T} = \left\{ \xi = x + iy : x \in \mathbb{M}^4,\ y \in \mathscr{V}^- \right\} \subset \mathbb{M}_{\mathbb{C}} = \mathbb{C}^4.$$

The function $w(\xi)$ is Lorentz invariant

$$w(\Lambda\xi) = w(\xi), \quad \Lambda \in \mathcal{L}^\uparrow \qquad (11.75)$$

in its domain of holomorphy.

We observe that the function (11.73) is the Fourier transform of a positive distribution, and this implies the following result.

Proposition 11.1.3 *The Wightman function $w_2(x, y) = \Delta_m^+(x - y)$ is a positive kernel function, that is,*

$$\int_{\mathbb{M}\times\mathbb{M}} w_2(x, y)\varphi(x)^*\varphi(y) d^4x d^4y \geq 0, \qquad (11.76)$$

for all test functions $\varphi \in \mathcal{S}(\mathbb{M})$.

Considering the definition (11.70), this inequality will then imply that $\|\Phi(\varphi)\Omega\|^2 = \langle \Phi(\varphi)\Omega \mid \Phi(\varphi)\,\Omega \rangle$ is not negative.

The distribution $\delta_+(k^2 - m^2) = \Theta(k^0)\delta(k^2 - m^2)$ is an element of $\mathcal{S}'(\widehat{\mathbb{M}})$. The Fourier transform

$$\int e^{-ik^0 t} \delta_+(k^2 - m^2) dk^0 = (2\omega(\mathbf{k}))^{-1} e^{-i\omega(\mathbf{k})t}$$

is a distribution in $\mathcal{S}'(\widehat{\mathbb{R}}^3)$ with parameter t. For any $f \in \mathcal{S}(\widehat{\mathbb{R}}^3)$, the integral $\int f(\mathbf{k})e^{-i\omega(\mathbf{k})t}\frac{d^3k}{2\omega(\mathbf{k})}$ has derivatives in t of arbitrary order. It therefore follows that $(2\omega(\mathbf{k}))^{-1} e^{-i\omega(\mathbf{k})t}$ and its Fourier transform $\Delta_m^+(t, \mathbf{x},)$ are distributions in $\mathcal{S}'(\mathbb{R}^3)$ that are arbitrarily often differentiable in $t \in \mathbb{R}$.

To evaluate the analytic function $w(\xi)$ explicitly, we take $\xi = (-is, \mathbf{0})$ in $\mathcal{T}$ with $s > 0$, such that

$$w(\xi) = \frac{1}{2}(2\pi)^{-3} g(s) \qquad (11.77)$$

with

$$\begin{aligned}
g(s) &= \int_{\mathbb{R}^3} e^{-s\sqrt{\mathbf{k}^2+m^2}} \frac{d^3k}{\sqrt{\mathbf{k}^2 + m^2}} \\
&= 4\pi \int_0^\infty \frac{\kappa^2 d\kappa}{\sqrt{\kappa^2 + m^2}} e^{-s\sqrt{\kappa^2+m^2}} = 4\pi m^2 \int_1^\infty e^{-tsm}\sqrt{t^2 - 1}\, dt \\
&= \begin{cases} 4\pi m s^{-1} K_1(ms) & \text{if } m > 0 \\ 4\pi s^{-2} & \text{if } m = 0. \end{cases} \qquad (11.78)
\end{aligned}$$

Thereby

$$K_1(s) = s \int_1^\infty e^{-st} \sqrt{t^2 - 1} \, dt = s e^{-s} \int_0^\infty e^{-su} \sqrt{u(u+2)} \, du \quad (11.79)$$

is the modified Bessel function of the second kind. The integrals in (11.79) are holomorphic for $s \in \mathbb{C}$ with $\operatorname{Re} s > 0$. By rotating the variable u in the range $-\frac{1}{2}\pi < \arg u < \frac{1}{2}\pi$, we can enlarge the domain of holomorphy to the complex plane with a cut along the negative real axis. Moreover, by rotating the variable u within the range $-\pi < \arg u < \pi$, an analytic continuation through the cut to the domain $\left\{ s = |s| \, e^{i\alpha} \mid s \neq 0, \ -\frac{3}{2}\pi < \alpha < \frac{3}{2}\pi \right\}$ is possible.

The function $K_1(s)$ has the representation, as shown in Magnus et al. (1966, sec. 3.2),

$$K_1(s) = s^{-1} + \left[\gamma + \log \frac{s}{2} \right] I_1(s) - \frac{1}{4}s + r_3(s). \quad (11.80)$$

In this formula we have Euler's constant γ, $r_3(s)$ is an entire analytic function with a zero of third order at $s = 0$, and $I_1(s)$ is the modified Bessel function of the first kind:

$$I_1(s) = -iJ_1(is) = \sum_{n=0}^\infty \frac{1}{n! \, (n+1)!} \left(\frac{s}{2} \right)^{2n+1} = \frac{1}{2}s + \frac{1}{16}s^3 + \cdots \quad (11.81)$$

In the neighborhood of $s = 0$, we obtain

$$K_1(s) = s^{-1} + sR(s) \quad (11.82)$$

with $R(s) = \frac{1}{2} \log \frac{s}{2} + \textit{locally bounded function}$. For large values of $|s|$ and $-\frac{3}{2}\pi < \arg s < \frac{3}{2}\pi$, the asymptotic behavior of $K_1(s)$ is, as shown in Magnus et al. (1966, sec. 3.14.1),

$$K_1(s) = \sqrt{\frac{\pi}{2s}} \, e^{-s} \left(1 + \frac{3}{8}s^{-1} + \mathcal{O}\left(s^{-2}\right) \right). \quad (11.83)$$

The function (11.78) has therefore the representation

$$g(s) = 4\pi \frac{K_1(ms)}{s} = \frac{4\pi}{s^2} + 4\pi m^2 R(ms), \quad (11.84)$$

where the last term vanishes for $m = 0$.

With $\xi = (-is, \mathbf{0}) \in \mathbb{C}^4$ such that $s = \sqrt{-\xi^2}$, the function $w(\xi)$ is

$$w(\xi) = \frac{m}{(2\pi)^2} \frac{K_1\left(m\sqrt{-\xi^2}\right)}{\sqrt{-\xi^2}}, \quad (11.85)$$

if $m > 0$, and $w(\xi) = (2\pi)^{-2} \xi^{-2}$ if $m = 0$. Due to the Lorentz invariance (11.75) and analyticity, this result is true for all $\xi \in \mathbb{M} + i\mathcal{V}^- \subset \mathbb{M}_\mathbb{C}$. In this domain, the sign of the square root satisfies $\mathrm{Re}\sqrt{-\xi^2} > 0$. In the limit $\xi = x - i(\varepsilon, \mathbf{0}), \varepsilon \to +0$, we obtain the value of w on the Minkowski space[3]

$$\Delta_m^+(x) = w(x^0 + i0, \mathbf{x}) = \frac{m}{(2\pi)^2} \frac{K_1\left(m\sqrt{-x^2 + i0x^0}\right)}{\sqrt{-x^2 + i0x^0}} \tag{11.86}$$

$$= \frac{1}{(2\pi)^2} \frac{1}{-x^2 + i0x^0} + \frac{m^2}{(2\pi)^2} R\left(m\sqrt{-x^2 + i0x^0}\right). \tag{11.87}$$

The function $\sqrt{-x^2 + i0x^0}$ is the limit of $\sqrt{-\xi^2}$ for $\mathrm{Im}\,\xi^0 \to -0$, that is,

$$\sqrt{-x^2 + i0x^0} = \begin{cases} \sqrt{|x^2|} & \text{if } x^2 < 0, \\ -i\sqrt{|x^2|} & \text{if } x^2 > 0,\ x^0 > 0, \\ +i\sqrt{|x^2|} & \text{if } x^2 > 0,\ x^0 < 0. \end{cases}$$

The explicit result (11.85) shows that the domain of analyticity of $w(\xi)$ is much larger than the tubular domain $\mathcal{T} = \mathbb{M} + i\mathcal{V}^-$. The function (11.85) is holomorphic for $\xi \in \mathbb{M}_{\mathbb{C}}$ if $\xi \neq 0$ and $-\pi < \arg\left(-\xi^2\right) < \pi$ (with $\mathrm{Re}\sqrt{-\xi^2} > 0$). This domain includes the domain of spacelike vectors of the Minkowski space and the *euclidean points* $\xi_E = (ix^4, x^1, x^2, x^3) \neq 0$ with $x = (x^1, x^2, x^3, x^4) \in \mathbb{E} = R^4$. At the euclidean points, the function w has the values

$$w(\xi_E) = (2\pi)^{-2} \frac{m}{|x|} K_1(m\,|x|), \tag{11.88}$$

with the euclidean length $|x| = \sqrt{\sum_{\mu=1}^4 (x^\mu)^2}$.

As a consequence of this result, the two-point Wightman function $w_2(x_1, x_2)$ at spacetime points $x_{1,2} \in \mathbb{M}$ is the boundary value of a function that is analytic in both variables ξ_1 and ξ_2. This function has an analytic continuation to the euclidean points $\xi_{jE} = (i x_j^4, x_j^1, x_j^2, x_j^3), j = 1, 2$, with $x_j = (x_j^1, x_j^2, x_j^3, x_j^4) \in \mathbb{E}$ and $x_1 \neq x_2$,

$$w_2(\xi_{1E}, \xi_{2E}) = (2\pi)^{-2} \frac{m}{|x_1 - x_2|} K_1(m\,|x_1 - x_2|). \tag{11.89}$$

Here $|x_1 - x_2| = \sqrt{\sum_{\mu=1}^4 \left(x_1^\mu - x_2^\mu\right)^2}$ is the euclidean distance.

The distribution $w(x)$ has no defined value at $x = 0$, and the vacuum expectation $(\Omega \mid \Phi(x)\Phi(y)\Omega) = \Delta_m^+(x - y)$ is not defined at coinciding points $x = y$.

---

[3] A general theory of Lorentz invariant distributions has been developed by Methée (1954); see section 3.3 of Bogoliubov et al. (1990). For the connection between boundary values of holomorphic functions and distributions, see Bremermann and Durand (1961) and chapter 2. of Streater and Wightman (1964).

Hence the product of fields $\Phi(x)\Phi(y)$ cannot be defined at coinciding points $x = y$. But for the free field, it turns out that the only ill-defined term comes from the vacuum expectation.

The **normal product** or Wick product of the field operators $\Phi(x)$ and $\Phi(y)$ is defined as

$$:\Phi(x)\Phi(y): \stackrel{\triangle}{=} \Phi(x)\Phi(y) - w_2(x,y)\, I. \qquad (11.90)$$

At the algebraic level, this is really just the usual process of expanding terms and then summarily moving all the creation operators to the left of all annihilation operators, ignoring the commutation relations. Indeed, the identity $\Phi(x)\Phi(y) - w_2(x,y)\, I = \Phi(x)\Phi(y) - \left[\Phi^{(+)}(x), \Phi^{(-)}(y)\right]$ implies this representation of the normal product by "normal ordering":

$$\begin{aligned} :\Phi(x)\Phi(y): = \; & \Phi^{(+)}(x)\Phi^{(+)}(y) + \Phi^{(-)}(y)\Phi^{(+)}(x) \\ & + \Phi^{(-)}(x)\Phi^{(+)}(y) + \Phi^{(-)}(x)\Phi^{(-)}(y). \qquad (11.91) \end{aligned}$$

The normal product transforms like

$$U(a, \Lambda) :\Phi(x)\Phi(y): U^*(a, \Lambda) = :\Phi(\Lambda x + a)\Phi(\Lambda y + a): \qquad (11.92)$$

under orthochronous Poincaré transformations $(a, \Lambda) \in \mathbb{M} \times \mathcal{L}^\uparrow$. This behavior follows from the transformation rules (11.62) and (11.72). Another important property of the normal product is its commutativity

$$:\Phi(x)\Phi(y): = :\Phi(y)\Phi(x): . \qquad (11.93)$$

This relation is an immediate consequence of (11.91) since creation operators commute with creation operators and annihilation operators commute with annihilation operators.

This normal product is defined as operator-valued distribution also at coincidence points

$$:\Phi^2(x): \stackrel{\triangle}{=} :\Phi(x)\Phi(x): . \qquad (11.94)$$

This singular operator is called the (second) **Wick power** of $\Phi(x)$. It is straightforward to prove that the inner product $\left(F \,|\!:\Phi^2(x)\!: \; G\right)$ is a continuous function of x if F and G are tensors in a suitable linear subset $\mathcal{D}$ of the Fock space as defined in Section 9.1.3. But it takes more effort to derive that $\int \varphi(x) :\Phi^2(x): \; d^4x$ is a well-defined operator on $\mathcal{D}$ for all test functions $\varphi \in \mathcal{S}(\mathbb{M})$, cf. Wightman and Garding (1964).

Commutator Function

The commutator of the field operators $\Phi(x)$ and $\Phi(y)$ is a c-number, which can be calculated from its vacuum expectation

$$-i\,\Delta_m(x-y) = \big\langle \Omega \mid [\Phi(x), \Phi(y)]\,\Omega \big\rangle = \Delta_m^+(x-y) - \Delta_m^+(y-x).$$

$$(11.95)$$

Taking into account the identity between Bessel functions

$$is\,(R(is) - R(-is)) = K_1(is) + K_1(-is) = -\pi J_1(s),\ s > 0,$$

the distribution $\Delta_m(x) = -i\big(\Delta_m^+(x) - \Delta_m^+(-x)\big)$ follows from (11.87) and (11.80) as

$$\Delta_m(x) = \frac{1}{2\pi}\operatorname{sign}(x^0)\delta(x^2) - \frac{m}{4\pi}\operatorname{sign}(x^0)\Theta(x^2)\frac{J_1\left(m\sqrt{x^2}\right)}{\sqrt{x^2}}.\quad (11.96)$$

The Fourier representation

$$\Delta_m(x) = i\,(2\pi)^{-3}\int e^{-ik\cdot x}\operatorname{sign}(k^0)\delta(k^2 - m^2)d^4k \qquad (11.97)$$

is an immediate consequence of (11.73). The generalized function $\Delta_m(x)$ is Lorentz invariant, $\Delta_m(\Lambda x) = \Delta_m(x), \Lambda \in \mathcal{L}^\uparrow$, and it is also antisymmetric, $\Delta_m(-x) = -\Delta_m(x)$. It vanishes therefore for spacelike x, as can also be seen from the representation (11.96).

We now investigate $\Delta_m(x) = \Delta_m(t, \mathbf{x})$ as distribution in $\mathbf{x}$ with parameter $t \in \mathbb{R}$. Choosing a test function $\varphi(\mathbf{x}) \in \mathcal{S}(\mathbb{R}^3)$, the most singular term $\frac{1}{2\pi}\operatorname{sign}(x^0)\delta(x^2)$ gives the contribution

$$\int \operatorname{sign}(t)\delta(t^2 - \mathbf{x}^2)\varphi(\mathbf{x})d^3x = \int_{|\mathbf{x}|=|t|} \frac{t^2}{2t}\varphi(\mathbf{x})d\Omega$$

$$= 2\pi t\,\varphi(0) + \mathcal{O}(t^3) \qquad (11.98)$$

if $t \neq 0$. The remaining contributions are of the order $\int_{|\mathbf{x}|\leq|t|} d^3x \sim |t|^3$ for small $|t|$. Hence $\Delta_m(t, \mathbf{x})$ is a solution of the Klein–Gordon equation

$$\left(\Box + m^2\right)\Delta_m(x) = 0 \qquad (11.99)$$

with the initial conditions

$$\Delta_m(t = 0, \mathbf{x}) = 0, \quad \frac{\partial}{\partial t}\Delta_m(t = 0, \mathbf{x}) = \delta^3(\mathbf{x}). \qquad (11.100)$$

The second derivative $\frac{\partial^2}{\partial t^2}\Delta_m(t = 0, \mathbf{x}) = 0$ vanishes at $t = 0$ as a consequence of (11.98).

Performing the k^0 integration in (11.97), we obtain the useful Fourier representation

$$\Delta_m(x) = (2\pi)^{-3}\int_{\mathbb{R}^3} \frac{\sin \omega(\mathbf{k})t}{\omega(\mathbf{k})}e^{i\mathbf{k}\mathbf{x}}d^3x. \qquad (11.101)$$

Retarded Propagator

The **retarded propagator** or **retarded Green's function** $\Delta_m^{ret}(x)$ is the fundamental solution of the Klein–Gordon equation

$$\left(\Box + m^2\right) \Delta_m^{ret}(x) = \delta^4(x) \tag{11.102}$$

that vanishes for $t < 0$. Its Fourier representation is

$$\Delta_m^{ret}(x) = (2\pi)^{-4} \int e^{-ik \cdot x} \left(m^2 - k^2 - i0k^0\right)^{-1} d^4k. \tag{11.103}$$

The distribution $\frac{1}{m^2 - k^2 - i0k^0}$ is the boundary value of the Lorentz invariant analytic function $\left(m^2 - w^2\right)^{-1}$ with vectors $w = k + iu$ in $\mathbb{C}^4$, $k \in \mathbb{R}^4$, $u \in \mathscr{V}^+$, for $u \to 0$. This function is holomorphic for $\text{Im } k^0 > 0$. Hence the distribution (11.103) vanishes for $t < 0$. The identity

$$\left(m^2 - k^2\right) \left(m^2 - k^2 - i0k^0\right)^{-1} = 1$$

implies the relation (11.102). The Green's function (11.103) is a well-defined Lorentz invariant distribution, $\Delta_m^{ret}(\Lambda x) = \Delta_m^{ret}(x)$, $\Lambda \in \mathcal{L}^\uparrow$, with a support in the closed forward cone $\overline{\mathscr{V}^+}$.

The **advanced propagator**

$$\Delta_m^{adv}(x) = \Delta_m^{ret}(-t, \mathbf{x}) = \Delta_m^{ret}(-x) \tag{11.104}$$

is another solution of the differential equation (11.102). It vanishes for $t > 0$. The relation $\Delta_m^{ret}(x) - \Delta_m^{ret}(-x) = \Delta_m(x)$ follows from the identity

$$\left(m^2 - k^2 - i0k^0\right)^{-1} - \left(m^2 - k^2 + i0k^0\right)^{-1} = 2\pi i \, \text{sign}(k^0) \, \delta(k^2 - m^2).$$

Then the support restrictions of $\Delta_m^{ret}(x)$ and $\Delta_m^{adv}(x)$ lead to the identification

$$\begin{aligned}
\Delta_m^{ret}(x) &= \Theta(x^0)\Delta_m(x) \\
&= i\Theta(x^0)\Delta_m^+(x) - i\Theta(x^0)\Delta_m^+(-x). \tag{11.105}
\end{aligned}$$

The last identity follows from (11.95).

Feynman Propagator and the Two-Point τ-Function

The **causal Green's function** or **Feynman propagator** $\Delta_m^c(x)$ is yet another a solution of the differential equation (11.102). It is defined as

$$\Delta_m^c(x) \triangleq \Delta_m^{ret}(x) + i\Delta_m^+(-x) = \begin{cases} i\Delta_m^+(x) & \text{if } x^0 > 0, \\ i\Delta_m^+(-x) & \text{if } x^0 < 0. \end{cases} \tag{11.106}$$

The second identification can be written as

$$\Delta_m^c(x) = i\Theta(x^0)\Delta_m^+(x) + i\Theta(-x^0)\Delta_m^+(-x);$$

it follows from (11.105). Its Fourier transform is

$$\tilde{\Delta}_m^c(k) = \int \Delta_m^c(x)e^{ik\cdot x}d^4x$$
$$= \left(m^2 - k^2 - i0k^0\right)^{-1} + 2\pi i\,\Theta(-k^0)\delta(k^2 - m^2)$$
$$= \left(m^2 - k^2 - i0\right)^{-1}. \tag{11.107}$$

The generalized functions $\tilde{\Delta}_m^c(k)$ and $\Delta_m^c(x)$ are invariant under Lorentz transformations of the full group $\mathcal{L}$. The propagator $i\Delta_m^c(x)$ and the distribution $\Delta_m^+(x)$ are boundary values of the analytic function $w(\xi)$; they coincide for spacelike x, a region that lies inside the analyticity domain, and for positive timelike x; but they are different boundary values for negative timelike x:

$$\begin{array}{llll}
-i\,\Delta_m^c(x) = & \Delta_m^+(x) = w(x) = w(-x) & \text{if} & x^2 < 0, \\
-i\,\Delta_m^c(x) = & \Delta_m^+(x) = w(x^0 - i0, \mathbf{x}) & \text{if} & x^2 > 0,\, x^0 > 0, \\
-i\,\Delta_m^c(x) = & w(x^0 + i0, \mathbf{x}) & \text{if} & x^2 > 0,\, x^0 < 0, \\
\Delta_m^+(x) = & w(x^0 - i0, \mathbf{x}) & \text{if} & x^2 > 0,\, x^0 < 0.
\end{array} \tag{11.108}$$

In analogy to (11.71), we define the two-point τ-function as

$$\tau(x_1, x_2) \overset{\triangle}{=} -i\Delta_m^c(x_1 - x_2)$$
$$= \Theta(x_1^0 - x_2^0)w_2(x_1, x_2) + \Theta(x_2^0 - x_1^0)w_2(x_2, x_1). \tag{11.109}$$

The second identity follows from (11.106). The two-point τ function is a well-defined generalized function in $\mathcal{S}'(\mathbb{M} \times \mathbb{M})$. The τ-function is symmetric in its arguments x_1 and x_2, and – as a consequence of the Lorentz invariance of $\Delta_m^c(x)$ – it has the invariance property

$$\tau(\Lambda x_1 + a, \Lambda x_2 + a) = \tau(x_1, x_2), \qquad a \in \mathbb{M}, \qquad \Lambda \in \mathcal{L}, \tag{11.110}$$

under Poincaré transformations.

This τ-function is related to the time-ordered product of field operators. The time-ordered product or T-product of the field operators $\Phi(x_1)$ and $\Phi(x_2)$ is formally introduced as

$$T\,\Phi(x_1)\Phi(x_2) = \Theta(x_1^0 - x_2^0)\Phi(x_1)\Phi(x_2) + \Theta(x_2^0 - x_1^0)\Phi(x_2)\Phi(x_1). \tag{11.111}$$

The right side of (11.111) does not depend on the Lorentz system, since the operators $\Phi(x_1)$ and $\Phi(x_2)$ commute for spacelike separation of the arguments. But it is only a formal definition since the multiplication of the generalized operator $\Phi(x_1)\Phi(x_2)$ with the step function $\Theta(x_1^0 - x_2^0)$ is so far only a formal device. As a consequence of (11.109), the vacuum expectation of (11.111) can be identified with the τ-function

$$\tau(x_1, x_2) = \langle \Omega \mid \mathcal{T} \Phi(x_1)\Phi(x_2)\,\Omega \rangle, \tag{11.112}$$

which is well defined. To proceed, we use the identity (11.90). The normal product $:\Phi(x_1)\Phi(x_2):$ is symmetric in the variables x_1 and x_2. The time-ordered prescription therefore does not affect this product, $\mathcal{T}\,(:\Phi(x_1)\Phi(x_2):)$ $=:\Phi(x_1)\Phi(x_2):$. The identities (11.90) and (11.109) then imply that (11.111) can be identified with the following well-defined product.

The **time-ordered product** of the field operators $\Phi(x_1)$ and $\Phi(x_2)$ is defined as

$$\mathcal{T} \Phi(x_1)\Phi(x_2) \triangleq \tau(x_1, x_2)\,I + \,:\Phi(x_1)\Phi(x_2):. \tag{11.113}$$

The product (11.113) is an operator-valued generalized function of the variables x_1 and x_2. It is symmetric in x_1 and x_2, and it transforms under orthochronous Poincaré transformations as

$$U(a, \Lambda)\,(\mathcal{T} \Phi(x)\Phi(y))\,U^*(a, \Lambda) = \mathcal{T}\,\Phi(\Lambda x + a)\Phi(\Lambda y + a). \tag{11.114}$$

This relation follows from (11.92) and (11.110).

Euclidean Propagator and Schwinger Function

The function (11.74) $w(\xi)$ can be analytically continued to the euclidean points $\xi_E = (ix_4, x_1, x_2, x_3)$ with $x = (x_1, x_2, x_3, x_4) \in \mathbb{E} = \mathbb{R}^4$, $x \neq 0$. The **euclidean propagator**

$$\Delta_m^E(x) \triangleq w(i\,x^4, \mathbf{x}) = (2\pi)^{-2}\,\frac{m}{|x|}\,K_1\,(m\,|x|) \tag{11.115}$$

is a function of the euclidean length $|x| = \sqrt{\sum_{\mu=1}^4 (x^\mu)^2}$ of the vector $x \in \mathbb{E} = \mathbb{R}^4$; see (11.88). From Section 11.1.5, we know that $w(i\,x^4, \mathbf{x})$ is an analytic continuation of the Feynman propagator $w(x) = -i\Delta_m^c(x)$, x spacelike. Starting from the Feynman propagator (11.107) in the momentum representation, the energy variable k^0 can be continued to the imaginary axis $k^0 \to ik^4$ with $k^4 \in \mathbb{R}$. This continuation is the **Wick rotation** $\mathbb{R} \ni k^0 \to e^{i\alpha}k^0, 0 \leq \alpha \leq \frac{\pi}{2}$, within the analyticity domain of $(m^2 - k^2 - i0)^{-1}$. Then the indefinite form $k^2 = (k^0)^2 - \mathbf{k}^2$ becomes the negative definite form $-(k^4)^2 - \mathbf{k}^2 = -|k|^2$, $k = (k^1, k^1, k^3, k^4) \in \mathbb{R}^4$. The euclidean propagator is therefore defined in the momentum variables as

$$\tilde{\Delta}_m^E(k) = \left(|k|^2 + m^2\right)^{-1}. \tag{11.116}$$

The functions (11.115) and (11.116) are locally integrable and decreasing for large arguments (including the case $m = 0$). They are well-defined generalized functions in $\mathcal{S}'(\mathbb{R}^4)$, which are related by the euclidean Fourier transform

$$\Delta_m^E(x) = (2\pi)^{-4} \int_M \left(|k|^2 + m^2\right)^{-1} \exp\left(-i\,kx\right) d^4x \tag{11.117}$$

(see, e.g., Gelfand and Shilov, 1964). Thereby kx is the euclidean inner product $kx = \sum_{\mu=1}^4 k_\mu x_\mu$. As already stated in Section 11.1.5, the Wightman function $w_2(x, y) = w(x - y)$ has an analytic continuation to $w_2(i\,x^4, \mathbf{x}, i\,y^4, \mathbf{y})$. The resulting function

$$S_2(x, y) = w_2(i\,x^4, \mathbf{x}, i\,y^4, \mathbf{y}) = \Delta_m^E(x - y) \tag{11.118}$$

is called the two-point **Schwinger function**.

The causal propagator $\Delta_m^c(x)$ has been introduced in quantum field theory by Stueckelberg and Rivier (1949) and by Feynman (1949). It is a basic ingredient of the standard perturbation theory. The transition from the Minkowski time variable to a variable on the imaginary axis and the corresponding transition of the energy variable by analytic continuation ("rotation") has been introduced by Wick (1954) as a method to simplify calculations. A systematic investigation of euclidean Green's functions started with the work of Schwinger (1958), Nakano (1959), and Symanzik (1966).

11.1.6 *N*-Point Functions and Normal Product

Wightman Functions

The vacuum expectation

$$w_n(x_1, \ldots, x_n) = \langle \Omega \mid \Phi(x_1) \ldots \Phi(x_n) \, \Omega \rangle \tag{11.119}$$

of the product of n field operators is called the n-point Wightman function. The field operators $\Phi(x_j) = \Phi^{(-)}(x_j) + \Phi^{(+)}(x_j)$ are sums of creation and annihilation operators, and the expectation (11.119) can be calculated as done in Section 9.3.1. The result yields again the Gaussian combinatorics of moments

$$w_{2k}(x_1, \ldots, x_{2k}) = \sum_{\sigma \in \mathfrak{S}_<(2k)} w_2(x_{\sigma(1)}, x_{\sigma(2)}) \ldots w_2(x_{\sigma(2k-1)}, x_{\sigma(2k)}), \tag{11.120}$$

with the odd terms vanishing. Recall that the two-point Wightman function is $w_2(x, y) = \Delta_m^+(x - y)$. Here $\mathfrak{S}_<$ is the set of all permutations σ of the index set $\{1, \ldots, 2k\}$ that respect the order relations $\sigma(1) < \sigma(3) < \cdots < \sigma(2m-1)$ and $\sigma(2j-1) < \sigma(2j)$, $j = 1, \ldots, k$. The Wightman functions are in general not invariant under an interchange of their arguments.

Let $\mathbf{X} = (x_1, \ldots, x_n)$ be a sequence of spacetime points, then we introduce the following notation for an ordered product of field operators,

$$\Phi(\mathbf{X}) = \Phi(x_1) \ldots \Phi(x_n),$$

and the **Wightman function** is then defined as

$$w(\mathbf{X}) = \langle \Omega \mid \Phi(\mathbf{X}) \, \Omega \rangle.$$

This is entirely consistent with (11.119).

The representation (11.120) and the analyticity properties of the two-point Wightman function together imply that the generalized function $w(x_1, \ldots, x_n)$ is the boundary value of an analytic function $w(\xi_1, \ldots, \xi_n)$. The domain of analyticity includes the region $(\xi_1, \ldots, \xi_n) \in \mathbb{C}^{4n}$ with the condition that $\mathrm{Re}\left(-(\xi_j - \xi_k)^2\right) > 0$, $1 \le j < k \le n$.

Normal (Wick) Product of Field Operators

Given a sequence $\mathbf{X} = (x_1, \ldots, x_n)$, we would like to obtain a simple way to describe the normal product or Wick product $:\Phi(\mathbf{X}):$. Evidently this should be the generalization of the identity (11.91), and therefore the same as taking $\Phi(x_1) \ldots \Phi(x_n)$, writing each term as $\Phi(x_k) = \Phi^{(-)}(x_k) + \Phi^{(+)}(x_k)$, expanding and then moving all annihilators to the left of creators, and ignoring the commutation relations. This procedure is called normal ordering or Wick ordering. For the given sequence $\mathbf{X} = (x_1, \ldots, x_n)$, we shall understand the equation $\mathbf{X}_1 + \mathbf{X}_2 = \mathbf{X}$ to mean a way of splitting $\mathbf{X}$ into a pair of disjoint subsequences, respecting the same order of terms; see Section 4.1.7. The null subsequence is allowed.

Given a sequence $\mathbf{X} = (x_1, \ldots, x_n)$, we define the **Wick product**, $:\Phi(x_1) \ldots \Phi(x_n):$, to be

$$:\Phi(\mathbf{X}): \triangleq \sum_{\mathbf{X}_1 + \mathbf{X}_2 = \mathbf{X}} \Phi^{(-)}(\mathbf{X}_1) \Phi^{(+)}(\mathbf{X}_2). \tag{11.121}$$

The Wick product (11.121) is an associative product, generated from the product (11.90) of two fields. The Wick product of field operators is commutative. The covariance properties (11.49) of $\Phi^{(\pm)}$ imply the covariance of the normal product

$$U(y, \Lambda) :\Phi(x_1) \ldots \Phi(x_n): U^*(y, \Lambda) = :\Phi(\Lambda x_1 + y) \ldots \Phi(\Lambda x_n + y): . \tag{11.122}$$

For $n = 1$, we obtain $:\Phi(x): \triangleq \Phi(x)$, and for $n = 2$ the identity (11.121) yields the normal product (11.91) of Section 11.1.5. More generally, the definition (11.121) leads to the following.

Proposition 11.1.4 *The sequentially ordered and Wick-ordered products of field operators are related by*

$$\Phi(\mathbf{X}) = \sum_{\mathbf{X}_1 + \mathbf{X}_2 = \mathbf{X}} w(\mathbf{X}_1) \; :\Phi(\mathbf{X}_2):, \tag{11.123}$$

$$:\Phi(\mathbf{X}): \; = \sum_{\mathbf{X}_1 + \mathbf{X}_2 = \mathbf{X}} (-1)^{\frac{1}{2} \# \mathbf{X}_1} w(\mathbf{X}_1) \; \Phi(\mathbf{X}_2). \tag{11.124}$$

Proof This is easily established by induction. These relations are true for $n = 1$ and $n = 2$. For larger n, they can be derived with the help of recurrence relations. The product $:\Phi(\mathbf{X}): \; \Phi(x_{n+1})$ may be calculated using (11.121), and it is straightforward to derive the recurrence relation

$$:\Phi(\mathbf{X}): \; \Phi(x_{n+1}) = \sum_{m \in \mathbf{N}} w_2(x_m, x_{n+1}) \; :\Phi(\mathbf{X}/x_m):$$
$$+ \; :\Phi(\mathbf{X} + x_{n+1}): \; .$$

The notation here is that $\mathbf{X}/x_m$ is the sequence $(x_1, \ldots, x_n)$ with x_m removed, and $\mathbf{X} + x_{n+1}$ means $(x_1, \ldots, x_n, x_{n+1})$. The sum (11.121) then leads to the recurrence relation The identities (11.123) and (11.124), which is Möbius inversion formula for sequences, are consequences of this relation. □

A crucially important remark is that the Wick product is symmetric under interchange of its arguments. Given $\mathbf{X} = (x_1, \ldots, x_n)$, we could take the corresponding set of elements $X = \{x_1, \ldots, x_n\}$, and since $:\Phi(\mathbf{X}):$ does not depend on the order of the elements of the sequence, we could think of it as $:\Phi(X):$ instead. That is, we may reconsider the Wick product as

$$:\Phi(X): \; \equiv \; :\prod_{x \in X} \Phi(x): \; .$$

In light of this, we may interpret equation (11.121) as a standard Wick product

$$:\Phi: \; \equiv \; \Phi^{(-)} \diamond \Phi^{(+)}, \tag{11.125}$$

where $\Phi^{(\pm)}(X) \triangleq \prod_{x \in X} \Phi^{(\pm)}(x)$. Note that the objects appearing in (11.125) are operators and that their order - the creators $\phi^{(-)}$ to the left of the annihilators $\Phi^{(+)}$ is essential.

The Wick product was introduced by Wick (1950). In this publication, Wick also derived the identities (11.123) and (11.124) (first Wick Theorem) and the subsequent relations (11.132) with the T-product (second Wick Theorem).

Time-Ordered Product and τ–Functions

The time-ordered product or T-product of the fields $\Phi(x_1), \ldots, \Phi(x_n)$ is formally defined as

$$
\mathcal{T}\,\Phi(x_1)\ldots\Phi(x_n) = \sum_{\sigma \in \mathfrak{S}(n)} \Theta(x^0_{\sigma(1)} > \cdots > x^0_{\sigma(n)})\, \Phi(x_{\sigma(1)})\ldots\Phi(x_{\sigma(n)}).
$$

$$(11.126)$$

The step function $\Theta(s_1 > \cdots > s_n)$ with $(s_1, \ldots, s_n) \in \mathbb{R}^n$ is defined as

$$
\Theta(s_1 > \cdots > s_n) = \begin{cases} 1, & \text{if } s_1 > \cdots > s_n, \\ 0, & \text{otherwise.} \end{cases}
$$

The sum in (11.126) extends over all permutations $\sigma \in \mathfrak{S}(n)$ of the index set $\{1, \ldots, n\}$. By formal arguments, one obtains the following properties, which should be satisfied by the final definition:

$$
\mathcal{T}\,\Phi(x_1)\ldots\Phi(x_n) = \Phi(x_1)\ldots\Phi(x_n) \text{ if } x^0_1 > \cdots > x^0_n, \tag{11.127}
$$

$$
\mathcal{T}\,\Phi(x_1)\ldots\Phi(x_n) = \mathcal{T}\,\Phi(x_{\sigma(1)})\ldots\Phi(x_{\sigma(n)}), \quad \sigma \in \mathfrak{S}(n), \tag{11.128}
$$

$$
U(y, \Lambda)\,\mathcal{T}\,\Phi(x_1)\ldots\Phi(x_n)\,U^*(y, \Lambda) = \mathcal{T}\,\Phi(\Lambda x_1 + y)\ldots\Phi(\Lambda x_n + y). \tag{11.129}
$$

In the last identity, the Poincaré transformations used were assumed to be orthochronous, that is, $(y, \Lambda) \in \mathbb{M} \times \mathcal{L}^\uparrow$. We note that the time-ordered product may be alternatively defined as

$$
\mathcal{T}\,\Phi(x_1)\ldots\Phi(x_n) = \Phi(x_{\pi(1)})\ldots\Phi(x_{\pi(n)}),
$$

where π is the permutation that leads to the time ordering $x^0_{\pi(1)} > \cdots > x^0_{\pi(n)}$. The vacuum expectation

$$
\tau_n(x_1, \ldots, x_n) = \langle \Omega \mid \mathcal{T}\,\Phi(x_1)\ldots\Phi(x_n)\,\Omega \rangle \tag{11.130}
$$

is called the n-**point** τ-**function** . Following (11.128), the τ-function is totally symmetric in the variables $x_1, \ldots, x_n$. It is therefore determined by its values modulo the simultaneous points. We note that we have

$$
\tau_n(x_1, \ldots, x_n) \equiv w_n(x_{\pi(1)}, \ldots, x_{\pi(n)})
$$

where π is the permutation putting the sequence $(x_1, \ldots, x_n)$ to chronological order – unfortunately, π depends on the sequence!

For definiteness, let us assume that $x^0_1 > \cdots > x^0_n$, then (11.130) coincides with the Wightman function (11.119), which in turn has the explicit representation (11.120) as sum over two-point Wightman functions. The corresponding sum over two-point τ-functions

$$\tau_{2k}(x_1,\ldots,x_{2m}) = \sum_{\sigma \in \mathfrak{S}_<} \tau_2(x_{\sigma(1)}, x_{\sigma(2)}) \ldots \tau_2(x_{\sigma(2m-1)}, x_{\sigma(2m)}),$$

$$(11.131)$$

with odd terms vanishing. These are distributions in $\mathcal{S}'(\mathbb{M}^n)$, which is a symmetric function of the variables $x_1,\ldots,x_n$ and which coincides with (11.119) for $x_1^0 > \cdots > x_n^0$. Hence the identities (11.131) give a well-defined meaning to the vacuum expectation of (11.126).

For given $\mathbf{X} = (x_1,\ldots,x_n)$, the time-ordered product $\mathcal{T}\Phi(\mathbf{X})$ will only depend on the set $X = \{x_1,\ldots,x_n\}$ since any permutation of the original sequence will lead to the same chronological reordering. It therefore makes sense to define

$$\mathcal{T}\Phi(X) = \mathcal{T}\prod_{x \in X}\Phi(x).$$

Likewise, the τ-functions are symmetric and we may write

$$\tau(X) \equiv \langle \Omega \mid \mathcal{T}\Phi(X)\,\Omega \rangle.$$

We immediately obtain the Gaussian decomposition as

$$\tau(X) \equiv \sum_{\mathcal{P} \in \mathrm{Pair}(X)} \tau_{\mathcal{P}},$$

where we use the same notation as in (6.15).

We can now give a precise meaning to the formal sum (11.126). If $x_1^0 > \cdots > x_n^0$ the T-product (11.126) coincides with the product (11.127), which can be expressed by Wightman functions and normal products

$$\mathcal{T}\Phi(x_1)\ldots\Phi(x_n) \overset{(11.124)}{=} \sum_{\mathbf{X}_1+\mathbf{X}_2=\mathbf{X}} w(\mathbf{X}_1)\,{:}\Phi(\mathbf{X}_2){:}\,.$$

Now as the subsequence $\mathbf{X}_1$ respects the correct chronological ordering, we may replace the Wightman $w(\mathbf{X}_1)$ by the $\tau(\mathbf{X}_1)$ leading to the identification

$$\mathcal{T}\Phi(x_1)\ldots\Phi(x_n) \triangleq \sum_{\mathbf{X}_1+\mathbf{X}_2=\mathbf{X}} \tau(\mathbf{X}_1)\,{:}\Phi(\mathbf{X}_2){:}\,. \qquad (11.132)$$

The main result is that (11.132) now yields a well-defined operator-valued generalized function that obviously satisfies (11.127) and (11.128). The identity (11.129) follows from the invariance $\tau_2(\Lambda x_1 + y, \Lambda x_2 + y) = \tau_2(x_1, x_2)$ of the two-point τ-function and the covariance (11.122) of the normal product.

For an arbitrary finite set $X = \{x_1,\ldots,x_n\}$, we may use the two previous lemmas to write (11.132) as $\mathcal{T}\Phi(X) = \sum_{X_1+X_2=X}\tau(X_1)\,{:}\Phi(X_2){:}$ where the decomposition is into disjoint subsets. This may be written in the more compact notation of (4.5):

$$\mathcal{T}\,\Phi \equiv \tau \diamond \,:\!\Phi\!:\,. \tag{11.133}$$

The **time-ordered exponential** of $\Phi\,[J] = \int J\,(x)\,\Phi\,(x)\,dx$ is defined to be

$$\mathcal{T}e^{\Phi[J]} = \sum_n \frac{1}{n!}\mathcal{T}\Phi\,[J]^n$$

$$= \sum_n \frac{1}{n!}\int J\,(x_1)\ldots J\,(x_n)\,\mathcal{T}\Phi\,(x_1)\ldots\Phi\,(x_n)\,dx_1\ldots dx_n$$

or in Guichardet notation,

$$\mathcal{T}e^{\Phi[J]} \equiv \int J^X \mathcal{T}\Phi\,(X)\,dX.$$

Setting $\Delta_n = \{(x_1,\ldots,x_n) : \mathrm{t}\,(x_1) > \cdots > \mathrm{t}\,(x_n)\}$, we see that the time-ordered exponential is

$$\mathcal{T}e^{\Phi[J]} = \sum_n \int_{\Delta_n} J\,(x_1)\ldots J\,(x_n)\,\mathcal{T}\Phi\,(x_1)\ldots\Phi\,(x_n)\,dx_1\ldots dx_n.$$

Now, we saw in (11.133) that $\mathcal{T}\Phi = \tau \diamond \,:\!\Phi\!:$, and by virtue of the Wick product we have that

$$\mathcal{T}e^{\Phi[J]} = Z_{\mathcal{T}\Phi} = Z_\tau\,[J]\,Z_{:\Phi:}\,[J]\,.$$

For the free particle case, the τ-functions are Gaussian, so that $Z_\tau\,[J] \equiv \exp\frac{1}{2}\int J\,(x)\,\tau\,(x,y)\,J\,(y)\,dxdy$, while we introduce the normal ordered exponential

$$Z_{:\Phi:}\,[J] = \int J^X \,:\!\Phi\,(X)\!:\,dX.$$

We also have (11.125), $:\Phi \triangleq \Phi^{(-)} \diamond \Phi^{(+)}$, so that

$$Z_{:\Phi:}\,[J] = Z_{\Phi^{(-)}}\,[J]\,Z_{\Phi^{(+)}}\,[J]\,.$$

Note that these generating functions do not commute! Therefore, we have in the free particle case

$$\mathcal{T}e^{\Phi[J]} = \exp\left(\frac{1}{2}\int J\,(x)\,\tau\,(x,y)\,J\,(y)\,dxdy\right)e^{\Phi^{(-)}[J]}e^{\Phi^{(+)}[J]},$$

which converts the time-ordered exponential into Wick-ordered form.

In perturbation theory, one needs a generalization of the T-product to time ordering of Wick products. Since the Wick product is commutative, we have

$$\mathcal{T}\,:\!\Phi\,(X)\!: \equiv\, :\!\Phi\,(X)\!:\,.$$

More generally, let $X_1, \ldots, X_m$ be disjoint finite subsets of $\mathbb{M}$. Then the time-ordered product $\mathcal{T} (:\Phi(X_1): \cdots :\Phi(X_m):)$ is again calculated with formula (11.132), but the τ-function may only contain contractions $\tau_2(x, y)$, where x and y belong to different sets X_k.

Schwinger Functions

The Wightman functions have an analytic continuation to the euclidean points $(\xi_1, \ldots, \xi_n) = (ix_1^0, \mathbf{x}_1, \ldots, ix_n^0, \mathbf{x}_n)$, and the n-point Schwinger function with arguments $(x_1, \ldots, x_n) \in \mathbb{E}^n = \mathbb{R}^{4n}$ is defined as

$$S_n (x_1, \ldots, x_n) = w_n \left(ix_1^0, \mathbf{x}_1, \ldots, ix_n^0, \mathbf{x}_n \right). \tag{11.134}$$

The two-point Schwinger function $S_2(x, y) = \Delta_m^E(x - y)$ is calculated in Section 11.1.5. Identities (11.120) imply the following explicit formula for the Schwinger functions, with odd terms vanishing:

$$S_{2m}(x_1, \ldots, x_{2m}) = \sum_{\sigma \in \mathfrak{S}_<} S_2(x_{\sigma(1)}, x_{\sigma(2)}) \ldots S_2(x_{\sigma(2m-1)}, x_{\sigma(2m)}),$$

The Schwinger functions are singular only at coinciding arguments.

The τ-functions and the Schwinger functions are symmetric Green's functions, $G_n(x_1, \ldots, x_n) = G_n(x_{\sigma(1)}, \ldots, x_{\sigma(n)})$, where σ is any permutation of $\{1, \ldots, n\}$. It is therefore possible to obtain them from a generating function $Z(\varphi)$ using the ideas introduced in Section 4.1. In the case of the free field, this generating function is of the Gaussian type. Let $G_n(x_1, \ldots, x_n)$ be the n-point τ-function or Schwinger function, then the following formula holds:

$$G_n(x_1, \ldots, x_n) = \frac{\delta^n}{\delta\varphi(x_1) \ldots \delta\varphi(x_n)} Z(\varphi).$$

It has the generating function

$$Z(\varphi) = \exp \left(\int_{\mathbb{M} \times \mathbb{M}} \varphi(x) G_2(x, y) \varphi(y) d^4x d^4y \right).$$

In this case, φ is a test function in $\mathcal{S}(\mathbb{M})$. The Schwinger functions (11.134) are the moments of a Gaussian measure. The elements of this measure space form a Markovian field; see Nelson (1973).

Symmetric Green's functions (distributions) are determined knowing them on $\{(x_1, \ldots, x_n) \in \mathbb{M}^n, x_j \neq x_k \text{ for } j \neq k\}$ (or respectively on the set $\{(x_1, \ldots, x_n) \in \mathbb{E}^n, x_j \neq x_k \text{ for } j \neq k\}$), with the possible exception of distributions localized at the coincidence points, that is, distributions of the form $P(\partial) \delta(x_i - x_j)$, where $P(\partial)$ is a polynomial of the derivatives $\frac{\partial}{\partial x_i}$. In the case of the free field, the τ-functions and the Schwinger functions are sufficiently well behaved at the coincidence points such that they have a unique extension to distributions on $\mathbb{M}^n$ or $\mathbb{E}^n$ by the demand of minimal singularity. These Green's functions can therefore be considered as functions on finite subsets X of $\mathbb{M}$ or $\mathbb{E}$ in the sense of Guichardet, and the summation techniques of Chapter 3. become applicable.

11.2 Canonical Operators for the Free Field

Canonical quantization has been an essential method at the beginning of quantum field theory, and it is still used to quantize classical models. Here we outline this method only for the free field (including a coupling to a classical current). The Lagrangian density of a classical scalar field $\phi(x)$ with mass m is

$$\mathcal{L}(x) \equiv \mathcal{L}\left(\phi(x), \partial_\mu \phi(x)\right)$$
$$= \frac{1}{2}(\partial_0 \phi(x))^2 - (\partial \phi(x))^2 - \frac{1}{2}m^2 \phi^2(x) + j(x)\phi(x). \quad (11.135)$$

Thereby we have introduced the coupling to an external current $j(x)$. The symbol ∂_μ is derivative $\partial_\mu = \partial/\partial x^\mu$, $\mu = 0, \dots, 3$, and ∂ means the 3-gradient $\partial = (\partial_1, \partial_2, \partial_3)$. The field equation derived from (11.135) is

$$\frac{\partial \mathcal{L}}{\partial \phi(x)} - \sum_\mu \partial_\mu \frac{\partial \mathcal{L}}{\partial\left(\partial_\mu \phi(x)\right)} = -m^2 \phi(x) - \Box \phi(x) + j(x) = 0.$$
$$(11.136)$$

The canonical momentum field is

$$\pi(x) \triangleq \frac{\partial \mathcal{L}}{\partial\left(\partial_0 \phi(x)\right)} = \partial_0 \phi(x) \quad (11.137)$$

and the Hamiltonian density follows as

$$\mathcal{H}(x) \triangleq \pi(x)\,\partial_0 \phi(x) - \mathcal{L}(x)$$
$$= \frac{1}{2}\pi^2(x) + (\partial \phi(x))^2 + m^2 \phi^2(x) - j(x)\phi(x). \quad (11.138)$$

In a canonical quantum field theory, these constructions are transferred to the corresponding operators. The one-particle Hilbert space is taken as $\mathcal{H}_m^+ = L(\mathbb{R}^3, \frac{d^3k}{2\omega(\mathbf{k})})$. The Hamiltonian of the free field is $H = d\Gamma(\widehat{M})$ with the one-particle Hamiltonian

$$(\widehat{M}f)(\mathbf{k}) = \omega(\mathbf{k})f(\mathbf{k}).$$

As in (11.137), the time derivative of the local field (11.58) is a good candidate for the canonical momentum field

$$\Pi(x) \triangleq \partial_0 \Phi(x) \overset{(9.63)}{=} i\,[H, \Phi(x)]$$
$$= iA^+\left((2\pi)^{-\frac{3}{2}}\omega(\mathbf{k})\,e^{ik\cdot x}\right) - iA^-\left((2\pi)^{-\frac{3}{2}}\omega(\mathbf{k})\,e^{ik\cdot x}\right),$$

$$(11.139)$$

where $e^{ik\cdot x}$ means $e^{-i\mathbf{k}\mathbf{x}+i\omega(\mathbf{k})t}$. The commutator $\left[\Phi(x), \Pi(y)\right]$ is a c-number, which can be calculated from the vacuum expectation

$$\left(\Omega \mid \left[\Phi(x), \Pi(y)\right] \Omega\right) = \frac{\partial}{\partial y^0}\left(\Omega \mid \left[\Phi(x), \Phi(y)\right] \Omega\right) \overset{(11.59)}{=} i\frac{\partial}{\partial y^0}\Delta_m(x - y).$$

Hence the commutator is

$$\left[\Phi(x), \Pi(y)\right] = -i\frac{\partial}{\partial x^0}\Delta_m(x - y)\,I. \qquad (11.140)$$

As already indicated in our remarks about singular arguments, the field operator $\Phi(x) = \Phi(t, \mathbf{x})$ and its time derivative $\Pi(t, \mathbf{x})$ have a well-defined meaning as operator-valued generalized functions in the variable $\mathbf{x}$ at a fixed time t. The *equal time commutator* of the operators Φ and Π follows from the initial conditions (11.100) of Δ_m as

$$\left[\Phi(t, \mathbf{x}), \Pi(t, \mathbf{y})\right] = i\delta^3\,(\mathbf{x} - \mathbf{y})\,I. \qquad (11.141)$$

The other equal time commutators vanish:

$$\left[\Phi(t, \mathbf{x}), \Phi(t, \mathbf{y})\right] = \left[\Pi(t, \mathbf{x}), \Pi(t, \mathbf{y})\right] = 0. \qquad (11.142)$$

Hence the (noncovariant) field $\Pi(x)$ is the **canonical momentum field**.

We give some formal arguments, which indicate the close connection of quantum field theory with classical Hamiltonian field theory. Knowing the field $\Phi(x)$ and the conjugate field $\Pi(x)$ at a time t, one can calculate the positive and negative frequency parts of the field operator. The momentum field (11.139) can be written as

$$\Pi(x) = iMA^+\left((2\pi)^{-\frac{3}{2}}e^{-i\mathbf{k}\mathbf{x}+i\omega(\mathbf{k})t}\right) - iMA\left((2\pi)^{-\frac{3}{2}}e^{-i\mathbf{k}\mathbf{x}+i\omega(\mathbf{k})t}\right),$$

where

$$M = \mathcal{F}_3^{-1}\widehat{M}\mathcal{F}_3 = \sqrt{-\Delta + m^2}$$

is the pseudodifferential operator of the one-particle energy. The creation and annihilation contributions to $\Phi(x)$ are therefore

$$\Phi^{(-)}(x) = 2^{-1}\left(\Phi(x) - iM^{-1}\Pi(x)\right),$$
$$\Phi^{(+)}(x) = 2^{-1}\left(\Phi(x) + iM^{-1}\Pi(x)\right). \qquad (11.143)$$

Using the Plancherel theorem and the Fourier representation (11.51) of the fields $\Phi^{(\pm)}(x)$, the Hamilton operator

$$H = d\Gamma(\widehat{M}) \overset{(9.54)}{=} \int \omega(\mathbf{k}) a_\mathbf{k}^+ a_\mathbf{k}^- \left(\frac{d^3 k}{2\omega(\mathbf{k})}\right) = \frac{1}{2} \int a_\mathbf{k}^+ a_\mathbf{k}^- d^3 k$$

is the integral

$$H = 2 \int_{\mathbb{R}^3} \Phi^{(-)}(x) \left(-\Delta + m^2\right) \Phi^{(+)}(x) d^3 x. \tag{11.144}$$

with a fixed value of $x^0 = t \in \mathbb{R}$. After insertion of (11.143) the integrand can be identified with $\frac{1}{2}\left(\Phi(x) - iM^{-1}\Pi(x)\right) M^2 \left(\Phi(x) + iM^{-1}\Pi(x)\right)$. After formal manipulations, which use that the vacuum expectation of H vanishes, we obtain

$$H = \frac{1}{2} \int_{\mathbb{R}^3} \left(:\Pi^2(x): + :\Phi(x)(-\Delta + m^2)\Phi(x):\right) d^3 x$$

$$\equiv \frac{1}{2} \int_{\mathbb{R}^3} \mathcal{H}(x) \, d^3 x, \tag{11.145}$$

where the operator of the energy density $\mathcal{H}(x) = :\Pi^2(x): + :(\partial\Phi(x))^2: + m^2 :\Phi^2(x):$ corresponds to the result (11.138) of the classical scalar field $\phi(x)$ with vanishing external current $j = 0$. The Hamiltonian equations of motion can be obtained from (11.145) as

$$\frac{\partial}{\partial t}\Phi(x) = i[H, \Phi(x)] = \Pi(x),$$

$$\frac{\partial}{\partial t}\Pi(x) = i[H, \Pi(x)] = -(-\Delta + m^2)\Phi(x) \tag{11.146}$$

using the equal time commutation relations (11.141) and 11.142). The equations (11.146) are equivalent to the Klein–Gordon equation (11.64) for the field operator $\Phi(x)$.

To derive these field equations with well-defined operations, we use the "smeared" fields $\Phi(t, \chi) = \int \Phi(t, \mathbf{x})\chi(\mathbf{x})d^3 x$ with test functions $\chi(\mathbf{x}) \in \mathcal{S}(\mathbb{R}^3)$ or $L^2(\mathbb{R}^3)$; cf. our previous remark about singular distributions as arguments. Instead of using $\chi(\mathbf{x})$, we can label the field operators by the Fourier transform of $\chi(\mathbf{x})$

$$f(\mathbf{k}) = (\mathcal{F}_3\chi)(\mathbf{k}) = (2\pi)^{-\frac{3}{2}} \int \chi(\mathbf{x}) e^{-i\mathbf{k}\mathbf{x}} d^3 x, \tag{11.147}$$

which is an element of the one-particle space $\mathcal{H}_m^+$. The function χ is real, if the function (11.147) satisfies the reality condition $f = f^*$, where

$$f(\mathbf{k}) \to f^*(\mathbf{k}) \overset{\triangle}{=} f(-\mathbf{k})^* \tag{11.148}$$

is an antiunitary involution of the space $\mathcal{H}_m^+$. This involution maps the linear subset $L^2(\widehat{\mathbb{R}}^3)$ into itself. The real subspace $\{f \in \mathcal{H}_m^+ : f = f^\star\}$ is denoted as $\mathcal{H}_{m,\mathbb{R}}^+$.

For the definition of the canonical field operators, it is convenient to start from the Segal field operator, cf. (9.36) in Section 9.1.2:

$$\hat{\Phi}(f) \triangleq A^+(f) + A^-(f), \qquad f \in \mathcal{H}_m^+. \tag{11.149}$$

The time evolution of the Segal field operator of a free field is, cf. (9.61),

$$\hat{\Phi}(f) \longrightarrow \Gamma\left(e^{i\widehat{M}t}f\right) \hat{\Phi}(f)\Gamma\left(e^{-i\widehat{M}t}f\right)$$
$$= A^+(e^{i\widehat{M}t}f) + A^-(e^{i\widehat{M}t}f). \tag{11.150}$$

The operator

$$\tilde{\Phi}_t(f) \triangleq A^+\left(e^{i\widehat{M}t}f\right) + A^-\left(e^{i\widehat{M}t}f^\star\right)$$
$$= \Gamma\left(e^{i\widehat{M}t}\right) \tilde{\Phi}_0(f)\Gamma\left(e^{-i\widehat{M}t}\right) \tag{11.151}$$

depends linearly on $f \in \mathcal{H}_m^+$ and coincides with the local field operator (11.69) $\Phi(t, \chi)$ if $f \in L^2(\widehat{\mathbb{R}}^3)$ and $\chi \in L^2(\mathbb{R}^3)$ are related by the Fourier transform (11.147). Formally, the operator $\tilde{\Phi}_t(f)$ can be written as

$$\tilde{\Phi}_t(f) = \int \tilde{\Phi}(t, \mathbf{k})f(-\mathbf{k})d^3k,$$

where $\tilde{\Phi}(t, \mathbf{k})$ is the Fourier transform of the local field operator $\Phi(t, \mathbf{x})$:

$$\tilde{\Phi}(t, \mathbf{k}) = (2\pi)^{-\frac{3}{2}} \int \Phi(t, \mathbf{x})e^{-i\mathbf{k}\mathbf{x}}d^3x. \tag{11.152}$$

If $\chi \in L^2(\mathbb{R}^3)$ is a real test function, the Fourier transform $f = \mathcal{F}_3\,\chi$ satisfies the reality condition $f = f^\star$, and the local field operator $\Phi(t, \chi) = \tilde{\Phi}_t(f)$ coincides with the Segal field operator $\hat{\Phi}(e^{i\widehat{M}t}f)$:

$$\Phi(t, \chi) = \tilde{\Phi}_t(f) = \hat{\Phi}(e^{i\widehat{M}t}f), \qquad f \in L^2(\widehat{\mathbb{R}}^3) \cap \mathcal{H}_{m,\mathbb{R}}^+. \tag{11.153}$$

The time derivative of $\Phi(t, \chi)$ is

$$\frac{\partial}{\partial t}\Phi(t, \chi) = \frac{\partial}{\partial t}\hat{\Phi}(e^{i\widehat{M}t}f) = \hat{\Phi}(i\widehat{M}e^{i\widehat{M}t}f). \tag{11.154}$$

This operator is the canonical momentum field operator $\Pi(t, \chi)$ (or $\tilde{\Pi}_t(f)$ using the Fourier transforms):

$$\Pi(t, \chi) = \tilde{\Pi}_t(f) \triangleq \hat{\Phi}(i\widehat{M}e^{i\widehat{M}t}f), \qquad f \in L^2(\widehat{\mathbb{R}}^3) \cap \mathcal{H}_{m,\mathbb{R}}^+. \tag{11.155}$$

Since $\omega(\mathbf{k})$ satisfies the identity $\omega(\mathbf{k}) = \omega(-\mathbf{k})^*$, the one-particle Hamilton operator $\widehat{M}$ is a real operator, $\left(\widehat{M}f\right)^\star = \widehat{M}f^\star$. Consequently, for $f \in L^2(\mathbb{R}^3) \cap$

$\mathcal{H}_{m,\mathbb{R}}^+$, the function $i\widehat{M}f = -(i\widehat{M}f)^\star$ is an "imaginary" element of $\mathcal{H}_m^+$. If the test function $\chi = \bar{\chi}$ is restricted to $\mathcal{S}(\mathbb{R}^3)$, the functions f and $\widehat{M}^n f$, $n \in \mathbb{N}$, are elements of $\mathcal{S}(\widehat{\mathbb{R}}^3)$, and the operators $\Phi(t, \chi)$ and $\Pi(t, \chi)$ are infinitely often differentiable with respect to t.

For the Segal field operators, the canonical commutation relations are summarized in the single formula (9.38):

$$\left[\hat{\Phi}(f), \hat{\Phi}(g)\right] = 2i \operatorname{Im}\langle f \mid g \rangle I, \ f, g \in \mathcal{H}_m^+. \tag{11.156}$$

Using the identities (11.153) and (11.155), we obtain from (11.156) the equal time commutators of the canonical fields

$$[\Phi(t, \chi_1), \Phi(t, \chi_2)] = [\Pi(t, \chi_1), \Pi(t, \chi_2)] = 0, \tag{11.157}$$

$$[\Phi(t, \chi_1), \Pi(t, \chi_2)] = 2i \langle f_1 \mid \widehat{M}f_2 \rangle$$

$$= i \int \chi_1(\mathbf{x}) \chi_2(\mathbf{x}) d^3 x, \tag{11.158}$$

with real functions $\chi_j \in L^2(\mathbb{R}^3)$, $j = 1, 2$, and $f_j = \mathcal{F}_3 \chi_j = f_j^\star$. These relations are the standard equal time commutators (11.141) and (11.142) of the canonical fields.

The canonical time zero fields

$$\tilde{\Phi}_0(f) = \hat{\Phi}(f), \qquad \tilde{\Pi}_0(f) = \hat{\Phi}(i\widehat{M}f) \tag{11.159}$$

are self-adjoint operators for $f \in L^2(\mathbb{R}^3) \cap \mathcal{H}_{m,\mathbb{R}}^+$. A vector $f \in \mathcal{H}_m^+$ can be decomposed into $f = h + ig$ with $f, g \in \mathcal{H}_{m,\mathbb{R}}^+$. The Segal field operator $\hat{\Phi}(f)$ is the sum of the canonical field operators

$$\hat{\Phi}(f) \overset{(9.39)}{=} \hat{\Phi}(h) + \hat{\Phi}(ig) = \tilde{\Phi}_0(h) + \tilde{\Pi}_0(\widehat{M}^{-1}g). \tag{11.160}$$

The time evolution of the free field exhibits an interesting correspondence to the equations of a classical harmonic oscillator. If $f \in \mathcal{H}_{m,\mathbb{R}}^+$, the vector $e^{i\widehat{M}t}f$ decomposes into $e^{i\widehat{M}t}f = (\cos \widehat{M}t)f + i(\sin \widehat{M}t)f$ with $(\cos \widehat{M}t)f \in \mathcal{H}_{m,\mathbb{R}}^+$ and $(\sin \widehat{M}t)f \in \mathcal{H}_{m,\mathbb{R}}^+$. Hence the fields (11.153) and (11.155) can be decomposed into

$$\tilde{\Phi}_t(f) = \tilde{\Phi}_0((\cos \widehat{M}t)f) + \tilde{\Pi}_0((\widehat{M}^{-1} \sin \widehat{M}t)f),$$

$$\tilde{\Pi}_t(f) = \tilde{\Pi}_0((\cos \widehat{M}t)f) - \tilde{\Phi}_0((\widehat{M} \sin \widehat{M}t)f). \tag{11.161}$$

These identities exactly correspond to the classical equations for position and momentum of a harmonic oscillator. The field equations follow by differentiation as $\frac{\partial}{\partial t} \tilde{\Phi}_t(f) = \tilde{\Pi}(t, f)$ and $\frac{\partial}{\partial t} \tilde{\Pi}(t, f) = -\tilde{\Phi}(t, \widehat{M}^2 f)$, $f \in \mathcal{S}(\widehat{\mathbb{R}}^3) \cap \mathcal{H}_{m,\mathbb{R}}^+$, or

$$\frac{\partial}{\partial t}\Phi(t,\chi) = \Pi(t,\chi),$$

$$\frac{\partial}{\partial t}\Pi(t,\chi) = -\Phi(t,(-\triangle + m^2)\chi). \tag{11.162}$$

Extracting the test function χ, we obtain the field equations (11.146).

The displacement relations (9.87) can be formulated for the canonical field operators (11.153) $\tilde{\Phi}_t(f)$ and (11.155) $\tilde{\Pi}_t(f)$, $f \in L^2(\widehat{\mathbb{R}}^3) \cap \mathcal{H}^+_{m,\mathbb{R}}$. With the Weyl operator $W(g) = \exp\left(i\,\hat{\Phi}(g)\right)$, $g \in \mathcal{H}^+_m$, we obtain

$$W(g)\tilde{\Phi}_t(f)W^{-1}(g) = \tilde{\Phi}_t(f) + i\left\langle e^{-i\widehat{M}t}g - e^{i\widehat{M}t}g^\star \mid f\right\rangle,$$

$$W(g)\tilde{\Pi}_t(f)W^{-1}(g) = \tilde{\Pi}_t(f) - \left\langle e^{-i\widehat{M}t}g + e^{i\widehat{M}t}g^\star \mid \widehat{M}f\right\rangle. \tag{11.163}$$

The identities (11.163) remain valid, if we substitute the vector g by a time-dependent vector function $g(t) \in \mathcal{H}^+_m$, if $t \mapsto g(t)$ is norm continuous.

12

Local Interacting Boson Fields

In the first part of this chapter, general properties of relativistic neutral scalar quantum fields and their n-point functions are discussed. The statements are based on general assumptions about the fields (Wightman axioms). Our presentation is incomplete, and technical details are omitted. For a full account (including charged fields and fields with spin), see, for example, Streater and Wightman (1964) or chapters 8 and 9 in Bogoliubov et al. (1990). This section is completed with a short overview of perturbation theory and constructive approaches.

The final section is devoted to the simplest interaction of a scalar particle: the interaction with a classical current. Though this model is quite simple, it allows an interesting study of causality within local quantum field theory that resolves problems with causality arising in a pure one-particle setting.

12.1 Interacting Neutral Scalar Fields

12.1.1 General Properties

(i) In the Hilbert space $\mathcal{H}$ of the theory, there exists a continuous unitary representation $U(y, \Lambda)$ of the proper orthochronous Poincaré group $\mathcal{P}_+^\uparrow$. There is a normalized vector Ω – called the vacuum vector – such that

$$U(y, \Lambda)\Omega = \Omega. \tag{12.1}$$

This vector is unique up to a constant phase factor. The abelian group of spacetime translations $U(y, \Lambda) = \exp(iP \cdot y)$ is generated by the energy momentum operator $P = (H, \mathbf{P})$. The spectrum of P is a subset of the closed forward cone $\overline{\mathcal{V}^+}$ (*spectral condition*).

(ii) The field $\Phi(x)$, $x \in \mathbb{M}$, is an operator-valued tempered distribution. The operators $\Phi(\varphi) = \int \Phi(x)\varphi(x)d^4x, \varphi \in \mathcal{S}(\mathbb{M})$ have a common dense domain $\mathcal{D} \subset \mathcal{H}$ that includes the vacuum. The field operators $\Phi(\varphi)$ and the unitary operators $U(y, \Lambda)$ map $\mathcal{D}$ into itself. Neutral scalar fields are hermitean:

$$\Phi^*(x) = \Phi(x). \tag{12.2}$$

More precisely, $\Phi^*(\varphi) = \Phi(\varphi^*)$ is true for all test functions $\varphi \in \mathcal{S}(\mathbb{M})$. As domain $\mathcal{D}$, we can choose the linear space spanned by the vacuum and by all vectors $\Phi(\varphi_1) \ldots \Phi(\varphi_n) \Omega$, $\varphi_j \in \mathcal{S}(\mathbb{M}), 1 \leq j \leq n = 1, 2, \ldots$, which are created by the field from the vacuum.

(iii) The transformation law under Poincaré transformations $(y, \Lambda) \in \mathcal{P}_+^\uparrow$ is

$$U(y, \Lambda)\Phi(x)U^*(y, \Lambda) = \Phi(\Lambda x + y). \tag{12.3}$$

(iv) The field is local

$$[\Phi(x), \Phi(y)] = 0 \qquad \text{if } (x - y)^2 < 0. \tag{12.4}$$

We notice a consequence of the spectral condition. The abelian translation group has the spectral representation $U(y, I) = \int e^{ip \cdot y} dE(p)$ with a spectrum concentrated inside $\overline{\mathcal{V}^+}$. The Fourier transform of the matrix elements $\langle F \mid U(y, I) G \rangle$ with vectors $F, G \in \mathcal{D}$ is the measure

$$\int e^{-ip \cdot y} \langle F \mid U(y, I) G \rangle \, d^4 y = (2\pi)^4 \, \langle F \mid dE(p) \, G \rangle$$

on the Minkowski space. This measure vanishes if $p \in \mathbb{M} \backslash \overline{\mathcal{V}^+}$, that is,

$$\int e^{-ip \cdot y} \langle F \mid U(y, I) G \rangle \, d^4 y = 0 \qquad \text{if } p \neq \overline{\mathcal{V}^+}. \tag{12.5}$$

12.1.2 N-Point Functions

Wightman Functions

The Wightman functions are the vacuum expectations of products of fields

$$w_n(x_1, \ldots, x_n) = \langle \Omega \mid \Phi(x_1)\Phi(x_2) \ldots \Phi(x_n) \, \Omega \rangle \, ; \tag{12.6}$$

they are distributions in $\mathcal{S}'(\mathbb{M}^n)$, $n \in \mathbb{N}$. A simple property of these distributions follows from the hermiticity of the field operators

$$w_n(x_1, \ldots, x_n)^* = w_n(x_n, \ldots, x_1). \tag{12.7}$$

The invariance (12.1) of the vacuum and the covariance (12.3) of the fields implies the invariance

$$w_n(\Lambda x_1 + y, \ldots, \Lambda x_n + y) = w_n(x_1, \ldots, x_n) \tag{12.8}$$

of the Wightman functions under inhomogeneous Lorentz transformations $(y, \Lambda) \in \mathcal{P}_+^\uparrow$. The invariance under translations implies that the n-point Wightman function is actually a generalized function of the difference variables $\xi_j = x_j - x_{j+1}, j = 1, \ldots, n - 1$.

Wightman has derived axioms for the n-point functions that are equivalent to the axioms for the fields (Wightman, 1956). Here we discuss only the two simplest cases.

Proposition 12.1.1 *The one-point function Wightman function is a real constant.*

Proof The invariance (12.8) under translations implies that
$w_1(x + y) = w_1(x)$ for $x, y \in \mathbb{M}$. Hence

$$w_1(x) = c_1 \overset{(12.7)}{=} \bar{c}_1 \tag{12.9}$$

is a real constant. $\qquad\square$

In the standard case of field theories the interacting field converges for large times weakly to a free field (Lehmann, Symanzik, Zimmermann – LSZ – asymptotic condition). Then $c_1 = \langle \Omega \mid \Phi(x) \Omega \rangle = 0$ has to vanish.

Proposition 12.1.2 *The two-point Wightman function admits the **Källén–Lehmann representation***

$$w_2(x_1, x_2) = c_2 + \int_0^\infty \Delta_m^+(x_1 - x_2) d\rho(m), \tag{12.10}$$

where $c_2 \geq 0$ is a positive constant, $\Delta_m^+(x)$ is the positive frequency Green's function (11.73) of a free field with mass $m \geq 0$, and $d\rho(m)$ is a positive polynomially bounded measure on $\mathbb{R}_+$.

Proof We indicate the derivation of this representation. The two-point function is a generalized function of the difference variable

$$w_2(x_1, x_2) = w(x_1 - x_2), \tag{12.11}$$

and this generalized function is a Lorentz invariant. From the positivity of the inner product of the Hilbert space follows that w is a function of positive type

$$\int w(x_1 - x_2)\varphi(x_1)^*\varphi(x_2) d^4x_1 d^4x_2 = \langle \Phi(\varphi) \mid \Phi(\varphi) \rangle \geq 0, \qquad \varphi \in \mathcal{S}(\mathbb{M}). \tag{12.12}$$

Then the Bochner–Minlos theorem implies that $w(x)$ is the Fourier transform of a positive tempered measure $\tilde{w}(k)$. The spectral condition (12.5) with the vectors $\Phi(\varphi_{1,2})\Omega$ yields the identity

$$\int e^{ip \cdot y} \langle \Omega \mid \Phi(\varphi_1^*) U(-y, I) \Phi(\varphi_2) \, \Omega \rangle \, d^4 y = 0 \qquad \text{if } p \neq \overline{\mathscr{V}^+},$$

that is, the two-point function (12.11) satisfies $\int e^{ip \cdot y} w(x_1 - x_2 + y) d^4 y = 0$ if $p \notin \overline{\mathscr{V}^+}$ for all $x_1, x_2 \in \mathbb{M}$. Hence the support of the Fourier transform $\tilde{w}(k)$ of $w(x)$ is a subset of $\overline{\mathscr{V}^+}$. The Fourier transform $\tilde{w}(k)$ is therefore a Lorentz invariant positive measure with support in $\mathscr{V}^+$. Such a measure has the representation

$$\tilde{w}(k) = (2\pi)^4 c_2 \, \delta^4(k) + 2\pi \int_0^\infty \Theta(k^0) \delta(k^2 - m^2) d\rho(m), \qquad (12.13)$$

where $c_2 \geq 0$ and $d\rho(m)$ are a polynomially bounded positive measure on $\mathbb{R}_+$. This relation implies the representation (12.10). $\qquad\qquad\qquad\square$

There is another property of the Wightman functions, which follows from the uniqueness of the vacuum: the cluster decomposition property. It says: If y is a spacelike vector, then $w_{m+n}(x_1, \ldots, x_m, x_{m+1} + \lambda y, \ldots, x_{m+n} + \lambda y) \to w_m(x_1, \ldots, x_m) w_n(x_{m+1}, \ldots, x_{m+n})$ is true for $\lambda \to \infty$ in the sense of convergence of distributions in $\mathcal{S}'(\mathbb{M}^{m+n})$. For the one- and two-point function, that means

$$c_2 = c_1^2. \qquad (12.14)$$

For the standard field theories, which satisfy the LSZ asymptotic condition, both constants c_1 and c_2 vanish.

Time-Ordered Product and τ-Functions

The time-ordered product or T-product of the fields $\Phi(x_1), \ldots, \Phi(x_n)$ is formally defined as in Section 11.1.6 by

$$\mathcal{T} \, \Phi(x_1) \ldots \Phi(x_n) = \sum_{\sigma \in \mathfrak{S}_n} \Theta(x_{\sigma(1)}^0 > \cdots > x_{\sigma(n)}^0) \Phi(x_{\sigma(1)}) \ldots \Phi(x_{\sigma(n)}),$$

$$(12.15)$$

where the sum extends over all permutations σ of the index set $\{1, \ldots, n\}$. Mathematically well-defined objects, which correspond to these formal products, might not exist in a Wightman quantum field theory. But it is possible to postulate the existence of such products as operator-valued generalized functions without contradiction to the Wightman axioms. By formal manipulation on expressions of the type (12.15), one obtains a list of properties these objects should have. These properties are then taken as defining assumptions ("axioms"); see, for example, Eckmann and Epstein (1979) or the extensive presentation in chapter 13 of Bogoliubov et al. (1990). We only mention two properties, which we already know from the free field: The T-product of arbitrary Wightman fields should satisfy the symmetry (11.128) and the covariance (11.129) under proper orthochronous Poincaré transformations.

The vacuum expectation of the T-product

$$\tau_n(x_1, \ldots, x_n) = \langle \Omega \mid T \, \Phi(x_1) \ldots \Phi(x_n) \, \Omega \rangle \qquad (12.16)$$

is a generalized function in $\mathcal{S}'(\mathbb{M}^n)$, and it is called n-point τ-function. It is invariant under exchange of its spacetime arguments, and it is invariant under proper orthochronous Poincaré transformations $\tau_n(\Lambda x_1 + y, \ldots, \Lambda x_n + y) = \tau_n(x_1, \ldots, x_n)$, $y \in \mathbb{M}$, $\Lambda \in \mathcal{L}_+^\uparrow$.

The τ-functions have a particular role in quantum field theory. All scattering matrix elements can be calculated from the τ-functions; and the standard perturbation theory has been developed for these functions; see, for example, Bogoliubov and Shirkov (1959, 1983), Itzykson and Zuber (1980), and Weinberg (1995).

Euclidean Green's Functions (Schwinger Functions)

The spectral condition and Lorentz invariance allows to derive analyticity domains for the Wightman functions $w_n(x_1, \ldots, x_n)$ that include the euclidean points $(\xi_1, \ldots, \xi_n) \in \mathbb{M}_\mathbb{C}^n$ with $\xi_j = (i x_j^4, \mathbf{x}_j)$, where $x_j = (x_j^1, x_j^2, x_j^3, x_j^4) \in \mathbb{E} = \mathbb{R}^4$, $j = 1, \ldots, n$, are vectors in the euclidean $\mathbb{R}^4$. Thereby, we have to exclude the "exceptional points" with $x_j = x_k$ for $j \neq k$. The n-point euclidean Green's functions or Schwinger functions are defined on the space

$$\mathbb{E}_{\neq}^n = \big\{ (x_1, \ldots, x_n) \in \mathbb{E}^n, \, x_j \neq x_k \qquad \text{for } j \neq k \big\} \qquad (12.17)$$

as

$$s(x_1, \ldots, x_n) = w(\xi_1, \ldots, \xi_n). \qquad (12.18)$$

The Schwinger functions are generalized functions in $\mathcal{S}'(\mathbb{E}_{\neq}^n)$, but it might not be possible to extend them to generalized functions in $\mathcal{S}'(\mathbb{E}^n)$. They are symmetric functions of their arguments, and they are invariant under euclidean transformations

$$s(Rx_1 + y, \ldots, Rx_n + y) = s(x_1, \ldots, x_n), \qquad R \in O_+(\mathbb{E}), \qquad y \in E.$$

Osterwalder and Schrader (1973, 1975) have derived a list of axioms for the Schwinger functions that follow from the axioms for the Wightman functions, and – assuming these axioms – the Wightman theory can be reconstructed. A comprehensive reference is chapter 9 in Bogoliubov et al. (1990); see also Zinoviev (1995). One can therefore obtain a Wightman field theory from Schwinger functions. But the Osterwalder–Schrader axioms are difficult to verify, and one usually starts from a more restrictive set of basic assumptions. The n-point Schwinger functions are taken as generalized functions in $S'(\mathbb{E}^n)$. Eckmann and Epstein (1979) have shown that this assumption is equivalent to the existence of well-defined T-products for the Wightman field theory. All successful constructions start from an additional very effective assumption: the Schwinger functions are taken as moments of a probability measure. The probabilistic approach is successful so far for field theories on a two- or a three-dimensional euclidean space. For higher spacetime dimensions, perturbation expansions and (formal) identities can be derived. The calculations are to some extend easier than those of the perturbation theory for τ-functions, since the singularity structure of euclidean Green's functions is simpler.

Connected Green's Functions

Let $G_n(\mathbf{X}) = G_n(x_1, \ldots, x_n)$, $n \in \mathbb{N}$, and be a family of Green's functions of quantum field theory (Wightman functions, τ-functions, or Schwinger functions). The symbol $\mathbf{X} = (x_1, \ldots, x_n)$ is a sequence of vectors x_j in $\mathbb{M}$ or $\mathbb{E}$, resp. Then connected Green's functions are defined by $K_1(x) = G_1(x)$ and the recursion

$$K_n(\mathbf{X}) = G_n(\mathbf{X}) - \sum_{\mathbf{X}_1 + \cdots + \mathbf{X}_p = \mathbf{X}} K(\mathbf{X}_1) \ldots K(\mathbf{X}_p), \qquad (12.19)$$

if $n \geq 2$. Here the summation is over all partitions of the sequence $\mathbf{X} = (x_1, \ldots, x_n)$ as detailed in Section 4.1.7. These Green's functions have been introduced by Haag (1958) for the Wightman functions and are called **truncated** Wightman functions. This notation persists in most books of mathematical physics. But standard books of quantum field theory, such as Itzykson and Zuber (1980), use the term **connected** Green's functions. At any rate, they are exactly the cumulant Green's functions already introduced in Section 4.1.7. In the case of free fields, the only nonvanishing connected Green's function is $K_2(x_1, x_2) = G_2(x_1, x_2)$. *Any interaction has to show up in the connected Green's functions!* Knowing the connected Green's functions, one obtains the ordinary ones from (4.16), namely $G(\mathbf{X}) = \sum K(\mathbf{X}_1) \ldots K(\mathbf{X}_m)$.

If the Green's functions are symmetric in their arguments – as in the case of τ-functions or Schwinger functions – the connected Green's functions are exactly the cumulants Green's functions of Section 4.1.4, and the techniques introduced there, in particular the use of generating functionals, apply.

12.1.3 Perturbation Expansions and Constructive Approaches

So far, the only example of a quantum field on four-dimensional Minkowski space, which satisfies all requirements, is the free field. But there is an involved theory of perturbation expansions for the Green's functions of interacting fields. Here we give only a rough outline with restriction to the formal expansion without regularization. A modern approach to renormalization of a self-interacting scalar field theory in four dimensions can be found in Gallavotti (1985). Constructions beyond perturbation theory have been successful for field theories on spacetime dimensions less than 4; see, for example, the extensive monograph Glimm and Jaffe (1987) and the recent review article Summers (2012).

Perturbation Expansions for τ-Functions

The field equation for a self-interacting scalar field $\Phi(x)$ of mass m is

$$\left(\Box + m^2\right) \Phi(x) = j(x). \tag{12.20}$$

Thereby, the current density $j(x)$ will be determined by the interaction Lagrangian $\mathcal{L}_{int}(x)$. For a polynomial interaction

$$\mathcal{L}_{int}(x) = -\lambda :P\left(\Phi(x)\right):$$

with the real polynomial $P(s)$ of the variable $s \in \mathbb{R}$, the relation for the source term is $j(x) = :P'\left(\Phi(x)\right):$ with $P'(s) = \frac{d}{ds}P(s)$. The powers of the field operator $\Phi(x)$ are usually understood as Wick powers $:\Phi^n(x):$. In standard quantum field theory, one assumes that the field $\Phi(x)$ approaches for $t \to \mp\infty$ free incoming/outgoing fields $\Phi^{in/out}(x)$ (LSZ asymptotic condition, Lehmann et al., 1955, 1957). This assumption can actually be derived from the Wightman axioms and the assumption that retarded products of field operators exist (Hepp, 1965; Steinmann, 1968). With the incoming asymptotic condition, the differential equation (12.20) can be written as the integral equation (Yang and Feldman, 1950)

$$\Phi(x) = \Phi^{in}(x) + \int \Delta_m^{ret}(x - y)j(y)d^4y, \tag{12.21}$$

with the retarded propagator (11.103). In this case, a closed formula for the τ-functions of the interacting field can be derived on a formal level: the *Gell-Mann and Low series* (see Gell-Mann and Low, 1951, and, e.g., sec. 6-1-1 of Itzykson and Zuber, 1980),

$$\tau(x_1, \ldots, x_n) = N^{-1} \sum_{p=0}^{\infty} \frac{i^p}{p!} \int d^4y_1 \ldots d^4y_p$$

$$\times \left\langle \Omega \mid \mathcal{T} \, \Phi^{in}(x_1) \ldots \Phi^{in}(x_n) \mathcal{L}_{int}(y_1) \ldots \mathcal{L}_{int}(y_p) \, \Omega \right\rangle, \tag{12.22}$$

where N is the normalization

$$N = \sum_{p=0}^{\infty} \frac{i^p}{p!} \int d^4y_1 \ldots d^4y_p \left\langle \Omega \mid \mathcal{T} \, \mathcal{L}_{int}(y_1) \ldots \mathcal{L}_{int}(y_p) \, \Omega \right\rangle.$$

In these formulas, the Lagrangian density $\mathcal{L}_{int}(y)$ is understood as functional of the incoming free field Φ^{in}. The series in (12.22) can be formally summed to the exponential $\exp\left(i \int d^4y\, \mathcal{L}_{int}(y)\right)$. The vacuum expectations in (12.22) are polynomials of integrals over products of two-point τ-functions of the field Φ^{in}. The euclidean version of this perturbation expansion and its formulation with Feynman diagrams is given in Section 6.2.3. The Gell-Mann and Low series (12.22) corresponds to equation (6.22) in that section.

Studying the combinatorics of all terms in (12.22), one sees that the denominator N cancels those factors in the nominator, which do not include an external x_j argument. The ratio (12.22) can therefore be written as the **linked cluster expansion**:

$$
\tau(x_1,\ldots,x_n) = \sum_{p=0}^{\infty} \frac{i^p}{p!} \int d^4y_1 \ldots d^4y_p
$$
$$
\times \left\langle \Omega \mid \mathcal{T}\,\Phi^{in}(x_1)\ldots\Phi^{in}(x_p)\mathcal{L}_{int}(y_1)\ldots\mathcal{L}_{int}(y_p)\,\Omega \right\rangle_L .
\tag{12.23}
$$

The subscript L means that in the evaluation of $\langle \Omega \mid \ldots \Omega \rangle$, only products of two-point functions are admitted, which are "linked" with one or more external x_j-arguments. The linked cluster expansion is derived in Section 6.2.3 for the euclidean (probabilistic) perturbation theory; see equation (6.25).

The Gell-Mann and Low series can be used to define a generating functional for the n-point functions. With a test function $J(x)$ in $\mathcal{S}(\mathbb{M})$, which can be interpreted as external classical current, we define the functional

$$
Z(J) = \frac{\left\langle \Omega \mid \mathcal{T}\, \exp i \int d^4y \left(\mathcal{L}_{int}(y) + J(y)\Phi^{in}(y)\right) \Omega \right\rangle}{\left\langle \Omega \mid \mathcal{T}\, \exp i \int d^4y \mathcal{L}_{int}(y)\, \Omega \right\rangle}.
\tag{12.24}
$$

Thereby the exponential in (12.24) is understood as formal power series. Then (12.22) is equivalent to

$$
\tau(x_1,\ldots,x_n) = \frac{\delta^n}{\delta J(x_1)\ldots\delta J(x_n)} Z(J)\,\big|_{J=0} .
\tag{12.25}
$$

The τ-functions (12.23) are generalized functions, and they have to be smeared with test functions in the external x_j variables. But also after these integrations with smooth and rapidly decreasing functions, severe problems with the d^4y integrals remain. Most of these integrals do not exist, for two reasons:

(i) Products of two point functions with arguments at the same point y_k occur. These local singularities are not integrable (ultraviolet divergence).

(ii) The integration over the Minkowski space leads to divergences at large distances (infrared divergence).

The main work of perturbation theory is therefore the renormalization of these singularities.

Euclidean Approaches

We give only a few details for a euclidean quantum field theory in $d \geq 2$ dimensions.

The euclidean field is denoted with $\phi(x)$, $x \in \mathbb{E} = \mathbb{R}^d$. The free euclidean field is a Gaussian field with covariance

$$\Delta_m^E(x - y) = (2\pi)^{-d} \int_{\mathbb{E}^d} \left(|k|^2 + m^2 \right)^{-1} \exp\left(-i\,k(x - y) \right) d^d x.$$

Let $d\mu_{\text{free}}(\phi)$ be the corresponding Gaussian measure on $\mathcal{S}'(\mathbb{E})$. Assume a polynomial interaction $\lambda P(\phi)$ where $P(t) = \sum_{j=0}^{2n} a_j t^j$, $a_{2n} > 0$ is a polynomial with lower bound, and $\lambda \in \mathbb{C}$ is a coupling constant. To define an interaction Lagrangian, the polynomial $P(\phi)$ is substituted by $:P(\phi):$ where the double dots indicate normal ordering with respect to the Gaussian measure $d\mu_m$. Moreover, the integration is restricted to a finite volume of $\mathbb{E}$ by introducing a cutoff function $g(x)$ that has compact support and satisfies $0 \leq g \leq 1$. Then a formal advice to calculate the n-point Schwinger functions is

$$s(x_1, \ldots, x_n) = \int \phi(x_1) \ldots \phi(x_n) d\mu_\lambda(\phi) \tag{12.26}$$

with the measure

$$d\mu_\lambda(\phi) = \lim_{g \to 1} N(\lambda; g)^{-1} \exp\left(-\lambda \int_{\mathbb{R}^n} :P\left(\phi(x) \right): g(x) d^2 x \right) d\mu_{\text{free}}(\phi), \tag{12.27}$$

where $N(\lambda; g) = \int \exp\left(-\lambda \int_{\mathbb{R}^n} :P\left(\phi(x) \right): g(x) d^2 x \right) d\mu_{\text{free}}(\phi)$ gives the probability normalization. But for $d \geq 4$, it is not yet clear whether such a probability measure exists. Nevertheless, these relations can be used to derive perturbation expansions and identities within perturbation theory. The perturbation expansion derived from (12.27) yields Schwinger functions, which are the analytic continuations of the τ-functions obtained from (12.23). An important advantage of the euclidean approach is that the methods introduced in Chapter 6 can be used.

In the case of $d = 2$ dimensions, the identities (12.26) and (12.27) have a well-defined mathematical meaning. For sufficiently small coupling $\lambda \geq 0$, the limit in (12.27) is a probability measure on $\mathcal{S}'(\mathbb{E})$, and the integration (12.26) leads to Schwinger functions, which satisfy the OS axioms. A complete presentation of this approach with the reconstruction of an interacting Wightman quantum field theory in two spacetime dimensions is given in Glimm and Jaffe (1987).

In the case of $d = 3$ dimensions, Feldman and Osterwalder (1976) have succeeded to construct the probability measure $d\mu_\lambda(\phi)$ for a ϕ^4 interaction. Also in this case, the moments (12.26) satisfy the OS axioms.

12.2 Interaction with a Classical Current

The simplest interaction of a scalar particle is the interaction with a classical current. Though this model is quite simple, it allows an interesting study of causality within local quantum field theory. The field equation is, as shown in the classical equation (11.136),

$$\left(\Box + m^2\right) \Phi(x) = j(x) I \tag{12.28}$$

with a real function $j(x)$. If $j(x) = j(\mathbf{x})$ does not depend on time, the model is called the van Hove model, which was first investigated by van Hove (1952). Further references are Cook (1961), Schweber (1961: sect. 12a), and Emch (1972: sect. 1.1.e).

For an investigation of the equation (12.28) with a time-dependent current $j(x) = j(t, \mathbf{x})$ we take functions that are square integrable

$$\int_M d^4x \, |j(x)|^2 < \infty, \tag{12.29}$$

and that have a compact support $G \subset M$

$$j(x) = 0 \qquad \text{if } x \notin G \subset [0, T] \times B_R \tag{12.30}$$

with a finite time $T > 0$ and a ball $B_R \subset \mathbb{R}^3$ of finite radius R.[1] The restrictions (12.29) and (12.30) allow the interpretation of $j(t, \mathbf{x})$ as a time-dependent one-particle vector. More precisely, the partial Fourier transform

$$\tilde{j}(t) = \tilde{j}(t, \mathbf{k}) = (2\pi)^{-\frac{3}{2}} \int j(t, \mathbf{x}) \, e^{-i\mathbf{k}\mathbf{x}} d^3k = \tilde{j}(t, -\mathbf{k})^* \tag{12.31}$$

is for almost all t a vector in $\mathcal{H}_m^+$ and the norm $\|\tilde{j}(t)\|$ is an integrable function of time

$$\int_\mathbb{R} \|\tilde{j}(t)\| \, dt < \sqrt{\frac{T}{2m} \int_M d^4x \, |j(x)|^2} < \infty. \tag{12.32}$$

---

[1] The calculations of this section can be performed with currents that have a finite norm (11.34). But the restriction to currents with a compact support is convenient for the discussion of causality.

The external current breaks the translation invariance. Haag's Theorem is therefore not applicable, and the interaction picture exists.

12.2.1 Solution of the Field Equation

Classical Fields

In the case of classical field equations, the inhomogeneous Klein–Gordon (KG) equation $\left(\Box + m^2\right)\phi(x) = -j(x)$ has the solution

$$\phi(x) = \phi_{in}(x) + \vartheta(x) \text{ with}$$

$$\vartheta(x) \triangleq \int_M \Delta_m^{ret}(x - y)j(y)d^4y. \tag{12.33}$$

Thereby $\phi_{in}(x)$ is a solution of the homogeneous KG equation, and the field $\vartheta(x)$ is generated by the current. The convolution kernel $\Delta_m^{ret}(x)$ is the retarded Green's function (11.101) of the Klein–Gordon equation. The support of $\Delta_m^{ret}(x)$ is restricted to the forward cone. The field $\vartheta(x) = \int \Delta_m^{ret}(x - y)j(y)d^4y$ therefore propagates into the causal shadow of the support of the current. If j has the compact support (12.30), the field $\vartheta(x)$ coincides for $x^0 > T$ with

$$\vartheta_\infty(x) \triangleq \int_M \Delta_m(x - y)j(y)d^4y, \tag{12.34}$$

which is a solution of the homogeneous Klein–Gordon equation.

Quantum Fields

The quantum field $\Phi(x)$ is a solution of the operator differential equation

$$\left(\Box + m^2\right)\Phi(x) = j(x)\,I. \tag{12.35}$$

For $t \to -\infty$, the field $\Phi(x)$ has to converge to the incoming free field $\Phi^{in}(x)$ according to the LSZ condition. The solution of (12.35), which satisfies this asymptotic condition, is

$$\Phi(x) = \Phi^{in}(x) + \int_M d^4y\,\Delta_m^{ret}(x - y)j(y)\,I = \Phi^{in}(x) + \vartheta(x)\,I. \tag{12.36}$$

The field $\Phi(x)$ obviously satisfies the commutation relation (11.63), which vanishes for spacelike separation of the arguments; hence (12.36) is a local field operator. Moreover, the field $\Phi(x)$ and the momentum field

$$\Pi(x) = \partial_0\Phi(x) = \Pi^{in}(x) + \partial_0\vartheta(x)\,I \tag{12.37}$$

are a canonical pair, which satisfies the canonical equal time commutation relations (11.141) and (11.142). The operators (12.36) and (12.37) are field operators in the Heisenberg picture. Thereby all operators and vectors are defined in the Fock space of the free field Φ^{in}. The one-particle Hilbert space of this field is $\mathcal{H}_m^{+,in} \simeq L^2(\mathbb{R}^3, \frac{d^3k}{2\omega(\mathbf{k})})$, and the Hilbert space of the full theory is the Fock space $\Gamma_{in}^+ = \Gamma^+(\mathcal{H}^{+,in})$. To obtain the Schrödinger picture, we have to calculate the unitary evolution operators $U(t_2, t_1)$, $t_{1,2} \in \mathbb{R}$, which have the properties

$$U(t_3, t_2)U(t_2, t_1) = U(t_3, t_1),$$

$$U(t_1, t_1) = I, \ U^*(t_2, t_1) = U^{-1}(t_2, t_1) = U(t_1, t_2),$$

$$U^{-1}(t_2, t_1)\Phi(t_1, \mathbf{x})U(t_2, t_1) = \Phi(t_2, \mathbf{x}),$$

$$U^{-1}(t_2, t_1)\Pi(t_1, \mathbf{x})U(t_2, t_1) = \Pi(t_2, \mathbf{x}), \qquad (12.38)$$

for arbitrary values of the time parameters $t_j \in \mathbb{R}$, $j = 1, 2, 3$. For this purpose, we use the smeared field operator $\tilde{\Phi}_t(f) = \int f(-\mathbf{k})\tilde{\Phi}(\mathbf{k}, t)d^3k$ from (11.153) with smooth test functions $f \in \mathcal{S}(\mathbb{R}^3) \cap \mathcal{H}_{m,\mathbb{R}}^+$. The field equation (12.35) is then given by

$$\frac{\partial^2}{\partial t^2}\tilde{\Phi}_t(f) = -\tilde{\Phi}_t\left(\widehat{M}^2 f\right) + 2\langle \tilde{j}(t) \mid \widehat{M}f\rangle.$$

Thereby, the operator $(\widehat{M}f)(\mathbf{k}) = \sqrt{m^2 + \mathbf{k}^2}f(\mathbf{k})$ is the Hamilton operator of a free particle. The inhomogeneous term appearing in (12.37) is

$$2\langle \tilde{j}(t) \mid \widehat{M}f\rangle = \int \overline{\tilde{j}(t, \mathbf{k})}f(\mathbf{k})d^3k = \int j(t, \mathbf{x})\chi(\mathbf{x})d^3x$$

with the test function $\chi = \mathcal{F}_3^{-1}f$. The retarded Green's function (11.101) is the integral kernel of the bounded operator

$$G(t) = \begin{cases} \widehat{M}^{-1}\sin\widehat{M}t & \text{if } t > 0, \\ 0 & \text{if } t < 0. \end{cases}$$

The derivative $\frac{d}{dt}G(t)$ is a bounded operator, and $\mathbb{R} \ni t \mapsto \frac{d}{dt}G(t)f \in \mathcal{H}_m^{+,in}$ is a norm continuous vector function for all $f \in \mathcal{H}_m^{+,in}$. Following (12.36), the solution of (12.37) is

$$\tilde{\Phi}_t(f) = \tilde{\Phi}_t^{in}(f) + 2\int \langle G(t - t')\tilde{j}(t') \mid \widehat{M}f\rangle dt'$$

$$= \tilde{\Phi}_t^{in}(f) + 2\int_{-\infty}^t \langle (\sin\widehat{M}(t - t'))\tilde{j}(t') \mid f\rangle dt'. \qquad (12.39)$$

The estimate (12.32) implies that the integral $\int_{-\infty}^t (\sin\widehat{M}(t - s))\tilde{j}(s)ds$ is a well-defined vector in $\mathcal{H}_m^{+,in}$, which is a differentiable function of t. The canonical momentum field follows by differentiation as

$$\tilde{\Pi}_t(f) = \tilde{\Pi}_t^{in}(f) - 2\int_{-\infty}^t \langle(\cos\widehat{M}(t-t'))\tilde{j}(t') \mid \widehat{M}f\rangle dt'. \qquad (12.40)$$

Comparing (12.39) and (12.40) with (11.163), we observe that the operator $\tilde{\Phi}_t(f)$ can be obtained from $\tilde{\Phi}_t^{in}(f)$ by a Weyl displacement. The vector

$$g(t) \triangleq \int_{-\infty}^t \exp(i\widehat{M}t')\tilde{j}(t')dt' \in \mathcal{H}_m^{+,in} \qquad (12.41)$$

is a continuous function of $t \in \mathbb{R}$, and the following identities are valid:

$$e^{-i\widehat{M}t}g - e^{i\widehat{M}t}g^\star = -2i\int_{-\infty}^t dt' \left(\sin\widehat{M}(t-t')\right)\tilde{j}(t'),$$

$$e^{-i\widehat{M}t}g + e^{i\widehat{M}t}g^\star = 2\int_{-\infty}^t dt' \left(\cos\widehat{M}(t-t')\right)\tilde{j}(t').$$

Hence we have derived the relation

$$W_t \tilde{\Phi}_t^{in}(f)W_t^{-1} = \tilde{\Phi}_t(f) \qquad (12.42)$$

with the Weyl operators

$$W_t = W(g(t)), \qquad t \in \mathbb{R}. \qquad (12.43)$$

The incoming field has the time evolution $\tilde{\Phi}_t^{in}(f) = U_{in}^{-1}(t)\tilde{\Phi}_0^{in}(f)U_{in}(t)$ with $U_{in}(t) = \Gamma\left(e^{-i\widehat{M}t}\right)$. The evolution operators (12.38) are therefore given by the unitary operators

$$U(t_2, t_1) = W_{t_1} U_{in}(t_2 - t_1)W_{t_2}^{-1}. \qquad (12.44)$$

12.2.2 S-Matrix

The assumptions (12.29) and (12.30) imply the identities

$$\Phi(x) = \begin{cases} \Phi^{in}(x) & \text{if } x^0 \le 0, \\ \Phi^{out}(x) & \text{if } x^0 \ge T \end{cases} \qquad (12.45)$$

with the incoming free field $\Phi^{in}(x)$ and the outgoing free field

$$\Phi^{out}(x) = \Phi^{in}(x) + \vartheta_\infty(x)I. \qquad (12.46)$$

The vector (12.41) has the properties $g(t) = 0$ if $t \le 0$, and $g(t) = g_\infty$ if $t \ge T$ with

$$g_\infty = \int_{\mathbb{R}} \exp(i\widehat{M}t')\tilde{j}(t')dt'. \qquad (12.47)$$

The vector (12.47) coincides with the one-particle vector derived from $j(x)$ by the mapping J of (11.31). The Weyl operator $W(g_\infty)$ is therefore the operator

$$W(g_\infty) = \exp\left(i\,\Phi^{in}(j)\right). \qquad (12.48)$$

The transition $\Phi^{in} \to \Phi^{out}$ can be achieved by the transform $\Phi^{out}(x) = \exp\left(i\,\Phi^{in}(j)\right)\Phi^{in}(x)\exp\left(-i\,\Phi^{in}(j)\right)$ of the in-field. In scattering theory, the unitary operator, which connects the infield with the outfield is called the S-matrix. In this notation, we write

$$\Phi^{out}(x) = S^{-1}\,\Phi^{in}(x)\,S, \qquad \text{with} \qquad S = \exp\left(-i\,\Phi^{in}(j)\right). \quad (12.49)$$

This result also easily follows from the displacement relation (11.68) of the local field and the identities (12.32) and (12.46).

12.2.3 Causality in the Schrödinger Picture

We now discuss the Schrödinger picture under the restrictions (12.29) and (12.30) on the source. We assume that the initial state at time $t \leq 0$ is the vacuum state $\Omega \in \Gamma^+_{in}$. The time evolution is given by the unitary operators (12.44), and the observables are built up from the field $\Phi(t = 0, \mathbf{x})$ and its derivatives at time $t = 0$. This state remains unchanged until the current is switched on, $U(t, 0)\Omega = U_{in}(t)\Omega = \Omega$ if $t \leq 0$. For $t \geq 0$, the classical current generates the coherent state

$$\Psi(t) = U(t, 0)\Omega = U_{in}(t)W\left(-g(t)\right)\Omega = W\left(-e^{-i\widehat{M}t}g(t)\right)\Omega.$$

If $t \geq T$, this vector becomes $\Psi(t) = U_{in}(t)S\Omega$ with the S-matrix (12.49). The coherent state $S\Omega = \exp\left(-i\,\Phi^{in}(j)\right)\Omega$ is the normalized exponential state (9.67) $N(h) = \exp\left(h - \frac{1}{2}\|h\|^2\right)$ of the one-particle vector $h = -i\,\Phi^{in}(j)\Omega = -ig_\infty \in \mathcal{H}^{+,in}_m$. This state propagates with the free evolution

$$\Psi(t) = U_{in}(t)N(-ig_\infty) = N(-ie^{-i\widehat{M}t}g_\infty), \qquad t \geq T. \quad (12.50)$$

The particle content of a coherent state is given in Section 9.2.1.. Here we discuss the causality of the propagation. In the Heisenberg picture, the contribution of the current to the field (12.36) obviously propagates into the causal shadow of the current. But for the coherent state (12.50), a causal propagation is not obvious, since the one-particle vector $e^{-i\widehat{M}t}g_\infty \in \mathcal{H}^{+,in}_m$ has no reasonable localization in the spacetime Minkowski space as discussed in Section 11.1.3. To resolve this apparent contradiction, we have to specify the observables, which are related to a measurement within a finite domain $\mathcal{O} \in \mathbb{R}^3$. In quantum field theory, such local observables are integrals of the type $\int_{\mathcal{O}} d^3x\,\sigma(\mathbf{x})A(\mathbf{x})$, where $\sigma(\mathbf{x}) \geq 0$ is a weight function (efficiency) and $A(\mathbf{x})$ is a Wick polynomial constructed from the local field $\Phi(t = 0, \mathbf{x})$ and its derivatives at $t = 0$. Take, for example, the field strength density with operator $\Phi(0, \mathbf{x}) = \Phi^{in}(0, \mathbf{x})$. The expectation of the field strength density in the state $\Psi(t)$ is

$$
\begin{aligned}
\langle \Psi(t) \mid \Phi(0, \mathbf{x})\Psi(t) \rangle &= \langle \Omega \mid U^*(t, 0)\Phi^{in}(0, \mathbf{x})U(t, 0)\Omega \rangle \\
&= \langle \Omega \mid \Phi(t, \mathbf{x})\Omega \rangle \overset{(12.36)}{=} \vartheta(t, \mathbf{x}).
\end{aligned}
\tag{12.51}
$$

The expectation of more general local observables – such as the energy density – can be calculated by the same method, and yields again the result obtained from the classical field $\vartheta(x)$. Hence effects of the created particles can only be observed within the causal shadow of the current. The noncausal parts of the one-particle wave function are not observable in the local quantum field theory.

13

Quantum Stochastic Calculus

In this chapter, we describe the quantum stochastic calculus developed by Hudson and Parthasarathy (1984). This is a branch of mathematics involving integration with respect to Fock space processes of creation, annihilation, and number, which generalizes the usual Itō theory. The motivations came from the desire to have a noncommutative theory of probability, and unitary dilations of quantum dynamical semigroups. However, the theory effectively starts with observations that Gaussian and Poissonian processes can naturally be constructed in a suitable Fock space, and that the Itō correction ought to be a manifestation of Wick ordering. As the title of the early paper, Hudson and Streater (1981), suggests, the Itō calculus should be just the ordinary calculus with Wick ordering. (See the account by Hudson, 2012, on the development of the theory.)

We will initiate this with a description of the *Maassen kernel calculus*, Maassen (1987), and its extension by Meyer (1993).

13.1 Operators on Guichardet Fock Space

We begin by displaying the creation, annihilation, and number operators in a more explicit form when the Fock space is a Guichardet Fock space of the form $L^2 (\mathrm{Power}(\mathfrak{X}), dX)$, which we have previously encountered. For ease of notation, we shall use the shorthand for this chapter

$$\mathfrak{P} \triangleq \mathrm{Power}(\mathfrak{X}).$$

Lemma 13.1.1 *The creation and annihilation operators on Guichardet Fock space may be defined by their actions on suitable* $\Psi \in L^2 (\mathfrak{P}, dX)$:

$$(A\,(g)\,\Psi)\,[X] \equiv \int_{\mathfrak{x}} dx\, g^*\,(x)\,\Psi\,[X+x]$$

$$\left(A^*\,(g)\,\Psi\right)[X] \equiv \sum_{x \in X} g\,(x)\,\Psi\,[X-x]. \tag{13.1}$$

The number operator is $(N\Psi)\,[X] \equiv \#X\,\Psi\,[X]$.

Proof Since $\exp\,(f)\,[X+x] = f\,(x)\exp\,(f)\,[X]$, we immediately obtain that exponential vectors are the eigenvectors for annihilation:

$$A\,(g)\,\exp(f) = \left(\int g^*(x)f(x)\,dx\right)\exp\,(f).$$

We see that $A^*\,(g)$ and $A\,(g)$ are indeed adjoint, since by (13.4),

$$\langle\Phi|A\,(g)\,\Psi\rangle = \int dX \int dx\, \Phi^*[X]g^*(x)\,\Psi[X+x]$$

$$= \int dZ \sum_{z \in Z} \Phi^*\,[Z-z]\,g^*\,(z)\,\Psi\,[Z]$$

$$= \langle A^*\,(g)\,\Phi|\Psi\rangle.$$

$\square$

Likewise, the commutation relations are apparent from looking at

$$\left(A\,(g)\,A^*\,(h)\,\Psi\right)[X] = \int dx\, g^*\,(x)\sum_{y \in X+x} h\,(y)\,\Psi\left[X+x-y\right].$$

The $y = x$ term in the sum yields $\int dx\, g^*\,(x)\,h\,(x)\,\Psi\,[X]$ while the remainder is readily seen to give $(A^*\,(h)\,A\,(g)\,\Psi)\,[X]$. Therefore, we obtain the CCR

$$\left[A\,(g)\,,A^*\,(h)\right] = \int g^*\,(x)\,h\,(x)\,dx. \tag{13.2}$$

13.1.1 Operator Densities

We may tentatively introduce an annihilator density $\mathfrak{a}_x$ so that formally

$$(\mathfrak{a}_x\Psi)\,[X] = \Psi\,[X+x],$$

and so that the annihilator may be written as

$$A\,(g) = \int_{\mathfrak{x}} g^*\,(x)\,\mathfrak{a}_x\,dx.$$

Likewise, we should define the creator density as

$$\left(\mathfrak{a}_x^*\Psi\right)[X] = \sum_{y \in X} \delta\,(x-y)\,\Psi\left[X-y\right]$$

so that

$$A^*(g) = \int_{\mathfrak{X}} g(x)\,\mathfrak{a}_x^*\,dx.$$

The annihilator density is, at least, defined almost everywhere; however, the creator density is clearly singular. We also have formally that the number operator is $N = \int_{\mathfrak{X}} \mathfrak{a}_x^* \mathfrak{a}_x\,dx$.

More generally, for a finite set X, we set

$$\mathfrak{a}_X = \prod_{x \in X} \mathfrak{a}_x \text{ and } \mathfrak{a}_X^* = \prod_{x \in X} \mathfrak{a}_x^*.$$

We obtain the elegant rule

$$(\mathfrak{a}_X \Psi)[Y] = \Psi[X + Y].$$

However, as the reader will quickly surmise, the creation analogue is a more ponderous expression involving multiple delta-functions.

Technically speaking, these are not operators on the Fock space; however, they are convenient objects to consider, and their formal manipulation reproduces the correct formulas for bona fide operators defined in the next subsection, which would otherwise be pure drudgery. The densities formally obey the singular canonical commutation relations

$$\left[\mathfrak{a}_x, \mathfrak{a}_y^*\right] = \delta(x - y).$$

Wick Ordering

It is convenient to give a rule for putting expressions like $\mathfrak{a}_X \mathfrak{a}_Y^*$ to Wick order. We first of all introduce the notion of a δ-function between sets: if $X = \{x_1, \ldots, x_p\}$ and $Y = \{y_1, \ldots, y_q\}$, then we set

$$\delta(X, Y) = \delta_{p,q} \sum_{\sigma \in \mathfrak{S}_p} \delta\left(x_1 - y_{\sigma(1)}\right) \ldots \delta\left(x_p - y_{\sigma(p)}\right).$$

Lemma 13.1.2 (Wick Ordering)

$$\mathfrak{a}_X \mathfrak{a}_Y^* = \sum_{\substack{X_1 + X_2 = X \\ Y_1 + Y_2 = Y}} \delta(X_2, Y_2)\,\mathfrak{a}_{Y_1}^* \mathfrak{a}_{X_1}.$$

Proof This is the result of repeated use of the commutation relations. Let us write $\mathfrak{a}_X \mathfrak{a}_Y^* = \mathfrak{a}_{x_p} \ldots \mathfrak{a}_{x_1} \mathfrak{a}_{y_1}^* \ldots \mathfrak{a}_{y_q}^*$, where $X = \{x_1, \ldots, x_p\}$ and $Y = \{y_1, \ldots, y_q\}$, and proceed to move all the creator symbols to the left, starting with $\mathfrak{a}_{y_1}^*$, then $\mathfrak{a}_{y_2}^*$, and so on, using the relations $\mathfrak{a}_x \mathfrak{a}_y^* = \mathfrak{a}_y^* \mathfrak{a}_x + \delta(x - y)$. The final result will be

$$a_X a_Y^* = \sum_{n=0}^{\max\{p,q\}} \sum_{\substack{1 \le j(1) < \cdots < j(n) \le p \\ 1 \le k(1) < \cdots < k(n) \le q}} \sum_{\sigma \in \mathfrak{S}_n}$$

$$\times \delta\left(x_{j(\sigma(1))} - y_{k(1)}\right) \ldots \delta\left(x_{j(\sigma(n))} - y_{k(n)}\right)$$

$$\times a_{Y/\{y_{k(1)},\ldots,y_{k(n)}\}}^* a_{X/\{x_{j(1)},\ldots,x_{j(n)}\}}.$$

The final expression has the desired form with $X_2 = \{x_{j(1)}, \ldots, x_{j(n)}\}$ and $Y_2 = \{y_{k(1)}, \ldots, y_{k(n)}\}$. $\qquad\square$

13.1.2 Emission-Absorption Kernels

We say that an operator K is represented by an *emission-absorption kernel* if there exists a measurable function

$$\mathcal{K} \colon \mathfrak{P} \times \mathfrak{P} \mapsto \mathbb{C} \colon (X, Y) \mapsto \mathcal{K}^{X,Y},$$

called the kernel function of the operator, such that

$$K = \int a_X^* \mathcal{K}^{X,Y} a_Y \, dX dY.$$

For a given operator, the kernel function may often be deduced by taking its matrix elements between exponential states to get

$$\frac{\langle \exp(f) | K \exp(g) \rangle}{\langle \exp(f) | \exp(g) \rangle} = \int f_X^* \mathcal{K}^{X,Y} g_Y \, dX dY.$$

The action of K on general Fock space vectors is readily deduced to be

$$(K\Psi)[X] = \int dY \sum_{X = X_1 + X_2} \mathcal{K}^{X_1, Y} \Psi[X_2 + Y].$$

Note that the identity operator on the Fock space has kernel

$$\mathcal{I}^{X,Y} = \delta_\emptyset(X) \delta_\emptyset(Y).$$

Weyl Operator Kernels

Proposition 13.1.3 *The Weyl operator $D(h)$ admits a kernel representation with the emission-absorption kernel function*

$$\mathcal{D}(h)^{X,Y} = e^{-\|h\|^2/2} h_X \left(-h^*\right)_Y.$$

Proof   Recall the Wick-ordered form $D(h) = e^{-\frac{1}{2}\|h\|^2} e^{A^*(h)} e^{-A(h)}$ of the Weyl operators. We have that

$$
\begin{aligned}
e^{A(g)} &= \sum_{n \geq 0} \frac{1}{n!} \int g^*(x_1) \ldots g^*(x_n) \, \mathfrak{a}_{x_1} \ldots \mathfrak{a}_{x_n} dx_1 \ldots dx_n \\
&\equiv \int (g^*)_X \, \mathfrak{a}_X \, dX,
\end{aligned}
$$

and similarly $e^{A^*(g)} \equiv \int \mathfrak{a}_X^* g_X \, dX$. This leads to the Wick-ordered form

$$
D(h) \equiv e^{-\frac{1}{2}\|h\|^2} \int \mathfrak{a}_X^* h_X \left(-h^*\right)_Y \mathfrak{a}_Y \, dXdY
$$

from which we read off the kernel. $\square$

It is worth writing the action of the Weyl displacement operator on a general Fock vector just to see how compact our shorthand notation is:

$$
\begin{aligned}
(D(h) \, \Psi)[X] = e^{-\|h\|^2/2} \int dY \prod_{y \in Y} \left(-h^*(y)\right) \\
\times \sum_{X_1 + X_2 = X} \prod_{x \in X_1} h(x) \, \Psi[X_2 + Y].
\end{aligned}
$$

Wick Ordering of Products

Lemma 13.1.4 *Let K and L be operators having emission-absorption kernels $\mathcal{K}^{X,Y}$ and $\mathcal{L}^{X,Y}$ respectively. Then KL also admits an emission-absorption kernel representation with kernel $\mathcal{K} \star \mathcal{L}$:*

$$
(\mathcal{K} \star \mathcal{L})^{X,Y} = \sum_{\substack{X_1 + X_2 = X \\ Y_1 + Y_2 = Y}} \int dZ \, \mathcal{K}^{X_1, Y_1 + Z} \, \mathcal{L}^{X_2 + Z, Y_2}.
$$

Proof   We have

$$
KL = \int dXdYdUdV \, \mathfrak{a}_X^* \mathcal{K}^{X,Y} \mathfrak{a}_Y \, \mathfrak{a}_U^* \mathcal{L}^{U,V} \mathfrak{a}_V,
$$

and using the Wick-ordering lemma, we write this as

$$
\int dXdYdUdV \sum_{\substack{Y_1 + Y_2 = Y \\ U_1 + U_2 = U}} \mathfrak{a}_{X+U_1}^* \mathcal{K}^{X,Y} \delta(U_2, Y_2) \mathcal{L}^{U,V} \mathfrak{a}_{V+Y_1}.
$$

Now set $M = Y_2$. We may use the $\pitchfork$ formula to write the integral as

$$\int dXdY_1 dU_1 dVdM\; \mathfrak{a}^*_{X+U_1} \mathcal{K}^{X,Y_1+M}\, \mathcal{L}^{U_1+M,V} \mathfrak{a}_{V+Y_1}$$

$$= \int dM \sum_{\substack{\tilde{X}_1+\tilde{X}_2=\tilde{X} \\ \tilde{Y}_1+\tilde{Y}_2=\tilde{Y}}} , \mathfrak{a}^*_{\tilde{X}} \mathcal{K}^{\tilde{X}_1,\tilde{Y}_1+M}\, \mathcal{L}^{\tilde{X}_2+M,\tilde{Y}_2} \mathfrak{a}_{\tilde{Y}},$$

where $\tilde{X}_1 = X$, $\tilde{X}_2 = U_1$, $\tilde{Y}_1 = Y_1$, $\tilde{Y}_2 = V$. $\qquad\qquad\square$

13.1.3 Kernels for Second Quantized Operators

Let U be a unitary operator on the one-particle space $L^2(\mathfrak{X})$ with kernel $U(x,y)$. That is, $(U\phi)(x) = \int U(x,y)\,\phi(y)\,dy$. We would like to display the second quantization $\Gamma(U)$ in Wick-ordered form. To this end, we have that

$$\frac{\langle \exp(f)\,|\,\Gamma(U)\exp(g)\rangle}{\langle \exp(f)\,|\,\exp(g)\rangle} = \exp\{\langle f|Ug\rangle - \langle f|g\rangle\}$$

$$\equiv \exp\left\{\iint f^*(x)\,U(x,y)\,g(y)\,dxdy - \int f^*(z)\,g(z)\,dz\right\}.$$

While we may expand the exponential to obtain multiple integrals (analogous to the Hermite-type expansion of an exponential of a quadratic), it is clear that we are not going to get a simple emission-absorption kernel representation as a rule.

As a special case, we assume that the unitary U is diagonal in the position representation, so that we may write

$$U(x,y) = e^{i\theta(x)}\,\delta(x,y)$$

for some real-valued function θ. In this case, we have

$$\frac{\langle \exp(f)\,|\,\Gamma(U)\exp(g)\rangle}{\langle \exp(f)\,|\,\exp(g)\rangle} = \exp\int f^*(x)\left(e^{i\theta(x)}-1\right)g(x)\,dx$$

$$\equiv \int (f^*)_X\left(e^{i\theta}-1\right)_X g_X\,dX,$$

where $\left(e^{i\theta}-1\right)_X$ of course means $\prod_{x\in X}\left(e^{i\theta(x)}-1\right)$. We may summarize as follows.

Proposition 13.1.5 *Suppose that $U \equiv e^{i\Theta}$ diagonal in the position representation so that $U(x,y) = e^{i\theta(x)}\,\delta(x,y)$, then the second quantization operator $\Gamma(U)$ takes the form*

$$\Gamma\left(e^{i\Theta}\right) = \int \mathfrak{a}^*_X\left(e^{i\theta}-1\right)_X \mathfrak{a}_X\,dX. \tag{13.3}$$

Taking θ to be a constant value u leads to the very special case

$$e^{iuN} \equiv \int \mathfrak{a}_X^* \left(e^{iu} - 1 \right)^{\#X} \mathfrak{a}_X \, dX.$$

13.1.4 Emission-Scattering-Absorption Kernels

These last examples have introduced a purely scattering form when we display in Wick integral form. We can combine all these forms by saying that an operator K is represented by an *emission–scattering-absorption kernel* if there exists a measurable function

$$\mathcal{K}: \mathfrak{P} \times \mathfrak{P} \times \mathfrak{P} \mapsto \mathbb{C}$$
$$: (X, Y, Z) \mapsto \mathcal{K}^{X,Y,Z}$$

such that

$$K = \int dX dY dZ \, \mathfrak{a}_{X+Y}^* \mathcal{K}^{X,Y,Z} \mathfrak{a}_{Y+Z}.$$

The indices X, Z indicate points where we have creation (emission) and annihilation (absorption), respectively, while the index Y are scattering points. The action on vectors is given by the formula

$$(K\Psi)[X] = \int dV \sum_{X=X_1+X_2+X_3} \mathcal{K}^{X_1,X_2,V} \, \Psi[X_2 + X_3 + V].$$

Lemma 13.1.6 *Let K and L be operators with kernels $\mathcal{K}^{X,Y,Z}$ and $\mathcal{L}^{X,Y,Z}$ respectively. Then KL is again represented by the emission-scattering-absorption kernel with*

$$(\mathcal{K} \star \mathcal{L})^{X,Y,Z} = \int dW \sum_{\substack{X_1+X_2+X_3=X \\ Y_1+Y_2+Y_3=Y \\ Z_1+Z_2+Z_3=Z}} \times \mathcal{K}^{X_1,X_2+Y_1+Y_2,Z_1+Z_2+W} \mathcal{L}^{X_2+X_3+W,Y_2+Y_3+Z_2,Z_3}.$$

A vector $\Psi \in L^2(\mathfrak{P}, dX)$ is called a **Maassen test vector** if it is localized in some compact region of $\mathfrak{X}$ and if there exists positive constants c, M such that

$$|\Psi[X]| \leq c M^{\#X}.$$

Furthermore, an operator K with a kernel representation is said to be Maassen regular if its kernel is localized in some common compact set in each of its arguments and if there exist positive constants c, M such that

$$\left| \mathcal{K}^{X,Y,Z} \right| \leq c M^{\#\{X+Y+Z\}}.$$

The Maassen test vectors include exponential vectors of the form $\exp(f)$, where f is bounded and has compact support, and they are clearly dense in the Fock space.

Theorem 13.1.7 *The Maassen test vectors are a stable domain of vectors under the action of operators with Maassen regular kernels.*

Proof Let R be a compact region localizing a Maassen test vector Ψ as well as the kernel of a Maassen regular operator K. Denoting the constants for Ψ as c_Ψ, M_Ψ and the constants for the kernel as c_K, M_K, we see that

$$|K\Psi[X]| \leq \int dW \sum_{X=X_1+X_2+X_3} c_\Psi M_\Psi^{\#\{X_1+X_2+W\}} c_K M_K^{\#\{X_2+X_3+W\}}$$

$$\leq \lambda c_\Psi c_K M^{\#X},$$

where $M = \max\{M_\Psi, M_K, M_\Psi M_K\}$ and $\lambda = \int dW M^{|W|} = e^{M \operatorname{vol}(R)}$. Therefore, $K\Psi$ is also a Maassen test vector. $\qquad\square$

13.1.5 Gradient and Cogradient

The following section borrows from the analysis of Lindsay and Maassen (1988). The Hilbert spaces $L^2(\mathfrak{X} \times \mathfrak{P})$ and $L^2(\mathfrak{X}) \otimes L^2(\mathfrak{P})$ are naturally isomorphic. Here we have measurable functions $F[\cdot,\cdot]$ on $\mathfrak{X} \times \mathfrak{P}$ with

$$\int_\mathfrak{X} dx \int_\mathfrak{P} dX \, |F[x,X]|^2 < \infty,$$

and the factoring functions are $f \otimes \xi[x,X] = f(x)\xi[X]$.

As we have mentioned, the density $\mathfrak{a}_x$ does not properly define an operator on the Fock space; however, it is reasonable to ask when the function: $(x,X) \mapsto (\mathfrak{a}_x \Psi)[X]$ will define an element of $L^2(\mathfrak{X} \times \mathfrak{P})$.

The **gradient operator** grad is defined to be the closure of linear operator

$$\operatorname{grad}: L^2(\mathfrak{P}) \mapsto L^2(\mathfrak{X} \times \mathfrak{P})$$

given by

$$\langle \operatorname{grad}\Psi | f \otimes \xi \rangle_{L^2(\mathfrak{X} \times \mathfrak{P})} \triangleq \langle A(f)\Psi | \xi \rangle,$$

and having domain $\{\Psi: \int_\mathfrak{X} \|\mathfrak{a}_x \Psi\|^2 \, dx < \infty\}$. Its adjoint is then

$$\operatorname{div}: L^2(\mathfrak{X} \times \mathfrak{P}) \mapsto L^2(\mathfrak{P}),$$

and this is called the **cogradient**, or divergence. We have the action

$$\operatorname{div}(f \otimes \xi) \equiv A^*(f)\xi.$$

The essential feature in the definition is that we have

$$(\mathrm{grad}\,\Psi)\,[x, X] \equiv (\mathfrak{a}_x \Psi)\,[X] = \Psi\,[X + x]$$

and

$$\mathrm{div} F\,[X] \equiv \sum_{x \in X} F\,[x, X - x]$$

for $\Psi \in L^2(\mathfrak{P})$ and $F \in L^2(\mathfrak{X} \times \mathfrak{P})$. It is easily verified that the number operator is then given by (a Laplacian!)

$$N = \mathrm{div}\,\mathrm{grad}.$$

Lemma 13.1.8 *For suitably integrable functions F,*

$$\int_{\mathfrak{P}} dX \int_{\mathfrak{X}} dx\, F\,[X, x] = \int_{\mathfrak{P}} dZ \sum_{z \in Z} F\,[Z - z, z]. \qquad (13.4)$$

Proof Let $\chi_1\,[\cdot]$ be the indicator function for the singletons $\mathfrak{P}_1$ on the power set $\mathfrak{P}$, then the left-hand side may be written as

$$\int_{\mathfrak{P}} dX \int_{\mathfrak{P}} dY\, F\,[X, Y]\,\chi_1\,[Y] = \int_{\mathfrak{P}} dZ \sum_{X + Y = Z} F\,[X, Y]\,\chi_1\,[Y]$$

and setting $Y = \{z\}$, $X = Z/\{z\}$ immediately gives the result. $\qquad \square$

Proposition 13.1.9 *The domain of* grad *coincides with that of* $\sqrt{N}$.

Proof We consider the operators $V_p \colon L^2(\mathfrak{P}) \mapsto L^2(\mathfrak{X} \times \mathfrak{P})$, for p real, defined by

$$(V_p \Psi)\,[x, X] \triangleq (\#X + 1)^p\,\Psi\,[X + x].$$

We then have

$$\begin{aligned}
\left\| V_p \Psi \right\|_{L^2(\mathfrak{X} \times \mathfrak{P})}^2 &= \int_{\mathfrak{X}} dx \int_{\mathfrak{P}} dX\, (\#X + 1)^{2p}\,|\Psi\,[X + x]|^2 \\
&= \int_{\mathfrak{P}} dZ\, (\#Z)^{2p+1}\,|\Psi\,[Z]|^2 \\
&= \left\| N^{2p+1} \Psi \right\|^2,
\end{aligned}$$

where we used the previous lemma. We have for instance $\mathrm{grad} \equiv V_0$. The case $p = -\frac{1}{2}$ is interesting, and here we must interpret N^0 as the orthogonal projection $P_{\geq 1}$ onto $L^2\left(\bigoplus_{n \geq 1} \mathfrak{P}_n\right)$, i.e. $0^0 = 0$, so $V_{-\frac{1}{2}}$ is a partial isometry. From the observation that $N^q V_p = V_{p+q}$, we therefore arrive at the polar decomposition for the gradient: $\mathrm{grad} = V_{-\frac{1}{2}} \sqrt{N}$. Thus $\mathrm{dom}\,\mathrm{grad} = \mathrm{dom}\,\sqrt{N}$. $\qquad \square$

Define the operators $\tilde{N}$ and $P_{\text{sym.}}$ on $L^2 (\mathfrak{X} \times \mathfrak{P})$ by

$$\left(\tilde{N}F\right) [x, X] \triangleq (\#X + 1)\, F\, [x, X]$$

and

$$P_{\text{sym.}} F\, [X] \triangleq \frac{1}{\#X} \sum_{y \in X + \{x\}} F\left[y, X + x - y\right].$$

Lemma 13.1.10 *The domain of* div *is* $\operatorname{dom}\sqrt{\tilde{N}} P_{\text{sym.}}$.

Proof For $\Psi \in L^2 (\mathfrak{P})$ and $F \in L^2 (\mathfrak{X} \times \mathfrak{P})$, we have

$$\langle F | V_p \Psi \rangle = \int_{\mathfrak{X}} dx \int_{\mathfrak{P}} dX\, F^* [x, X]\, (\#X + 1)^p\, \Psi\, [X + x]$$

$$= \int_{\oplus_{n \geq 1} \mathfrak{P}_n} dZ\, (\#Z)^p \sum_{z \in Z} F^* [z, Z - z]\, \Psi\, [Z]$$

and so

$$\left(V_p^* F\right) (Z) \equiv \chi_{\geq 1} (Z)\, (\#Z)^p \sum_{z \in Z} F [z, Z - z]$$

with $\chi_{\geq 1} (Z) = 1$ if $\#Z \geq 1$, and $= 0$ if $Z = \emptyset$. We therefore see that div $= \sqrt{N} V^*_{-\frac{1}{2}}$.

Now

$$\int dZ \left| \sum_{z \in Z} F [z, Z - z] \right|^2 = \int dZ\, (\#Z) \sum_{z \in Z} |F [z, Z - z]|^2$$

$$= \int_{\mathfrak{X}} dx \int_{\mathfrak{P}} dX\, (N + 1) \left| F_{\text{sym.}} [X + x] \right|^2,$$

giving the result. $\qquad\qquad\square$

Lemma 13.1.11 *For* Ψ *measurable on* $\mathfrak{P}$, *let* $\Psi^{(r)}$ *be the measurable function on* $\mathfrak{P}^r$ *given by* $\Psi^{(r)} (X_1, \ldots, X_r) = \Psi\, [X_1 + \cdots + X_r]$. *Then* $\Psi^{(r)} \in L^2 (\mathfrak{P}^r)$ *if and only if* $\Psi \in \operatorname{dom}\sqrt{r}^N$.

Proof To see this, take $c > 0$, then

$$\int_{\mathfrak{P}^r} dX_1 \ldots dX_r\, c^{\#X_1 + \cdots + \#X_r} \left| \Psi^{(r)} (X_1, \ldots, X_r) \right|^2$$

$$= \int_{\mathfrak{P}} dX\, (rc)^{\#X} |\Psi\, [X]|^2 = \left\| \sqrt{rc}^N \Psi \right\|^2.$$

Setting $c = 1$, we get $\left\| \Psi^{(r)} \right\| = \left\| \sqrt{r}^N \Psi \right\|$. $\qquad\qquad\square$

13.2 Wick Integrals

Let $K = K_x$ be a measurable map from $\mathfrak{X}$ to the linear operators on $L^2(\mathfrak{P})$. We define four operators $\Lambda^{\alpha\beta}(K)$ on $L^2(\mathfrak{P})$, which we term **Wick integrals**, by

$$\Lambda^{\alpha\beta}(K) \triangleq \int_{\mathfrak{X}} dx \left[\mathfrak{a}_x^*\right]^\alpha K_x \left[\mathfrak{a}_x\right]^\beta ,$$

where α and β take the values zero and one, and where we use the convention that $[\mathfrak{x}]^0 = 1$ and $[\mathfrak{x}]^1 = \mathfrak{x}$ for any object $\mathfrak{x}$. The appropriate domains are dom $\Lambda^{\alpha\beta}(K)$ consisting of all $\Psi \in L^2(\mathfrak{P})$ such that

i) $\mathfrak{a}_x^\beta \Psi \in \text{dom } K_x$ for almost all $x \in \mathfrak{X}$;

ii) if $\alpha = 0$, then: $x \mapsto K_x [\mathfrak{a}_x]^\beta \Psi$ is norm integrable;

iii) if $\alpha = 1$, then: $x \mapsto K_x [\mathfrak{a}_x]^\beta \Psi$ is square integrable;

iv) if $\beta = 1$, then $\Psi \in \text{dom } \sqrt{N}$.

Here a measurable function: $x \mapsto \xi_x$ taking values in some fixed Hilbert space is deemed to be norm integrable, also known as **Bochner integrable**, if $\int_{\mathfrak{X}} \|\xi_x\| \, dx <\in \infty$. It is square integrable, also known as **Skorohod integrable**, if $\int_{\mathfrak{X}} \|\xi_x\|^2 \, dx <\in \infty$. The requirement when $\beta = 1$ is just the condition that Ψ is in the domain of the gradient operator.

In more detail, these operators are given by

$$\Lambda^{00}(K) \Psi : X \mapsto \int_{\mathfrak{X}} dx \, (K_x \Psi)[X] ,$$

$$\Lambda^{01}(K) \Psi : X \mapsto \int_{\mathfrak{X}} dx \, (K_x \mathfrak{a}_x \Psi)[X] ,$$

$$\Lambda^{10}(K) \Psi : X \mapsto \sum_{x \in X} (K_x \Psi)[X - x] ,$$

$$\Lambda^{11}(K) \Psi : X \mapsto \sum_{x \in X} (K_x \mathfrak{a}_x \Psi)[X - x] . \tag{13.5}$$

As the notation suggests, when we restrict to the appropriate domains, we have

$$\Lambda^{\alpha\beta}(K)^* \subset \Lambda^{\beta\alpha}(K^*) ,$$

and we verify this explicitly for $\Lambda^{10}(K)$:

$$\left\langle \Lambda^{01}(K^*) \Psi | \xi \right\rangle = \int_{\mathfrak{X}} dx \, \langle K_x^* \mathfrak{a}_x \Psi | \xi \rangle = \int_{\mathfrak{X}} dx \, \langle \mathfrak{a}_x \Psi | K_x \xi \rangle$$

$$= \int_{\mathfrak{P} \times \mathfrak{X}} dX dx \, \Psi^*[X + x] \, (K_x \xi)[X]$$

$$= \int_{\mathfrak{P}} dZ \, \Psi^*[Z] \sum_{z \in Z} (K_z \xi)[Z - z]$$

$$\equiv \left\langle \Psi | \Lambda^{10}(K) \xi \right\rangle .$$

The other identities are established in a similar manner.

Let us make some further remarks. If matrix elements exist between exponential states, then these are given by

$$\langle \exp(f) | \Lambda^{\alpha\beta}(K) \exp(g) \rangle =$$
$$\int_{\mathfrak{X}} dx \quad [f^*(x)]^\alpha \langle \exp(f) | K_x \exp(g) \rangle [g(x)]^\beta.$$

Proposition 13.2.1 (The Skorohod Equality) *Provided that we have both*

$$\int dx \int dX \, (\#X + 1) \, |K_x \Psi[X]|^2 < \infty$$

and

$$\int dx \int dX \, (\#X + 1) \, |L_x \xi[X]|^2 < \infty,$$

then

$$\left\langle \Lambda^{10}(K) \Psi | \Lambda^{10}(L) \xi \right\rangle = \int_{\mathfrak{X}\times\mathfrak{X}} dxdy \, \langle \mathfrak{a}_y K_x \Psi | \mathfrak{a}_x L_y \xi \rangle$$
$$+ \int_{\mathfrak{X}} dx \, \langle K_x \Psi | L_x \xi \rangle .$$

To see this, we use the singular CCR to write formally

$$\left\langle \Psi | \Lambda^{10}(K^*) \Lambda^{10}(L) \xi \right\rangle = \int_{\mathfrak{X}\times\mathfrak{X}} dxdy \, \left\langle \Psi | K_x^* \mathfrak{a}_x \mathfrak{a}_y^* L_y \xi \right\rangle$$
$$= \int_{\mathfrak{X}\times\mathfrak{X}} dxdy \, \left\langle \Psi | K_x^* \left[\mathfrak{a}_y^* \mathfrak{a}_x + \delta(x - y) \right] L_y \xi \right\rangle.$$

It should be emphasized that the term $\left\langle \Psi | K_x^* \mathfrak{a}_y^* \mathfrak{a}_x L_y \xi \right\rangle$ inside the final expression is not really in Wick order yet, as we still have to move $\mathfrak{a}_y^*$ to the left of K_x^* and $\mathfrak{a}_x$ to the right of L_y: we will need additional assumptions on K and L to achieve that. (Although our demonstration was formal, a correct proof of the Skorohod equality is easily given using the $\mathfrak{L}$ formula.)

13.2.1 Wick Integrals of Kernel Represented Operators

The choice of the operators K has so far been quite general. If we restrict to those having a kernel representation, say with kernel $K[\cdot, \cdot, \cdot]$, then the Wick integrals take the form

$$\Lambda^{\alpha\beta}(K) = \int_{\mathfrak{X}} dx \int_{\mathfrak{P}^3} dY_1 dY_2 dY_3$$
$$\times [\mathfrak{a}_x^*]^\alpha \, \mathfrak{a}_{Y_1+Y_2}^* \mathcal{K}_x[Y_1, Y_2, Y_3] \, \mathfrak{a}_{Y_2+Y_3} [\mathfrak{a}_x]^\beta.$$

The expression is again in Wick order and it is possible to represent each of the four $\Lambda^{\alpha\beta}(K)$ using kernels. We have

$$\ker \Lambda^{00}(K)[X_1, X_2, X_3] = \int_{\mathcal{X}} dx\, \mathcal{K}_x[X_1, X_2, X_3],$$

$$\ker \Lambda^{10}(K)[X_1, X_2, X_3] = \sum_{x \in X_1} \mathcal{K}_x[X_1 - x, X_2, X_3],$$

$$\ker \Lambda^{11}(K)[X_1, X_2, X_3] = \sum_{x \in X_2} \mathcal{K}_x[X_1, X_2 - x, X_3],$$

$$\ker \Lambda^{01}(K)[X_1, X_2, X_3] = \sum_{x \in X_3} \mathcal{K}_x[X_1, X_2, X_3 - x].$$

These are very easily established. For instance,

$$\Lambda^{11}(K) = \int dx dY_1 dY_2 dY_3\, \mathfrak{a}^*_{\{x\}+Y_1+Y_2} \mathcal{K}_x[Y_1, Y_2, Y_3]\, \mathfrak{a}_{\{x\}+Y_2+Y_3}$$

and we use the rule (13.4) to write

$$\int dx dY_2\, F[x, Y_2] = \int_{\mathfrak{P}} dX_2 \sum_{x \in X_2} F[x, X_2 - x]$$

yielding

$$\Lambda^{11}(K) = \int_{\mathfrak{P}^3} dY_1 dX_2 dY_3$$

$$\times \sum_{x \in X_2} \mathfrak{a}^*_{Y_1+X_2} \mathcal{K}_x[Y_1, X_2 - x, Y_3]\, \mathfrak{a}_{X_2+Y_3},$$

from which we may read off the kernel.

13.3 Chronological Ordering

Let us now suppose that there is a time-coordinate $t(\cdot)$ on the space $\mathcal{X}$ and that we can also apply a simultaneity-free principle. That is, we have a time-coordinate function $t\colon \mathcal{X} \mapsto \mathbb{R}$, and, for a typical sequence $(x_1, \ldots, x_n)$, we may assume that no simultaneities exist. In particular, we will have a permutation σ such that $t\left(x_{\sigma(n)}\right) > t\left(x_{\sigma(n-1)}\right) > \cdots t\left(x_{\sigma(1)}\right)$. We take $t_{\min}(X)$ and $t_{\max}(X)$ to be the earliest and latest times, respectively occurring in a set X. We then set

$$\Delta_n(\mathcal{X}) = \{(x_n, \ldots, x_1) \in \mathcal{X}^n : t(x_n) > \cdots > t(x_1)\}. \tag{13.6}$$

The notion of localization and chronological order concepts can now be combined. For each time t, we may write

$$\mathcal{X} = \mathcal{X}_{\leq t} + \mathcal{X}_{>t}, \tag{13.7}$$

where $\mathfrak{X}_{\leq t}$ is the set of all points $x \in \mathfrak{X}$ with $t(x) \leq t$, while $\mathfrak{X}_{>t}$ is its complement. Similarly, every subset X of $\mathfrak{X}$ can be decomposed as the disjoint union $X_{\leq t} + X_{>t}$, with $X_{>t}$ being the set of elements of X that are later than time t. We then have $L^2(\mathfrak{X}) = L^2(\mathfrak{X}_{\leq t}) \oplus L^2(\mathfrak{X}_{>t})$, and we denote the orthogonal projection from $L^2(\mathfrak{X})$ onto $L^2(\mathfrak{X}_{\leq t})$ by $\Pi_{t]}$.

We then have the tensor product decomposition at time t

$$L^2(\mathfrak{P}) \cong L^2(\mathfrak{P}_{\leq t}) \otimes L^2(\mathfrak{P}_{>t}), \tag{13.8}$$

splitting up into a past and future factor of Fock space respectively. Thus, every vector in the Fock space can be approximated by a linear combination of vectors of the form $\Psi = \Psi_{\leq t} \otimes \Psi_{>t}$, and in Guichardet notation this means

$$\Psi_{\leq t} \otimes \Psi_{>t}[X] = \Psi_{\leq t}[X_{\leq t}] \Psi_{>t}[X_{>t}]. \tag{13.9}$$

An operator is said to be trivial after time t if it has trivial action on the future Fock space $L^2(\mathfrak{P})$. That is, K is trivial after time t if it takes the form

$$K \equiv K_{\leq t} \otimes 1_{>t} \tag{13.10}$$

with respect to the tensor product decomposition at time t. (Here $1_{>t}$ is the identity on the future Fock space.) We say that a measurable function $K = K_x$ taking values in the operators on Fock space is adapted if, for each x, we have that K_x is trivial after time $t(x)$.

If $K = K_x$ is adapted, then we have the commutation relations

$$[K_y, A^{\#}(g)] = 0,$$

whenever g is localized in $\mathfrak{X}_{>t}$. We may also consider the more formal relations $[K_y, a_x^{\#}] = 0$, whenever $t(x) > t(y)$.

For a given $K = K_x$, we can construct an operator $\Lambda_t^{\alpha\beta}[K]$ by taking

$$\Lambda_t^{\alpha\beta}[K] \triangleq \int_{t(x)<t} dx \, [a_x^*]^{\alpha} K_x [a_x]^{\beta}.$$

The family $\left\{ \Lambda_{t(x)}^{\alpha\beta}(K) : x \in \mathfrak{X} \right\}$ will then be adapted.

13.3.1 Wick Products

The product $\Lambda^{\alpha\beta}(K) \Lambda^{\mu\nu}(L)$ of two Wick integrals will require an application of the canonical commutation relations in order to convert into Wick-ordered form. This expression is

$$\Lambda^{\alpha\beta}(K) \Lambda^{\mu\nu}(L) = \int_{\mathfrak{X}} dx \int_{\mathfrak{X}} dy \, [a_x^*]^{\alpha} K_x [a_x]^{\beta} [a_y^*]^{\mu} L_y [a_y]^{\nu}$$

and we wish to discuss the problem of how we would set about putting it into Wick-ordered form again. The first part of the maneuver, putting $[\mathfrak{a}_x]^\beta \left[\mathfrak{a}_y^*\right]^\mu$ to Wick order, only applies when neither β or μ are zero. This gives

$$\int_{\mathfrak{X}\times\mathfrak{X}} dxdy \left[\mathfrak{a}_x^*\right]^\alpha K_x \left[\mathfrak{a}_y^*\right]^\mu [\mathfrak{a}_x]^\beta L_y [\mathfrak{a}_y]^\nu + \hat{\delta}^{\beta\mu} \int_{\mathfrak{X}} dx \left[\mathfrak{a}_x^*\right]^\alpha K_x L_x [\mathfrak{a}_x]^\nu ,$$

where we have introduced the **Evans–Hudson** δ symbol

$$\hat{\delta}^{\alpha\beta} = \begin{cases} 1, & \alpha = \beta \neq 0; \\ 0, & \text{otherwise.} \end{cases}$$

The second integral is already in Wick order. In the first integral, we still have some work to do because, in general, we cannot assume that $\mathfrak{a}_y^*$ commutes with K_x or $\mathfrak{a}_x$ commutes with L_y. (This is as far as we can go with just the Skorohod equality alone!)

To proceed further, we now assume that both K and L are adapted. We then split the region of integration in the first integral up into $t(x) < t(y)$ and $t(x) > t(y)$. For $t(x) > t(y)$, the integral will be

$$\int_{t(x)>t(y)} dxdy \left[\mathfrak{a}_x^*\right]^\alpha K_x \left[\mathfrak{a}_y^*\right]^\mu L_y [\mathfrak{a}_y]^\nu [\mathfrak{a}_x]^\beta =$$
$$\int_{\mathfrak{X}} dx \left[\mathfrak{a}_x^*\right]^\alpha K_x \Lambda_{t(x)}^{\mu\nu} (L) [\mathfrak{a}_x]^\beta$$

since $\mathfrak{a}_x^*$ will now commute with the earlier L_y. Likewise, if $t(x) < t(y)$, then $\mathfrak{a}_y^*$ will commute with earlier K_x and so

$$\int_{t(x)<t(y)} dxdy \left[\mathfrak{a}_y^*\right]^\mu [\mathfrak{a}_x^*]^\alpha K_x [\mathfrak{a}_x]^\beta L_y [\mathfrak{a}_y]^\nu =$$
$$\int_{\mathfrak{X}} dy \left[\mathfrak{a}_y^*\right]^\mu \Lambda_{t(y)}^{\alpha\beta} (K) L_y [\mathfrak{a}_y]^\nu .$$

We may therefore write all of this as

$$\Lambda_t^{\alpha\beta} (K) \Lambda_t^{\mu\nu} (L) = \Lambda_t^{\alpha\beta} \left(K\Lambda_{t(\cdot)}^{\alpha\beta} (L)\right) + \Lambda_t^{\mu\nu} \left(\Lambda_{t(\cdot)}^{\alpha\beta} (K) L\right)$$
$$+ \hat{\delta}^{\beta\mu} \Lambda_t^{\alpha\nu} (KL). \tag{13.11}$$

Here we understand KL to be the functions: $x \mapsto K_x L_x$, and $K\Lambda_{t(\cdot)}^{\alpha\beta} (L)$ to be: $x \mapsto K_x \Lambda_{t(x)}^{\alpha\beta} (L)$, and so on. The first two terms in (13.11) are Wick integrals of Wick integrals and are therefore automatically Wick ordered. The final term is the correction from the canonical commutation relations and arises only when we need to reorder the operator densities (i.e., $\beta = \mu = 1$): this term is also explicitly a Wick integral.

Note that the adaptedness of K and L was essential to get to (13.11).

13.4 Quantum Stochastic Processes on Fock Space

For the remainder of this chapter, we are going to restrict our discussion to the special case $\mathfrak{X} = \mathbb{R}^+$, which we shall interpret as the positive time axis. The one-particle space is $L^2\left(\mathbb{R}^+\right)$, and we shall denote the Fock space over it by $\mathfrak{F}$. For I, a subset of $\mathbb{R}^+$, we shall denote the space of square integrable functions over it by $L^2\left(I\right)$, with the corresponding Fock space being written as $\mathfrak{F}_I$. As before, we have the tensor product decompositions

$$\mathfrak{F} = \mathfrak{F}_{[0,t]} \otimes \mathfrak{F}_{[t,\infty)} \tag{13.12}$$

for each $t > 0$. We next consider a family of operators $K = \{K_t : t \geq 0\}$ on $\mathfrak{F}$ and define the four Wick integrals as before:

$$\Lambda_t^{\alpha\beta}[K] \triangleq \int_{[0,t]} ds \left[\mathfrak{a}_s^*\right]^\alpha K_s \left[\mathfrak{a}_s\right]^\beta . \tag{13.13}$$

We shall generally refer to families of operators on a fixed Hilbert space as quantum stochastic processes. The issue of domains will be postponed until later, and for now we shall concentrate on algebraic features. An important first observation is that the Wick integrals of an adapted process K will again be adapted processes $\left\{\Lambda_t^{\alpha\beta}[K] : t \geq 0\right\}$.

13.4.1 The Fundamental Quantum Stochastic Processes

To give a concrete example of a process, let us take the simplest situation where $K = 1$, the identity on Fock space. We shall denote the corresponding processes as

$$A_t^{\alpha\beta} = \Lambda_t^{\alpha\beta}(I) \equiv \int_{[0,t]} ds \left[\mathfrak{a}_s^*\right]^\alpha \left[\mathfrak{a}_s\right]^\beta . \tag{13.14}$$

They will be automatically adapted and explicitly, for each time $t \geq 0$, they are the following four operators on $\mathfrak{F}$:

$$A_t^{00} = t \quad \text{(time)},$$
$$A_t^{01} = A\left(\chi_{[0,t]}\right) \quad \text{(annihilation)},$$
$$A_t^{00} = A^*\left(\chi_{[0,t]}\right) \quad \text{(creation)},$$
$$A_t^{00} = d\Gamma_+\left(\Pi_{[0,t]}\right) \quad \text{(scattering)}.$$

Here $\chi_{[0,t]}$ is the characteristic function of the interval $[0,t]$ and $\Pi_{[0,t]}$ is orthogonal projection[1] from the one particle space $L^2\left(\mathbb{R}^+\right)$ onto the subspace $L^2\left([0,t]\right)$.

---

[1] That is, $\Pi_{[0,t]}f\left(s\right) = \chi_{[0,t]}\left(s\right)f\left(s\right)$.

13.4.2 Quantum Stochastic Integrals

We now consider four adapted processes $\{X_{\alpha\beta}(t) : t \geq 0\}$ and construct the process

$$X(t) = \sum_{\alpha=0,1} \sum_{\beta=0,1} \int_0^t ds \, [a_s^*]^\alpha \, X_{\alpha\beta}(s) \, [a_s]^\beta$$

$$\equiv \int_0^t ds \, [a_s^*]^\alpha \, X_{\alpha\beta}(s) \, [a_s]^\beta ,$$

with the summation convention that repeated Greek indices are to be summed over the values 0 and 1. Matrix elements $\langle \exp(f) | X_t \exp(g) \rangle$ between exponential vectors then take the form

$$\int_0^t ds \, [f(s)^*]^\alpha \, \langle \exp(f) | X_{\alpha\beta}(s) \exp(g) \rangle \, [g(s)]^\beta .$$

We will also make use of a fluxion notation

$$\dot{X}(t) \triangleq [a_t^*]^\alpha \, X_{\alpha\beta}(t) \, [a_t]^\beta ,$$

and an (Itō) integral notation

$$X(t) = \int_0^t X_{\alpha\beta}(s) \, dA_s^{\alpha\beta},$$

with differential form

$$dX(t) = X_{\alpha\beta}(t) \, dA_t^{\alpha\beta}.$$

13.4.3 The Quantum Itō Formula

Let $X(t)$ be a quantum stochastic integral as in the preceding, and suppose that we have a second one $Y(t)$ with $\dot{Y}(t) = [a_t^*]^\mu \, Y_{\mu\nu}(t) \, [a_t]^\nu$ with the $Y_{\mu\nu}(\cdot)$ likewise adapted.

According to formula (13.11), we will have

$$X(t)\,Y(t)$$

$$= \int_0^t ds \, [a_s^*]^\alpha \, X_{\alpha\beta}(s) \, Y(s) \, [a_s]^\beta + \int_0^t ds \, [a_s^*]^\mu \, X(s) \, Y_{\mu\nu}(s) \, [a_s]^\nu$$

$$+ \hat{\delta}^{\beta\mu} \int_0^t ds \, [a_s^*]^\alpha \, X_{\alpha\beta}(s) \, Y_{\mu\nu} \, [a_s]^\nu$$

$$= \int_0^t ds \, [a_s^*]^\alpha \, \left\{ x_{\alpha\beta}(s) \, Y(s) + X(s) \, Y_{\alpha\beta}(s) + X_{\alpha 1}(s) \, Y_{1\beta}(s) \right\} [a_s]^\beta .$$

$$\tag{13.15}$$

The formula (13.15) can then be put in the form

$$X(t)\,Y(t) = \int_0^t dX(t)\,Y(t) + \int_0^t X(t)\,dY(t) + \int_0^t dX(t)\,dY(t)$$

$$(13.16)$$

under the convention that

$$dX(t)\,Y(t) \triangleq X_{\alpha\beta}(t)\,Y(t)\,dA_t^{\alpha\beta},$$ (13.17)

$$X(t)\,dY(t) \triangleq X(t)\,Y_{\alpha\beta}(t)\,dA_t^{\alpha\beta}$$ (13.18)

$$dX(t)\,dY(t) \triangleq X_{\alpha\beta}(t)\,Y_{\mu\nu}(t)\,dA_t^{\alpha\beta}\,dA_t^{\mu\nu},$$ (13.19)

and that the second-order product of differentials obeys the rules

$$dA_t^{\alpha\beta}\,dA_t^{\mu\nu} = \hat{\delta}^{\beta\mu}\,dA_t^{\alpha\nu}.$$ (13.20)

We refer to (13.20) as the Itō formula and (13.21) as an Itō table, as tabulated in the following:

$\times$	dA^{00}	dA^{01}	dA^{11}	dA^{10}
dA^{00}	0	0	0	0
dA^{01}	0	0	dA^{01}	dA^{00}
dA^{11}	0	0	dA^{11}	dA^{10}
dA^{10}	0	0	0	0

$$(13.21)$$

13.5 Quantum Stochastic Calculus

We now describe the Hudson–Parthasarathy theory of quantum stochastic processes and their extension of the Itō calculus to Bosonic processes. We have done much of the groundwork in the earlier sections, so now we can spend some time dealing with analytic issues.

We begin by fixing a subset $\mathcal{A}$ of $L^2(\mathbb{R}^+)$, which we call a space of **admissible functions**: it is required to have the properties of being a dense linear manifold in $L^2(\mathbb{R}^+)$ and being invariant under the projections $f \mapsto \chi_{[0,t]}f$ for each $t \geq 0$. As a generalization of our earlier constructions, Wick integrals will now be allowed to take values as operators in a fixed Hilbert space $\mathfrak{h}_0$, called the initial algebra. The action shall take place on the Hilbert space

$$\mathfrak{H} \triangleq \mathfrak{h}_0 \otimes \mathfrak{F}.$$ (13.22)

As before, $\mathfrak{F} = \Gamma_+\left(L^2(\mathbb{R}^+)\right)$ and, for each $t > 0$, we have the decomposition $\mathfrak{H} = \mathfrak{H}_{[0,t]} \otimes \mathfrak{H}_{[t,\infty)}$ with $\mathfrak{H}_{[0,t]} = \mathfrak{h}_0 \otimes \mathfrak{F}_{[0,t]}$ and $\mathfrak{H}_{[t,\infty)} = \mathfrak{F}_{[t,\infty)}$.

Let $\mathfrak{D}_0$ be a fixed linear manifold in $\mathfrak{F}$. A family $X = (X_t)_{t \geq 0}$ of operators on $\mathfrak{H}$, with the domain contained in $\mathfrak{D}_0 \otimes \exp(\mathcal{A})$ is said to be a quantum stochastic process based on $(\mathfrak{D}_0, \mathcal{A})$. If X_t has trivial action on the factor $\mathfrak{H}_{[t,\infty)}$ for the continuous tensor decomposition at time t, for each $t > 0$, then we say that the process is adapted. If there exists an increasing sequence $\{t_n\}_{n=0}^{\infty}$ with $t_0 = 0$ with $t_n \uparrow \infty$ such that X_t takes a constant (operator) value y_n on the interval $t_n \leq t < t_{n+1}$, then the process is simple: in addition, it will be adapted if the y_n act trivially on the future decomposition factor $\mathfrak{H}_{[t_n,\infty)}$, for each n. If the map $t \mapsto X_t u \otimes \exp(f)$ is strongly continuous for all $u \in \mathfrak{D}_0, f \in \mathcal{A}$ and $t > 0$, the process is said to be continuous, while if the map is strongly measurable with $\int_{T_1}^{T_2} \|X_s u \otimes \exp(f)\|^2 \, ds < \infty$, for all $u \in \mathfrak{D}_0, f \in \mathcal{A}$, then X is said to be locally square-integrable over $[T_1, T_2]$.

We now show that the Wick integral $X_t = \int_0^t [a_s^*]^\alpha X_{\alpha\beta}(s) [a_s]^\beta$, or equivalently as $X_t = \int_0^t X_{\alpha\beta}(s) \, dA_s^{\alpha\beta}$, where the $X_{\alpha\beta}(\cdot)$ are four locally square-integrable, adapted processes based on $(\mathfrak{D}_0, \mathcal{A})$, and can be obtained as a limit of finite sum approximations similar to the usual Itō construction. We begin by fixing an appropriate notion of convergence:

Let $X(\cdot)$ and $X_n(\cdot)$, for each $n = 1, 2, \ldots$, be locally square integrable processes based on $(\mathfrak{D}_0, \mathcal{A})$; then we say that the $X_n(\cdot)$ converge to $X(\cdot)$ over $[T_1, T_2]$ if

$$\lim_{n \to \infty} \int_{T_1}^{T_2} \|(X(t) - X_n(t)) u \otimes \exp(f)\|^2 \, dt = 0$$

for all $u \in \mathfrak{D}_0$ and $f \in \mathcal{A}$. We also say that the $X_n(\cdot)$ converge to $X(\cdot)$ locally if we have this convergence for every finite subinterval $[T_1, T_2]$.

Here the strategy is similar to how ordinary stochastic integrals of Itō type are constructed. We may smooth a locally square-integrable process to get a continuous process in a manner that preserves adaptedness, and conversely always construct an approximating sequence of continuous processes to any locally square-integrable process. In turn, we may discretize any continuous process to obtain a simple process, again maintaining adaptedness, and conversely approximate any adapted continuous process by a sequence of simple ones. The proof of this is given in Hudson and Parthasarathy (1984) as Proposition 3.2, but is almost identical to the standard construction – see for instance Øksendal (1992) – and we omit it.

Given for simple adapted processes $X_{\alpha\beta}(\cdot)$, we define their quantum stochastic integral to be

$$\int_{T_1}^{T_2} X_{\alpha\beta}(s) \, dA_s^{\alpha\beta} = \sum_{j=0}^{n-1} X_{\alpha\beta}(t_j) \left(A_{t_{j+1}}^{\alpha\beta} - A_{t_j}^{\alpha\beta} \right) \tag{13.23}$$

where $T_1 = t_0 < t_1 < \cdots < t_n = T_2$ is a partition of $[T_1, T_2]$ that includes all discontinuity points of each of the four processes. Let us introduce the notation

$$\bar{h}\Big|_{t_1}^{t_2} = \frac{1}{t_2 - t_1} \int_{t_1}^{t_2} h(t)\, dt$$

for the time average of a function over an interval $[t_1, t_2]$.

Lemma 13.5.1 *Let* $X_{\alpha\beta}(\cdot)$ *be* $Y_{\alpha\beta}(\cdot)$ *adapted simple processes based on* $(\mathfrak{D}_0, \mathcal{A})$ *and let* $X(t) = \int_0^t X_{\alpha\beta}(s)\, dA_s^{\alpha\beta}$, $Y(t) = \int_0^t Y_{\mu\nu}(s)\, dA_s^{\mu\nu}$; *then, for all* $u, v \in \mathfrak{D}_0$ *and* $f, g \in \mathcal{A}$,

$$\left\langle u \otimes \exp(f) \,\Big|\, \int_{T_1}^{T_2} X_{\alpha\beta}(s)\, dA_s^{\alpha\beta}\, v \otimes \exp(g) \right\rangle$$

$$= \int_{T_1}^{T_2} ds\, [f^*(s)]^\alpha \left\langle u \otimes \exp(f) \,|\, X_{\alpha\beta}(s)\, v \otimes \exp(g) \right\rangle [g(s)]^\beta \quad (13.24)$$

and

$$\langle X(t)\, u \otimes \exp(f) \,|\, Y(t)\, v \otimes \exp(g) \rangle$$

$$= \int_0^t ds\, [f^*(s)]^\alpha \Big\{ \langle X_{\alpha\beta}(s)\, u \otimes \exp(f) \,|\, Y(s)\, v \otimes \exp(g) \rangle +$$

$$+ \langle X(s)\, u \otimes \exp(f) \,|\, Y_{\alpha\beta}(s)\, v \otimes \exp(g) \rangle$$

$$+ \langle X_{1\alpha}(s)\, u \otimes \exp(f) \,|\, Y_{1\beta}(s)\, v \otimes \exp(g) \rangle \Big\} [g(s)]^\beta. \quad (13.25)$$

Proof Substituting the expression (13.23) into the left-hand side of (13.24), we see that it equals

$$\sum_j \delta t_j\, \overline{[f^*]^\alpha [g]^\beta}\Big|_{t_j}^{t_{j+1}} \langle u \otimes \exp(f) \,|\, X_{\alpha\beta}(t_j)\, v \otimes \exp(g) \rangle,$$

where we set $\delta t_j = t_{j+1} - t_j$. This can be rewritten as the right-hand side of (13.24) by virtue of the piecewise continuous nature of the process.

Next we take a partition of $[T_1, T_2] = [0, t]$ that includes all discontinuity points of each of $X_{\alpha\beta}$ and $Y_{\alpha\beta}$. The right-hand side of (13.25) can then be written as

$$\sum_j \delta t_j\, \overline{[f^*]^\alpha [g]^\beta}\Big|_{t_j}^{t_{j+1}} \Big\{ \langle X_{\alpha\beta}(t_j)\, u \otimes \exp(f) \,|\, Y(t_j)\, v \otimes \exp(g) \rangle$$

$$+ \langle X(t_j)\, u \otimes \exp(f) \,|\, Y_{\alpha\beta}(t_j)\, v \otimes \exp(g) \rangle$$

$$+ \langle X_{1\alpha}(t_j)\, u \otimes \exp(f) \,|\, Y_{1\beta}(t_j)\, v \otimes \exp(g) \rangle \Big\},$$

and similarly we recover the required integral form by inspection. $\qquad\square$

These are of course expressions we have already seen before when formally manipulating Wick integrals. The following lemma helps us to show that the formulas (13.24) and (13.25) hold true also for adapted locally square-integrable processes.

Lemma 13.5.2 *Suppose that $\mathcal{A}$ consists of locally bounded functions; then for $T_1 \leq t \leq T_2$, $u \in \mathfrak{D}_0$ and $f \in \mathcal{A}$, e have*

$$\|X(t)\,u \otimes \exp(f)\|^2 \leq 6\,C^2 \int_{T_1}^{T_2} dr\, e^{r-T_2} \sum_{\alpha,\beta} \|X_{\alpha\beta}(r)\,u \otimes \exp(f)\|^2,$$

(13.26)

where $C = C(f, T_1, T_2) = \max_{p=0,2} \sup_{T_1 \leq t \leq T_2} |f(t)|^{p/2}$.

Proof　Formula (13.25) allows us to deduce that

$$\|X(t)\,u \otimes \exp(f)\|^2 = \int_{T_1}^{t} ds\, \left[f^*(s)\right]^{\alpha} \left[f(s)\right]^{\beta}$$
$$\times 2\mathrm{Re} \left\langle X(s)\,u \otimes \exp(f) \,|\, X_{\alpha\beta}(s)\,u \otimes \exp(f) \right\rangle$$
$$+ \int_{T_1}^{t} ds\, \|(X_{11}(s)f(s) + X_{10})\,u \otimes \exp(f)\|^2.$$

The Hilbert space inequalities $2\,|\langle \xi | \eta \rangle| \leq \|\xi\|^2 + \|\eta\|^2$ and $\left\| \sum_{j=1}^{n} \xi_j \right\|^2 \leq n \sum_{j=1}^{n} \|\xi_j\|^2$ then imply

$$\|X(t)\,u \otimes \exp(f)\|^2 \leq \int_{T_1}^{t} ds\, \Big\{ \|X(s)\,u \otimes \exp(f)\|^2$$
$$+ 6C(f, T_1, T_2)^2 \sum_{\alpha,\beta} \|X_{\alpha\beta}(s)\,u \otimes \exp(f)\|^2 \Big\}.$$

By monotonicity, we may differentiate wrt. t to get

$$\frac{d}{dt} \|X(t)\,u \otimes \exp(f)\|^2 \leq \|X(t)\,u \otimes \exp(f)\|^2$$
$$+ 6C(f, T_1, T_2)^2 \sum_{\alpha,\beta} \|X_{\alpha\beta}(t)\,u \otimes \exp(f)\|^2.$$

The result follows from applying the integrating factor e^{-t} to both sides and integrating up.　$\square$

Whenever $X_{\alpha\beta}(\cdot)$ are adapted square integrable processes based on $(\mathfrak{D}_0, \mathcal{A})$, with $\mathcal{A}$ consisting of locally bounded functions, then there exists an approximating family $X_{\alpha\beta}^{(n)}(\cdot)$ of adapted simple processes based on $(\mathfrak{D}_0, \mathcal{A})$, and as a corollary to the previous lemma we have that $X^{(n)}(t) \, u \otimes \exp(f) = \int_0^t X^{(n)}_{\alpha\beta}(s) \, dA_s^{\alpha\beta} \, u \otimes \exp(f)$ converges in $\mathfrak{F}$, for every $u \in \mathfrak{D}_0$ and $f \in \mathcal{A}$. (Apply the lemma to $X^{(n)} - X^{(m)}$!) The limit vector will be independent of the approximating sequence and can be denoted as $\int_0^t X_{\alpha\beta}(s) \, dA_s^{\alpha\beta} \, u \otimes \exp(f)$. This serves as the definition of the quantum stochastic integral $X(t) = \int_0^t X_{\alpha\beta}(s) \, dA_s^{\alpha\beta}$ on the domain $\mathfrak{D}_0 \otimes \mathcal{A}$. By construction, $X(\cdot)$ will be adapted and the formulas (13.24), (13.25), and (13.26) apply to such integrals.

13.5.1 QSDEs

Let $G_{\alpha\beta}$ be four operators on the initial space $\mathfrak{h}_0$ with common invariant domain $\mathfrak{D}$. We wish to show the existence and uniqueness of solutions to the quantum stochastic differential equation (QSDE):

$$dU_t = G_{\alpha\beta} U_t dA_t^{\alpha\beta}, \qquad U_0 = 1.$$

This can be equivalently be considered as the integro-differential equation

$$U_t = 1 + \int_0^t G_{\alpha\beta} U_s dA_s^{\alpha\beta},$$

and the standard technique for dealing with such problems is Picard iteration. We define a sequence $\{U^{(n)} : n = 0, 1, 2, \ldots\}$ of processes by setting $U^{(0)} = 1$ and then

$$U_t^{(n)} = 1 + \int_0^t G_{\alpha\beta} U_s^{(n-1)} dA_s^{\alpha\beta}.$$

We may write $U_t^{(n)} = \sum_{m=0}^n K_t^{(n)}$, where

$$K_t^{(n)} \equiv G_{\alpha_n\beta_n} \ldots G_{\alpha_1\beta_1} \otimes \int_{\Delta_n(t)} dA_{t_n}^{\alpha_n\beta_n} \ldots dA_{t_1}^{\alpha_1\beta_1}$$

(implied summation over 4^n terms!) and we recall that $\Delta_n(t)$ is the simplex $t > t_n > \cdots > t_1 > 0$. In the following, we will again take $\mathcal{A}$ to consist locally bounded functions.

Lemma 13.5.3 (Hudson–Parthasarathy) *For fixed $u \in \mathfrak{D}$, let us set $G(u, n) = \max_{\alpha_n,\beta_n,\ldots,\alpha_1,\beta_1} \left\| G_{\alpha_n\beta_n} \ldots G_{\alpha_1\beta_1} u \right\|$. Let $f \in \mathcal{A}$, and $T > 0$; we have for each $t \in [0, T]$*

$$\left\| K_t^{(n)} u \otimes \exp(f) \right\|^2 \le e^{(T+\|f\|^2)} G(u, n)^2 \frac{1}{n!} \left(24T \, C(f, 0, T)^2 \right)^n.$$

Proof Let us set $V_t^{(n)} = \sum_{\alpha_n,\beta_n,\ldots,\alpha_1,\beta_1} \int_{\Delta_n(t)} dA_{t_n}^{\alpha_n\beta_n} \ldots dA_{t_1}^{\alpha_1\beta_1}$, then

$$\left\| K_t^{(n)} u \otimes \exp(f) \right\| \le G(u,n) \left\| V_t^{(n)} \exp(f) \right\|.$$

Now $V_t^{(n)} = \int_0^t V_t^{(n-1)} dN_t$, where $N_t = A_t^{11} + A_t^{10} + A_t^{01} + A_t^{00}$, and from the previous lemma we find that

$$\left\| V_t^{(n)} \exp(f) \right\|^2 \le 6C(f,0,T)^2 \int_0^t e^{s-t} 4 \left\| V_s^{(n-1)} \exp(f) \right\|^2 ds$$

and so, by induction, we get the required estimate

$$\left\| K_t^{(n)} u \otimes \exp(f) \right\|^2 \le \left(24C(f,0,T)^2 \right)^n G(u,n)$$
$$\times e^T \|\exp(f)\|^2 \operatorname{vol}(\Delta_n(T)).$$

$\square$

(Note that $V_t^{(n)} \equiv \binom{N_t}{n}$, although we didn't make us of this fact.) From the lemma, we see that

$$\int_0^T dt \left\| U_t^{(n)} u \otimes \exp(f) \right\|^2 \le n \sum_{m=0}^n \int_0^T dt \left\| K_t^{(m)} u \otimes \exp(f) \right\|^2 < \infty$$

and so $U^{(n)}{}_t$ is locally square-integrable. By construction, it also is an adapted process.

At this stage, we should give a domain for the $U^{(n)}$. Let us set

$$\mathfrak{D}_0 = \left\{ u \in \mathfrak{D} : \sum_n \frac{1}{\sqrt{n!}} G(u,n) \tau^n < \infty, \text{ for all } \tau > 0 \right\}.$$

Now $\left\| K_t^{(n)} u \otimes \exp(f) \right\| \le e^{(T+\|f\|^2)/2} G(u,n) \frac{1}{\sqrt{n!}} \left(\sqrt{24T} C(f,0,T) \right)^n$ and so

$$\sum_n \sup_{0 \le t \le T} \left\| K_t^{(n)} u \otimes \exp(f) \right\| < \infty$$

and so we see that $U_t^{(n)} u \otimes \exp(f) = \sum_{m=0}^n K_t^{(n)} u \otimes \exp(f)$ converges as $n \to \infty$ in a manner that is uniform in t. The limit can be denoted $U_t u \otimes \exp(f)$ and clearly yields a quantum stochastic process U based on $(\mathfrak{D}_0, \mathcal{A})$. The uniformity of the convergence coupled with the estimates in the lemma allow us to take strong limits on $\mathfrak{D}_0 \otimes \mathcal{A}$ to obtain

$$U_t = s{-}\lim_{n\to\infty} U_t^{(n)} = 1 + \int_0^t G_{\alpha\beta} s{-}\lim_{n\to\infty} U_s^{(n-1)} dA_s^{\alpha\beta}$$
$$= 1 + \int_0^t G_{\alpha\beta} U_s dA_s^{\alpha\beta},$$

and so U satisfies the QSDE and is consequently continuous. Of course, it is also locally square-integrable.

To establish uniqueness of solution, we note that if U and U' are both solutions, then $W = U - U'$ satisfies $W_t = \int_0^t G_{\alpha\beta} W_s dA_s^{\alpha\beta}$. Comparing with the relation $K_t^{(n)} = \int_0^t G_{\alpha\beta} K_s^{(n-1)} dA_s^{\alpha\beta}$ in the previous lemma, we likewise arrive at the estimate

$$\| W_t u \otimes \exp{(f)} \|^2 \leq e^{(T+\|f\|^2)} G\,(u,n)^2 \frac{1}{n!} \left(24 T \alpha\,(f,0,T)^2 \right)^n$$

for $t \in [0, T]$ and so, taking $n \to \infty$, we find that U and U' agree on $\mathfrak{D}_0 \otimes \mathcal{A}$.

13.5.2 Unitarity

We now seek conditions on the coefficients $G_{\alpha\beta}$ such that the process U will be unitary. For simplicity, we take them to be unitary and so we can set $\mathfrak{D} = \mathfrak{h}$. The isometry condition $U_t^* U_t = 1$ implies that

$$0 = d\left(U_t^* U_t \right) = \left(dU_t^* \right) U_t + U_t^* \left(dU_t \right) + \left(dU_t^* \right) \left(dU_t \right)$$
$$= U_t^* \left(G_{\beta\alpha}^* + G_{\alpha\beta} + G_{1\alpha}^* G_{1\beta} \right) U_t\, dA_1^{\alpha\beta},$$

while the co-isometry condition $U_t U_t^* = 1$ implies that

$$0 = d\left(U_t U_t^* \right) = \left(dU_t \right) U_t^* + U_t \left(dU_t^* \right) + \left(dU_t \right) \left(dU_t^* \right)$$
$$= \left(G_{\alpha\beta} + G_{\beta\alpha}^* + G_{1\alpha} G_{1\beta}^* \right) dA_1^{\alpha\beta},$$

leading us to the conditions

$$G_{\beta\alpha}^* + G_{\alpha\beta} + G_{1\alpha}^* G_{1\beta} = 0 = G_{\alpha\beta} + G_{\beta\alpha}^* + G_{1\alpha} G_{1\beta}^*.$$

The general solution to these equations is

$$G_{11} = S - 1,$$
$$G_{10} = L,$$
$$G_{01} = -SL^*,$$
$$G_{00} = -\frac{1}{2} LL^* - iH,$$

where S is unitary, L bounded but otherwise arbitrary, and H is self-adjoint.

For X, a bounded operator on $\mathfrak{h}$, we set $J_t(X) = U_t^*(X \otimes 1) U_t$, and from the quantum Itō formula, we find

$$dJ_t(X) = J_t\left(\mathcal{L}_{\alpha\beta}(X) \right) dA_t^{\alpha\beta},$$

where we introduce the bounded linear maps

$$\mathcal{L}_{\alpha\beta}(X) = XG^*_{\beta\alpha} + G_{\alpha\beta}X + G^*_{1\alpha}XG_{1\beta}.$$

These maps each have the algebraic property that

$$\mathcal{L}_{\alpha\beta}(1) = 0,$$

which just follows from the unitary requirement on the $G_{\alpha\beta}$.

We see that

$$\| U_t u \otimes \exp(f) \| \le e^{-\|f\|^2/2} \sum_n \left(\sqrt{24T} \alpha(f,0,T) \max_{\alpha\beta} \| G_{\alpha\beta} \| \right)^n \frac{\|u\|}{\sqrt{n!}},$$

for $t \in [0,T]$. Therefore, for fixed $f,g \in \mathcal{A}$, we have a well-defined bounded operator R_t acting on the initial space such that

$$\langle u | R_t v \rangle = \langle U_t u \otimes \exp(f) | U_t v \otimes \exp(g) \rangle$$

for all $u,v \in \mathfrak{h}$. R will be locally bounded, and we see that

$$\frac{dR_t}{dt} = [f^*(t)]^\alpha \mathcal{L}_{\alpha\beta}(1) [g(t)]^\beta = 0.$$

We see that for R_t is the constant $R_0 = \langle \exp(f) | \exp(g) \rangle$. Therefore, $\langle U_t u \otimes \exp(f) | U_t v \otimes \exp(g) \rangle = \langle u \otimes \exp(f) | v \otimes \exp(g) \rangle$. It follows that U_t is an isometry.

13.6 Quantum Stratonovich Integrals

Given a pair of quantum stochastic integral processes determined by $\dot{X} = [\mathfrak{a}^*]^\alpha X_{\alpha\beta} [\mathfrak{a}]^\beta$ and $\dot{Y} = [\mathfrak{a}^*]^\alpha Y_{\alpha\beta} [\mathfrak{a}]^\beta$, their quadratic covariation is given by

$$[[X,Y]]_t = \int_0^t dX dY \equiv \int_0^t [\mathfrak{a}^*]^\alpha X_{\alpha 1} Y_{1\beta} [\mathfrak{a}]^\beta .$$

As we are dealing with noncommutative process, the covariation will generally be nonsymmetric. The quantum Itō rule may then be written in the form

$$d(XY) = (dX)Y_- + X_-(dY) + d[[X,Y]].$$

The analogue of the Stratonovich integral is to define $\int X\delta Y$ as the limit of finite sum approximations using a midpoint rule, first shown by Chebotarev (1997) (see also Gough, 1997): $\sum_k X_{t_k^*}(Y_{t_{k+1}} - Y_{t_k})$, with $t_k^* = \frac{1}{2}(t_{k+1} + t_k)$. This turns out to be the same as

$$\int X\delta Y \equiv \int X(dY) + \frac{1}{2}[[X,Y]]$$

or in differential language

$$X(\delta Y) = X(dY) + \frac{1}{2}dXdY.$$

Similarly, we have $\int (\delta X) Y \equiv \int (dX)Y + \frac{1}{2}[[X, Y]]$ and

$$(\delta X) Y = (dX) Y + \frac{1}{2}dXdY.$$

The quantum Itô rule then implies the Stratonovich rule

$$d(XY) \equiv (\delta X) Y_- + X_- (\delta Y), \tag{13.27}$$

which restores the Leibniz form.

Remarkably, the Stratonovich differentials take the form

$$X(\delta Y) \equiv \left(X + \frac{1}{2}dX\right) dY, \text{ and } (\delta X) Y \equiv dX \left(Y + \frac{1}{2}dY\right).$$

As such, the Itô correction (covariation) is shared equally between the two differentials. When we originally derived the Wick product formula, we used the singular commutation relations first and separated off the correction, then split the remainder into two time-ordered integrals.

13.7 The Quantum White Noise Formulation

Let us repeat, for clarity, the arguments in the setting of quantum stochastic integral processes. We have $X_t Y_t = \int_0^t ds \int_0^t dr \, \dot{X}_s \dot{Y}_r$ with

$$\dot{X}_s \dot{Y}_r = \left[a_s^*\right]^\alpha X_{\alpha\beta}(s) \left[a_s\right]^\beta \left[a_r^*\right]^\mu Y_{\mu\nu}(r) \left[a_r\right]^\nu. \tag{13.28}$$

We know that $X_t Y_t \equiv \int_0^t (\delta X) Y + \int_0^t X (\delta Y)$, while at the same time we would expect, from a reasonable definition of fluxions, to have

$$X_t Y_t = \int_0^t \dot{X} Y + \int_0^t X \dot{Y}. \tag{13.29}$$

Let us therefore do the obvious thing and define

$$\int_0^t \dot{X}_s Y_s ds \triangleq \int (\delta X_s) Y_s, \qquad \int_0^t X_s \dot{Y}_s \triangleq \int_0^t X_s (\delta Y_s), \tag{13.30}$$

so that we have

$$\dot{X}_t Y_t \equiv \left[a_t^*\right]^\alpha X_{\alpha\beta}(t) Y(t) \left[a_t\right]^\beta + \frac{1}{2}\left[a_t^*\right]^\alpha X_{\alpha 1}(t) Y_{1\beta}(t) \left[a_t\right]^\beta. \tag{13.31}$$

However, in a reasonable definition of fluxions, we might expect that

$$\dot{X}_t Y_t \equiv \left[a_t^* \right]^{\alpha} X_{\alpha\beta} (t) [a_t]^{\beta} Y (t) \tag{13.32}$$

Evidently the equations (13.31) and (13.32) will be consistent if

$$[a_t]^{\beta} Y (t) = Y (t) [a_t]^{\beta} + \frac{1}{2} \hat{\delta}_{\beta\mu} Y_{\mu\nu} (t) [a_t]^{\nu},$$

that is,

$$[a_t, Y (t)] = \frac{1}{2} Y_{1\nu} (t) [a_t]^{\nu}. \tag{13.33}$$

We arrived at this equation by asking for some reasonable algebraic rules for the fluxions. So far, nothing is rigorous; however, we have been able at least to express otherwise undefined objects in terms of well-defined Wick-ordered ones. We may justify (13.33) on the following formal manipulations

$$\left[a_t, Y(t^-) \right] = \left[a_t, \int_0^{t^-} \dot{Y}_s ds \right]$$

$$= \int_0^{t^-} ds \left[a_t, \dot{Y}_s \right]$$

$$= \int_0^{t^-} ds \, \delta (t - s) Y_{1\nu} (s) [a_s]^{\nu}.$$

Here we have dropped the term $\left[a_t, Y_{\mu\nu} (s) \right]$ since this should vanish for $t > s$ as the integrands are adapted by assumption. We get the answer we want if we adopt the rule that the $\int_0^{t^-} ds \, \delta (t - s) f (s) \equiv \frac{1}{2} f (t^-)$. A similar set of manipulations would suggest that $\left[X_t, a_t^* \right] = \frac{1}{2} \left[a_t^* \right]^{\alpha} X_{\alpha 1} (t)$, and so

$$X_t \dot{Y}_t = X_t \left[a_t^{\alpha} \right] Y_{\alpha\beta} (t) [a_t]^{\beta}$$

$$= \left[a_t^* \right]^{\alpha} X_t Y_{\alpha\beta} (t) [a_t]^{\beta} + \frac{1}{2} \left[a_t^* \right]^{\alpha} X_{\alpha 1} (t) Y_{1\beta} (t) [a_t]^{\beta}$$

and therefore

$$\dot{X}_t Y_t + X_t \dot{Y}_t = \left[a_t^* \right]^{\alpha} X_{\alpha\beta} (t) Y (t) [a_t]^{\beta} + \left[a_t^* \right]^{\alpha} X_t Y_{\alpha\beta} (t) [a_t]^{\beta}$$

$$+ \left[a_t^* \right]^{\alpha} X_{\alpha 1} (t) Y_{1\beta} (t) [a_t]^{\beta},$$

which is precisely the quantum Itō product rule we desire.

So far, we have used the symbols a_t and a_t^* purely as a notation for terms under the integral sign in quantum stochastic integrals. However, this is motivated by the following: whenever we multiply these integrals together, we encounter these operator densities out of normal (Wick) order, and the upshot of the preceding is that in the process of putting these to normal order, we end up with the Itō correction!

One ends up with a technique that extends to multiple integrals where one can essentially ignore the diagonal terms provided you give a rule for what to do when you encounter a delta-function supported at the boundary of a simplex. A straightforward way of tackling this is to take the commutation relations to be (see Accardi et al., 2002; Gough, 1997)

$$\left[a_t, a_s^* \right] = \eth_* (t - s),$$ (13.34)

where we have

$$\eth_* (t - s) = \frac{1}{2} \eth_+ (t - s) + \frac{1}{2} \eth_- (t - s)$$ (13.35)

with the rules that

$$\int f (s) \, \eth_\pm (t - s) = f \left(t^\pm \right).$$ (13.36)

This brings us back to our introductory comments. A fully rigorous theory has been developed known as *quantum white noise* analysis, Accardi et al. (2000), and is a natural generalization of the Hida theory of (classical) white noise, as in Obata (1994). We should mention that an extension to higher-order processes has been given by Accardi et al. (1996); see also Accardi et al. (2002).

13.8 Quantum Stochastic Exponentials

As in the classical case, we are faced with several possible mechanisms to define a stochastic exponential of a quantum stochastic integral process. Filling the role of the diagonal free exponential is the Itō time-ordered exponential.

The **quantum Itō time-ordered exponential** is denoted as

$$U_t = \vec{T}_I e^{X_t}$$

and defined as the solution to the QSDE

$$dU_t = (dX_t) \, U_t, \qquad U_0 = I.$$

This admits the series expansion

$$\vec{T}_I e^{X_t} = \sum_n \int_{\Delta_n(t)} dX_{t_n} \dots dX_{t_1}.$$

Note that the increments do not necessarily commute – even for different times!

An alternative is the Holevo time-ordered exponential introduced in Holevo (1992).

The **Holevo time-ordered exponential** is denoted as

$$V_t = \vec{T}_H e^{Y_t}$$

and defined by

$$dV_t = \left(e^{dY_t} - 1\right) V_t, \qquad V_0 = I.$$

If the increments commuted, we could reconstitute this as $\vec{T}_H e^{Y_t} = e^{Y_t}$, but this would be an exceptional situation.

The time-ordered exponential is to be understood as the "Trotterization"

$$\vec{T}_H e^{Y_t} = \lim_{n \to \infty} \left(e^{Y(t_n) - Y(t_{n-1})} - 1\right) \cdots \left(e^{Y(t_1) - Y(t_0)} - 1\right),$$

where $t = t_n > t_{n-1} > \cdots > t_0 = 0$ is a grid with $\max_k (t_k - t_{k-1}) \to 0$ in the limit. Holevo (1992, 1996) establishes convergence on the domain of exponential vectors for stochastic integrals $Y_t = \int_0^t Y_{\alpha\beta} dA^{\alpha\beta}$ under the condition that the coefficients are ultrastrongly admissible, that is, there is a sequence $\left(Y_{\alpha\beta}^{(n)}\right)$ of adapted simple processes such that $|||Y^{(n)}|||_t < \infty$ and $|||Y - Y^{(n)}|||_t \to 0$, where

$$|||Y|||| \triangleq \text{ess} - \sup_{t \le s \le 0} \sup_{i,j} \|Y_{ij}(s)\| + \sqrt{\int_0^t \sum_i \|Y_{i0}(s)\|^2 \, ds}$$

$$+ \sqrt{\int_0^t \sum_j \|Y_{0j}(s)\|^2 \, ds} + \int_0^t \|Y_{00}(s)\| \, ds.$$

In particular, the number of noise channels may be countable. The Holevo time-ordered exponential arises naturally as the continuous limit of discrete time open quantum dynamics; see, for instance, Attal and Pautrat (2006) and Gough (2004).

To determine the relationship between the two, we need to work out the integrals $X_t^{[n]} \equiv \int_0^t (dX)^n$.

Lemma 13.8.1 *Let X be a quantum stochastic integral determined by $\dot{X} = [a_t]^\alpha X_{\alpha\beta}(t) [a_t]^\beta$; then (for $n \ge 2$)*

$$\dot{X}_t^{[n]} \equiv [a_t^*]^\alpha X_{\alpha 1}(t) X_{11}(t)^{(n-2)} X_{1\beta}(t) [a_t]^\beta. \tag{13.37}$$

*Let f be an analytic function, say $f(z) = \sum_{n \ge 0} \frac{1}{n!} f_n z^n$, then the kth **decapitated version** of f is the analytic function:*

$$f_k(z) \triangleq \sum_{n\geq 0} \frac{1}{(n+k)!} f_{n+k} z^n = \frac{f(z) - \sum_{n=0}^{k-1} \frac{1}{n!} f_n z^n}{z^k}.$$

Proposition 13.8.2 *We have the identification* $\vec{T}_I e^{X_t} = \vec{T}_H e^{Y_t}$ *with the quantum stochastic integrals related by*

$$X_{\alpha\beta} = Y_{\alpha\beta} + Y_{\alpha 1} \exp_2(Y_{11}) Y_{1\beta},$$
$$Y_{\alpha\beta} = X_{\alpha\beta} + X_{\alpha 1} \varphi_2(X_{11}) X_{1\beta},$$

where $\varphi(z) = \ln(1+z)$.

Note that $\exp_2(z) = \frac{e^z - 1 - z}{z^2}$ and $\varphi_2(z) = \frac{\ln(1+z) - z}{z^2}$. The result is immediate from the Itō rule, and the proposition is a consequence of the relations $X_t = \sum_{n\geq 1} \frac{1}{n!} Y_t^{[n]}$ and $Y_t = \sum_{n\geq 1} \frac{(-1)^{n+1}}{n} X_t^{[n]}$, which we encountered in Theorem 2.5.1, and which also holds in the noncommutative case. In particular, note that

$$X_{11} = e^{Y_{11}} - I, X_{10} = \exp_1(Y_{11}) Y_{10},$$
$$X_{01} = Y_{01} \exp_1(Y_{11}), X_{00} = Y_{00} + Y_{01} \exp_2(Y_{11}) Y_{10},$$
$$Y_{11} = \ln(I + Y_{11}), Y_{10} = \varphi_1(X_{11}) X_{10},$$
$$Y_{01} = X_{01} \varphi_1(X_{11}), Y_{00} = X_{00} + X_{01} \varphi_2(X_{11}) X_{10}.$$

Yet another possibility is afforded by the Stratonovich time-ordered exponential. We define

$$W_t = \vec{T}_S e^{Z_t}$$

as the solution to the QSDE

$$dW_t = (\delta Z_t) W_t, \qquad W_0 = I.$$

This is the same formal definition as the time-ordered Stratonovich exponential (2.27), though the specification that the Stratonovich increment appears on the left is now a crucial aspect.

This admits the series expansion

$$\vec{T}_S e^{Z_t} = \sum_n \int_{\Delta_n(t)} \delta Z_{t_n} \ldots \delta Z_{t_1}.$$

Proposition 13.8.3 *We have* $\vec{T}_S e^{Z_t} = \vec{T}_I e^{X_t}$, *where*

$$X_{\alpha\beta} = Z_{\alpha\beta} + Z_{\alpha 1} \frac{1}{I - \frac{1}{2} Z_{11}} Z_{1\beta},$$

$$Z_{\alpha\beta} = X_{\alpha\beta} + X_{\alpha 1} \frac{1}{I + \frac{1}{2} X_{11}} X_{1\beta}.$$

Proof As in Proposition 2.6.1, the consistency condition is $dX = dZ + \frac{1}{2}dZdX$; however, this time the order of the product is important. Comparing coefficients of the fundamental noise processes shows that

$$X_{\alpha\beta} = Z_{\alpha\beta} + \frac{1}{2}Z_{\alpha 1}X_{1\beta},$$

and in particular $X_{11} = \frac{Z_{11}}{1 - \frac{1}{2}Z_{11}}$ and $Z_{11} = \frac{X_{11}}{1 + \frac{1}{2}X_{11}}$. The other terms are readily worked out through simple algebra. $\qquad\square$

13.9 The Belavkin–Holevo Representation

The Hudson–Parthasarathy theory leads to a quantum Itō calculus where we have the product rule

$$d(X_t.Y_t) = dX_t.Y_t + X_t.dY_t + dX_t.dY_t. \tag{13.38}$$

Here the Itō correction may be thought of as a being due to Wick ordering. It is also possible to give a representation, akin to the matrix representation of the the Heisenberg group, where correction is accounted for by having the ordinary product but with higher dimensional matrices. This was given by Belavkin (1988) and Holevo (1989).

13.9.1 Matrix Algebra Notation

Let $\mathfrak{A}$ be a fixed algebra and let $\mathfrak{A}^{n \times m}$ denote the set of $n \times m$ arrays with entries in $\mathfrak{A}$. Given $X \in \mathfrak{A}^{n \times m}$ and $Y \in \mathfrak{A}^{m \times r}$, we say that the pair (X, Y) is composable and define their product $XY \in \mathfrak{A}^{n \times r}$ to be the usual matrix product.

Taking $n \geq 1$ to be a fixed dimension, we consider arrays of the form

$$X_{00} \in \mathfrak{A}^{1 \times 1} \equiv \mathfrak{A},$$
$$X_{0\ell} = [X_{01}, \ldots, X_{0n}] \in \mathfrak{A}^{1 \times n},$$
$$X_{\ell 0} = \begin{bmatrix} X_{10} \\ \vdots \\ X_{n0} \end{bmatrix} \in \mathfrak{A}^{n \times 1},$$
$$X_{\ell\ell} = \begin{bmatrix} X_{11} & \cdots & X_{1n} \\ \vdots & & \vdots \\ X_{n1} & \cdots & X_{nn} \end{bmatrix} \in \mathfrak{A}^{n \times n}.$$

We may assemble these components into the following square matrices

$$\mathbf{X} = \begin{bmatrix} X_{00} & X_{0\ell} \\ X_{\ell 0} & X_{\ell\ell} \end{bmatrix} \in \mathfrak{A}^{(1+n) \times (1+n)} \tag{13.39}$$

and

$$\mathbb{X} = \begin{bmatrix} 0 & X_{0\ell} & X_{00} \\ 0 & X_{\ell\ell} & X_{\ell 0} \\ 0 & 0 & 0 \end{bmatrix} \in \mathfrak{A}^{(1+n+1)\times(1+n+1)}. \tag{13.40}$$

We shall write $\mathbf{X} \longleftrightarrow \mathbb{X}$ whenever $\mathbf{X}$ and $\mathbb{X}$ take the forms (13.39) and (13.40) respectively with the same component entries $\{X_{\alpha\beta}\}$. Important special cases are the following:

$$\mathbf{P} \triangleq \begin{bmatrix} 0 & 0 \\ 0 & I_n \end{bmatrix}, \qquad \mathbb{P} \triangleq \begin{bmatrix} 0 & 0 & 0 \\ 0 & I_n & 0 \\ 0 & 0 & 0 \end{bmatrix},$$

$$\mathbb{I} \triangleq \begin{bmatrix} 1 & 0 & 0 \\ 0 & I_n & 0 \\ 0 & 0 & 1 \end{bmatrix}, \qquad \mathbb{J} \triangleq \begin{bmatrix} 0 & 0 & 1 \\ 0 & I_n & 0 \\ 1 & 0 & 0 \end{bmatrix},$$

where I_n is the $n \times n$ identity matrix.

Lemma 13.9.1 *We have the following identifications:*

$$\mathbf{X}^\dagger \longleftrightarrow \mathbb{X}^* \triangleq \mathbb{J}\mathbb{X}^\dagger\mathbb{J},$$
$$\mathbf{XPY} \longleftrightarrow \mathbb{X}\mathbb{Y},$$
$$\mathbf{XY} \longleftrightarrow \mathbb{X}\mathbb{J}\mathbb{Y}.$$

It is instructive to check these relations. To begin with, we easily see that

$$\mathbb{X}^\star = \begin{bmatrix} 0 & (X_{\ell 0})^\dagger & (X_{00})^\dagger \\ 0 & (X_{\ell\ell})^\dagger & (X_{0\ell})^\dagger \\ 0 & 0 & 0 \end{bmatrix},$$

which is the required expression. We shall refer to $\mathbb{X}^\star$ as the *twisted involution* of $\mathbb{X}$, and we indeed have the properties $(\mathbb{X}^\star)^\star = \mathbb{X}$, $(\alpha\mathbb{X} + \beta\mathbb{Y})^\star = \bar\alpha\mathbb{X}^\star + \bar\beta\mathbb{Y}^\star$. Next we compare $\mathbf{XPY} = \begin{bmatrix} X_{0\ell}Y_{\ell 0} & X_{0\ell}Y_{\ell\ell} \\ X_{\ell\ell}Y_{\ell 0} & X_{\ell\ell}Y_{\ell\ell} \end{bmatrix}$ with

$$\mathbb{X}\mathbb{Y} = \begin{bmatrix} 0 & X_{0\ell} & X_{00} \\ 0 & X_{\ell\ell} & X_{\ell 0} \\ 0 & 0 & 0 \end{bmatrix} \begin{bmatrix} 0 & Y_{0\ell} & Y_{00} \\ 0 & Y_{\ell\ell} & Y_{\ell 0} \\ 0 & 0 & 0 \end{bmatrix}$$

$$= \begin{bmatrix} 0 & X_{0\ell}Y_{\ell\ell} & X_{0\ell}Y_{\ell 0} \\ 0 & X_{\ell\ell}Y_{\ell\ell} & X_{\ell\ell}Y_{\ell 0} \\ 0 & 0 & 0 \end{bmatrix}. \tag{13.41}$$

The last relation we leave as an exercise.

It is convenient to the following vector notations on $\mathfrak{A}^{1+n+1}$:

$$
\phi = \begin{bmatrix} \phi_{-1} \\ \phi_1 \\ \vdots \\ \phi_n \\ \phi_0 \end{bmatrix}, \qquad \phi^\star = \phi^\dagger \mathbb{J} = \left[\phi_0^\dagger | \phi_1^\dagger, \ldots, \phi_n^\dagger | \phi_{-1}^\dagger \right].
$$

13.9.2 Quantum Stochastic Integrals

Take $\{X_{\alpha\beta}(t) : t \geq 0\}$ to be a family of adapted quantum stochastic processes, with quantum stochastic integral $X_t = \int_0^t X_{\alpha\beta}(s) \, dA_s^{\alpha\beta}$, where the differentials are understood in the Itō sense.

The coefficients $\{X_{\alpha\beta}(t)\}$ may be assembled into a matrix $\mathbf{X}_t$, as in the preceding, which we call the *Itō matrix* for the process, and also into a matrix $\mathbb{X}_t$, as in the preceding, and we shall refer to this as the *Belavkin–Holevo matrix* for the process.

The Itō matrix for a product $X_t Y_t$ of quantum Itō integrals will then have entries $\{X_{\alpha\beta}Y + XY_{\alpha\beta} + X_{\alpha k}Y_{k\beta}\}$ and is therefore given by $\mathbf{X}Y + X\mathbf{Y} + \mathbf{XPY}$. We now see the importance of Lemma 13.9.1: the Itō correction is described by the Itō matrix $\mathbf{XPY}$, but this is equivalent to just the ordinary product $\mathbb{X}\mathbb{Y}$ of the Belavkin–Holevo matrices.

In particular, the Belavkin–Holevo matrix for the product $X_t Y_t$ can be written as

$$
\mathbb{X}Y + X\mathbb{Y} + \mathbb{X}\mathbb{Y} = (\mathbb{X}\mathbb{I} + X)(Y\mathbb{I} + \mathbb{Y}) - XY\mathbb{I}. \tag{13.42}
$$

We recover the fluxion differentials from either the Itō or Belavkin–Holevo matrices via the formulas

$$
\frac{dX}{dt} = \mathfrak{a}^\dagger \mathbf{X} \mathfrak{a} = \mathrm{a}^\star \mathbb{X} \mathrm{a}
$$

with

$$
\mathfrak{a} = \begin{bmatrix} a_0 \\ a_1 \\ \vdots \\ a_n \end{bmatrix}, \qquad \mathrm{a} = \begin{bmatrix} 0 \\ a_1 \\ \vdots \\ a_n \\ a_0 \end{bmatrix}.
$$

If we wish to use the Itō differential notation, then we must use the slightly more cumbersome matrix formula

$$dX = tr\left\{\mathbb{X}d\widetilde{\mathbb{A}}\right\},$$

where (with $\prime$ denoting the usual transpose for arrays)

$$d\widetilde{\mathbb{A}} \triangleq \begin{bmatrix} 0 & 0 & 0 \\ \left(dA^{0\ell}\right)' & \left(dA^{\ell\ell}\right)' & 0 \\ dA^{00} & \left(dA^{\ell 0}\right)' & 0 \end{bmatrix}.$$

Note here that the matrix $\mathbb{X}d\widetilde{\mathbb{A}}$ then takes the form $\begin{bmatrix} \# & \# & 0 \\ \# & \# & 0 \\ 0 & 0 & 0 \end{bmatrix}$, where $\#$

indicates the nonzero entries, and it is easy to see that $tr\left\{\mathbb{X}d\widetilde{\mathbb{A}}\right\}$ takes the form

$$X_{0\ell}\left(dA^{0\ell}\right)' + X_{00}\left(dA^{00}\right)' + tr_d\left\{X_{\ell\ell}\left(dA^{\ell\ell}\right)' + X_{\ell 0}\left(dA^{\ell 0}\right)'\right\}$$

$$= X_{0j}dA^{0j} + X_{00}dA^{00} + X_{ij}dA^{ij} + X_{i0}dA^{i0}.$$

Putting these observations together, we get the following lemma.

Lemma 13.9.2 *Let X_t and Y_t be quantum stochastic integrals, then the quantum Itō product rule may be written as*[2]

$$\frac{d}{dt}(X_tY_t) = \mathsf{a}^\star(t)\left[(X_t\mathbb{I} + \mathbb{X}_t)(Y_t\mathbb{I} + \mathbb{Y}_t) - (X_tY_t)\mathbb{I}\right]\mathsf{a}(t). \quad (13.43)$$

The multiple version of the quantum Itō product rule is

$$d(X_tY_t\ldots Z_t) = (X_t + dX_t)(Y_t + dY_t)\cdots(Z_t + dZ_t) - (X_tY_t\ldots Z_t),$$

which follows from the product rule by basic induction. Putting the multiple product rule into Itō matrix notation involves some rather unwieldy expressions. In contrast, this is handled very efficiently in terms of Belavkin–Holevo matrices as the lemma generalizes immediately.

Corollary 13.9.3 *Let $X_t, Y_t, \ldots, Z_t$ be quantum stochastic integrals, then the multiple quantum Itō product rule may be written as*[3]

$$\frac{d}{dt}(X_tY_t\ldots Z_t) = \mathsf{a}^\star(t)\Bigg((X_t\mathbb{I} + \mathbb{X}_t)(Y_t\mathbb{I} + \mathbb{Y}_t)\cdots(Z_t\mathbb{I} + \mathbb{Z}_t)$$

$$- (X_tY_t\ldots Z_t)\,\mathbb{I}\Bigg)\mathsf{a}(t).$$

Another straightforward corollary of the lemma is the differential rule for functions of quantum integral processes.

---

[2] Alternatively, $d(X_tY_t) = tr\left\{\left[(X_t\mathbb{I} + \mathbb{X}_t)(Y_t\mathbb{I} + \mathbb{Y}_t) - (X_tY_t)\mathbb{I}\right]d\widetilde{\mathbb{A}}_t\right\}.$

[3] Or as $d(X_tY_t\ldots Z_t) = tr\left\{\left[(X_t\mathbb{I} + \mathbb{X}_t)(Y_t\mathbb{I} + \mathbb{Y}_t)\cdots(Z_t\mathbb{I} + \mathbb{Z}_t) - (X_tY_t\ldots Z_t)\mathbb{I}\right]d\widetilde{\mathbb{A}}_t\right\}.$

Corollary 13.9.4 *Let X_t be a quantum stochastic integral process and let f be analytic, then the process $f(X_t)$ has differential*

$$\frac{d}{dt} f(X_t) = \mathsf{a}^*(t) \left[f(X_t \mathbb{I} + \mathbb{X}_t) - f(X_t) \mathbb{I} \right] \mathsf{a}(t). \qquad (13.44)$$

To appreciate just how compact the expression (13.44) is, we give the explicit form of the differential when $f(x) = \sum_n f_n x^n$:

$$df(X_t) = f_{0\ell} \left(dA^{0\ell} \right)' + f_{00} \left(dA^{00} \right)' + tr_d \left\{ f_{\ell\ell} \left(dA^{\ell\ell} \right)' + f_{\ell 0} \left(dA^{\ell 0} \right)' \right\},$$

where

$$f_{0\ell} = \sum_n f_n \sum_{\substack{p+q=n-1 \\ p,q \geq 0}} X^p X_{0\ell} (X \mathbb{I}_d + X_{\ell\ell})^q,$$

$$f_{00} = \sum_n f_n \sum_{\substack{p+q=n-1 \\ p,q \geq 0}} X^p X_{00} X^q$$

$$+ \sum_n f_n \sum_{\substack{p+q+r=n-2 \\ p,q,r \geq 0}} X^p X_{0\ell} (X \mathbb{I}_d + X_{\ell\ell})^q X_{\ell 0} X^r,$$

$$f_{\ell\ell} = \left[f(X \mathbb{I}_d + X_{\ell\ell}) - f(X \mathbb{I}_d) \right],$$

$$f_{\ell 0} = \sum_n f_n \sum_{\substack{p+q=n-1 \\ p,q \geq 0}} (X \mathbb{I}_d + X_{\ell\ell})^p X_{\ell 0} X^q.$$

This leads us to the second key observation about the Belavkin–Holevo matrix notation: *The differential rule for quantum processes follows immediately from the Poisson–Itō differential rule.*

The expected value of $X_t = \int_0^t X_{\alpha\beta}(s) \, dA_s^{\alpha\beta}$ in the Fock vacuum state is given by $E_0[X_t] = \int_0^t X_{00}(s) \, ds$, as each of the $A^{\alpha\beta}$ are martingales for this state, with the exception of time. Let us denote the 00-coordinate map for Belavkin–Holevo matrices by ρ_0: that is, in terms of our previous notation, $\rho_0(\mathbb{X}) = X_{00}$. We need to be able to pull out the top-right component of the matrix, and the mapping ρ_0 that achieves this is

$$\rho_0(\mathbb{X}) \equiv \Omega_0^\star \mathbb{X} \Omega_0,$$

where we introduce the vector

$$\Omega_0 \triangleq \begin{bmatrix} 0 \\ 0 \\ 1 \end{bmatrix}.$$

13.9.3 Quantum Itō Algebras

It is of interest to consider abstract spaces of Itō algebras. Let us think of $\mathbb{X}$ as an "infinitesimal generator" of a quantum stochastic process $X_t = \Lambda_t(\mathbb{X})$. We then wish to consider the algebraic properties that would be desired for the set $\mathfrak{a}$ of these generators, as well as the possible representations for such sets. We begin by detailing some well-known examples.

Weiner–Itō Algebra

We consider the Wiener–Itō SDE $dX_t = v(X_t)\,dt + \sigma(X_t)\,dW_t$, where W_t is a Wiener process. In this case, the $d = 1$ representation can be used. The Wiener process may be described as $A_t^{10} + A_t^{01}$ and we have the Belavkin–Holevo matrix $\mathbb{X} = \sigma\mathbb{W} + v\mathbb{T}$, where we introduce the matrices

$$\mathbb{W} = \begin{bmatrix} 0 & 1 & 0 \\ 0 & 0 & 1 \\ 0 & 0 & 0 \end{bmatrix}, \qquad \mathbb{T} = \begin{bmatrix} 0 & 0 & 1 \\ 0 & 0 & 0 \\ 0 & 0 & 0 \end{bmatrix}.$$

The vector space $\mathfrak{w}$ spanned by $\{\mathbb{W}, \mathbb{T}\}$ is closed under matrix multiplication and has the product table

$\times$	$\mathbb{W}$	$\mathbb{T}$
$\mathbb{W}$	$\mathbb{T}$	0
$\mathbb{T}$	0	0

making $\mathfrak{w}$ a matrix algebra. We readily see that for f analytic

$$f(X\mathbb{I} + \sigma\mathbb{W} + v\mathbb{T}) = f(X)\mathbb{I} + \sigma f'(X)\mathbb{W} + \left(vf'(X) + \frac{1}{2}\sigma^2 f''(X)\right)\mathbb{T}.$$

From this we deduce the Wiener–Itō differential formula

$$df(X_t) = \sigma f'(X)\,dW_t + \left(vf'(X) + \frac{1}{2}\sigma^2 f''(X)\right)dt.$$

Poisson–Itō Algebra

Likewise, we consider the Poisson–Itō SDE $dX_t = v(X_t)\,dt + \sigma(X_t)\,dN_t$, where N_t is a Poisson process, here described as $A_t^{11} + A_t^{10} + A_t^{01} + A_t^{00}$. This time, we work with the Belavkin–Holevo matrix $\mathbb{X} = \sigma\mathbb{N} + v\mathbb{T}$, where we introduce the new matrices

$$\mathbb{N} = \begin{bmatrix} 0 & 1 & 1 \\ 0 & 1 & 1 \\ 0 & 0 & 0 \end{bmatrix}.$$

The matrix algebra $\mathfrak{n}$ generated by $\{\mathbb{N}, \mathbb{T}\}$ again closes and we have the product table

×	$\mathbb{N}$	$\mathbb{T}$
$\mathbb{N}$	$\mathbb{N}$	0
$\mathbb{T}$	0	0

and, for f analytic, we find

$$f(X\mathbb{I} + \sigma\mathbb{N} + v\mathbb{T}) = f(X)\mathbb{I} + \left[f(X+\sigma) - f(X)\right]\mathbb{N} + vf'(X)\mathbb{T}.$$

This implies the Poisson–Itō differential formula

$$df(X_t) = \left[f(X+\sigma) - f(X)\right]dN_t + vf'(X)dt.$$

Hudson–Parthasarathy Algebra

The matrices $\mathbb{W}$ and $\mathbb{N}$ do not commute, however, and by inspection one finds that the algebra generated by $\{\mathbb{T}, \mathbb{W}, \mathbb{N}\}$ is actually a four-dimensional algebra that is spanned by the elements $\mathbb{T}, \mathbb{A}, \mathbb{A}^\star$ and $\mathbb{S}$, where

$$\mathbb{A} \triangleq \begin{bmatrix} 0 & 0 & 0 \\ 0 & 0 & 1 \\ 0 & 0 & 0 \end{bmatrix}, \quad \mathbb{A}^\star \triangleq \begin{bmatrix} 0 & 1 & 0 \\ 0 & 0 & 0 \\ 0 & 0 & 0 \end{bmatrix}, \quad \mathbb{S} \triangleq \begin{bmatrix} 0 & 0 & 0 \\ 0 & 1 & 0 \\ 0 & 0 & 0 \end{bmatrix}.$$

In particular, $\mathbb{W} = \mathbb{A} + \mathbb{A}^\star$ and $\mathbb{N} = \mathbb{S} + \mathbb{A} + \mathbb{A}^\star + \mathbb{T}$. This algebra is called the Hudson–Parthasarathy algebra and is denoted as $\mathfrak{hp}$.

General Theory

Let us denote the space of matrices $\begin{bmatrix} 0 & X_{0\ell} & X_{00} \\ 0 & X_{\ell\ell} & X_{\ell 0} \\ 0 & 0 & 0 \end{bmatrix}$, considered as a sub-set of $\mathfrak{A}^{(1+n+1)\times(1+n+1)}$ by $\mathfrak{hp}(n)$. This is becomes a nonunital $\star$-algebra when we specify the twisted involution $\star$. We may extend to a unital algebra by adding the identity $\mathbb{I}$.

An example of some elements are the following matrices determined from their Itō coefficients $X_{\alpha\beta}$: for fixed $1 \leq i \leq d$,

$$\mathbb{T}: \quad X_{\alpha\beta} = \delta_{\alpha 0}\delta_{\beta 0}$$
$$\mathbb{A}_i: \quad X_{\alpha\beta} = \delta_{\alpha 0}\delta_{\beta i}$$
$$\mathbb{S}_{ij}: \quad X_{\alpha\beta} = \delta_{\alpha i}\delta_{\beta j},$$

along with $\mathbb{W}_i = \mathbb{A}_i + \mathbb{A}_i^\star$, and $\mathbb{N}_i = \mathbb{S}_{ii} + \mathbb{A}_i + \mathbb{A}_i^\star + \mathbb{T}$. These elements are all self-adjoint with respect to the twisted involution. We consider the following matrix $\star$-subalgebras of $\mathfrak{hp}(n)$:

$$t = \text{span}\{\mathbb{T}\}$$
$$\mathfrak{w}_i = \text{span}\{\mathbb{T}, \mathbb{W}_i\}$$
$$\mathfrak{n}_i = \text{span}\{\mathbb{T}, \mathbb{N}_i\}$$
$$\mathfrak{qd}_i = \text{span}\{\mathbb{A}_i, \mathbb{A}_i^*, \mathbb{T}\}$$
$$\mathfrak{hp}_i = \text{span}\{\mathbb{S}_{ii}, \mathbb{A}_i, \mathbb{A}_i^*, \mathbb{T}\}$$

We shall refer to this matrix representation as the *canonical representation* of the Itō algebra $\mathfrak{hp}(n)$ and its subalgebras. Here t is the Newtonian algebra for a deterministic time variable. $\mathfrak{w}_i$ is the Wiener–Itō algebra for the ith channel and $\mathfrak{n}_i$ is the Poisson–Itō algebra for the channel. Next we have the quantum diffusion algebra for the channel $\mathfrak{qd}_i$, and we write $\mathfrak{qd}(n)$ for the full algebra of quantum diffusions for all n channels. The latter consists of all matrices of the

form $\mathbb{X} = \begin{bmatrix} 0 & X_{0\ell} & X_{00} \\ 0 & 0 & X_{\ell 0} \\ 0 & 0 & 0 \end{bmatrix}$ and we have the property that $\mathfrak{qd}(n).\mathfrak{qd}(n) = t$, that is, $\mathbb{X}\mathbb{Y}$ is a multiple of $\mathbb{T}$ whenever $\mathbb{X}, \mathbb{Y} \in \mathfrak{qd}(n)$. Finally $\mathfrak{hp}_i$ is the Hudson–Parthasarathy algebra for the ith channel, and it is interesting to note that it is the algebra generated by the elements $\{\mathbb{N}_i, \mathbb{W}_i, \mathbb{T}\}$.

In general, we have an identification $dX_t = X_{\alpha\beta}(t)\, dA_t^{\alpha\beta} \equiv d\Lambda_t(\mathbb{X})$ between a quantum stochastic integral and its Belavkin–Holevo matrix process $\{\mathbb{X}_s : s \geq 0\}$. We can write out the basic axiomatic requirements for an abstract Itō algebra $\mathfrak{a}$ for stochastic integrals with expectation E:

- $\alpha d\Lambda_t(\mathbb{X}) + \beta d\Lambda_t(\mathbb{Y}) = d\Lambda_t(\alpha\mathbb{X} + \beta\mathbb{Y})$
- $d\Lambda_t(\mathbb{X}).d\Lambda_t(\mathbb{Y}) = d\Lambda_t(\mathbb{X}\mathbb{Y})$
- $d\Lambda_t(\mathbb{X})^\dagger = d\Lambda_t(\mathbb{X}^*)$
- $d\{\Lambda_t(\mathbb{X})\Lambda_t(\mathbb{Y})\} = (\Lambda_t(\mathbb{X})\mathbb{I} + \mathbb{X})(\Lambda_t(\mathbb{Y})\mathbb{I} + \mathbb{Y}) - \Lambda_t(\mathbb{X})\Lambda_t(\mathbb{Y})\mathbb{I}$
- There exists a unique self-adjoint $\mathbb{T} \in \mathfrak{a}$ such that $\mathbb{T}\mathbb{X} = \mathbb{X}\mathbb{T} = 0$ for all $\mathbb{X} \in \mathfrak{a}$.
- There exists a linear functional ρ on $\mathfrak{a}$ such that

$$E[\Lambda_t(\mathbb{X})] = \int_0^t \rho(\mathbb{X}_s)\, ds.$$

In particular, $\mathfrak{a}$ must be a $\star$-algebra. The element $\mathbb{T}$ is self-adjoint ($\mathbb{T} = \mathbb{T}^\star$) but also nilpotent ($\mathbb{T}^2 = 0$), and this means that we cannot have a representation of $\mathfrak{a}$ in terms of matrices with the ordinary involution of Hermitean conjugation: hence the necessity of a twisted involution. The functional ρ is said to be faithful on $\mathfrak{a}$ if the only element, $\mathbb{Z}$, such that $\rho(\mathbb{Z}) = \rho(\mathbb{X}\mathbb{Z}) = \rho(\mathbb{Z}\mathbb{Y}) = \rho(\mathbb{X}\mathbb{Z}\mathbb{Y}) = 0$ for all $\mathbb{X}, \mathbb{Y} \in \mathfrak{a}$ is $\mathbb{Z} = 0$. Under conditions of faithfulness, it can be shown that any finitely generated Itō algebra can be canonically represented.

13.9.4 Evolutions and Dynamical Flows

Hudson and Parthasarathy show that the QSDE

$$\frac{dU}{dt} = \mathrm{a}^{\dagger}\mathbb{G}\mathrm{U}\mathrm{a} \quad U_0 = 1$$

has a unique solution for a given constant Belavkin–Holevo matrix $\mathbb{G}$ on coefficients on $\mathfrak{B}(\mathfrak{h}_0)$, the bounded operators on $\mathfrak{h}_0$. Necessary and sufficient conditions for unitarity are then given by

$$(\mathbb{I}+\mathbb{G})(\mathbb{I}+\mathbb{G})^{\star} = \mathbb{I} = (\mathbb{I}+\mathbb{G})^{\star}(\mathbb{I}+\mathbb{G}),$$

that is, $\mathbb{I}+\mathbb{G}$ is (twisted) unitary on $\mathfrak{B}(\mathfrak{h}_0)^{(1+n+1)\times(1+n+1)}$.

The next result shows how we may arrive at the Holevo time-ordered exponential coefficients and the Stratonivich coefficients.

Proposition 13.9.5 *Let $\mathbb{H}$ and $\mathbb{E}$ be (twisted) Hermitean matrices, that is, $\mathbb{H} = \mathbb{H}^{\star}$ and $\mathbb{E} = \mathbb{E}^{\star}$, then either prescription*

$$\mathbb{I}+\mathbb{G} = e^{-i\mathbb{H}}, \text{ or } \mathbb{I}+\mathbb{G} = \frac{\mathbb{I}-\frac{i}{2}\mathbb{E}}{\mathbb{I}+\frac{i}{2}\mathbb{E}}$$

determines a (twisted) unitary.

This is relatively easy to see, as the first expression is evidently unitary while the second is a Cayley transform. Conversely, given a twisted unitary $\mathbb{I}+\mathbb{G}$, we refer to $\mathbb{H} = i\ln(\mathbb{I}+\mathbb{G})$ as the generator matrix for the Holevo time-ordered exponential (logarithm may not be unique!) and $\mathbb{E} = -i\left(\mathbb{I}-\frac{i}{2}\mathbb{G}\right)^{-1}\mathbb{G}$ as the Stratonovich generator matrix.

We remark that for quantum diffusions, these notions coincide since, if $\mathbb{X} \in \mathfrak{qd}(n)$, then $\mathbb{X}^2 \equiv \xi_X \mathbb{T}$ for some operator ξ_X, and we have that

$$e^{-i\mathbb{X}} = \frac{\mathbb{I}-\frac{i}{2}\mathbb{X}}{\mathbb{I}+\frac{i}{2}\mathbb{X}} = \mathbb{I}-i\mathbb{X}-\frac{1}{2}\xi_X\mathbb{T}.$$

The expressions generally differ, however, when scattering is involved.

Let us explain briefly the origin of the Stratonovich generator. We may define the scalar product of a process X_t, having Belavkin–Holevo matrix $\mathbb{X}_t$, with a second Belavkin–Holevo matrix $\mathbb{Y}_t$ to be

$$\mathbb{X}_t \circ \mathbb{Y}_t \triangleq \left(X_t\mathbb{I}+\frac{1}{2}\mathbb{X}_t\right)\mathbb{Y}_t$$

and similarly $\mathbb{X}_t \circ Y_t \triangleq X_t \left(Y_t \mathbb{I} + \frac{1}{2} \mathbb{Y}_t \right)$, then

$$\frac{d}{dt} (X_t Y_t) = \frac{dX_t}{dt} \circ Y_t + X_t \circ \frac{dY_t}{dt},$$

where $\frac{dX_t}{dt} \circ Y_t = a^\star (t) (\mathbb{X}_t \circ Y_t) a(t)$, and so on. Formally, we have the Wick ordering rules

$$a^\star (t) (\mathbb{X}_t \circ Y_t) a(t) = a^\star (t) \mathbb{X}_t a(t) \circ Y_t,$$
$$a^\star (t) (X_t \circ \mathbb{Y}_t) a(t) = X_t \circ a^\star (t) \mathbb{Y}_t a(t).$$

For the QSDEs

$$\frac{dU_t}{dt} = -i a^\star (t) \mathbb{E} a(t) \circ U_t \text{ and } \frac{dU_t}{dt} = a^\star (t) \mathbb{G} U_t a(t)$$

to be equivalent, we need the consistency condition $\mathbb{G} = -i\mathbb{E} - \frac{i}{2}\mathbb{E}\mathbb{G}$, and this is precisely the relation introduced in the preceding.

14

Quantum Stochastic Limits

In this chapter, we look at the problem of quantum stochastic processes as approximations to physical models. Here, in contrast to the usual approximations for open systems that restrict their attention to just the reduced model, we are interested in a limit that captures both the system and its environment, with the latter taking a limit form of the Fock space of the Hudson–Parthasarathy quantum stochastic processes. The earliest formulation of this is by Accardi et al. (1989), which was formulated as a weak coupling limit resulting in a quantum diffusive evolution (Accardi et al., 1990). Subsequently, this was extended to include low-density limits that involved the Poissonian processes in the limit. For a detailed account, we refer to the book on the *Quantum Stochastic Limit* (Accardi et al., 2002). In this chapter, we formulate a general problem leading to a mixed Gaussian–Poissonian limit (Gough, 2005).

14.1 A Quantum Wong Zakai Theorem

We will consider limit problems of the following type. First fix an initial space $\mathfrak{h}_S$ describing the Hilbert space of a quantum mechanical systems, and Bose Fock space $\mathfrak{F}_R$ to model the reservoir (the environment of the system). On the joint space $\mathfrak{h}_S \otimes \mathfrak{F}_R$, we consider the time-dependent dynamics due to a unitary family $U_t^{(\lambda)}$, where $t \geq 0$ is time and $\lambda > 0$ is a parameter, with the associated Schrödinger equation

$$\frac{\partial}{\partial t} U_t^{(\lambda)} = -i \,\Upsilon_t(\lambda)\, U_t^{(\lambda)}, \tag{14.1}$$

where the time-dependent Hamiltonian $\Upsilon_t(\lambda)$ is assumed to take the form

$$\begin{aligned}
\Upsilon_t^{(\lambda)} &= E_{11} \otimes a_t^+(\lambda) a_t^-(\lambda) + E_{10} \otimes a_t^+(\lambda) + E_{01} \otimes a_t^-(\lambda) + E_{00} \otimes 1 \\
&= E_{\alpha\beta} \otimes [a_t^+(\lambda)]^\alpha [a_t^-(\lambda)]^\beta.
\end{aligned} \tag{14.2}$$

Here the operators $a_t^{\pm}(\lambda)$ are creation $(+)$ and annihilation $(-)$ operators on the Bose Fock space $\mathfrak{F}_R$. Our main assumption is that the reservoir is in its vacuum state, which we denote by Ω, and that the creation/annihilation operators in the Hamiltonian (14.2) satisfy commutation relations of the form

$$\left[a_t^-(\lambda), a_s^+(\lambda)\right] = C_\lambda(t - s). \tag{14.3}$$

We shall then assume that the two-point function will converge to a singular delta-function as $\lambda \to 0$. For definiteness, we assume that there exists a continuous integrable function C with the properties that

$$\int_{-\infty}^0 C(t)\, dt = \frac{1}{2} = \int_0^\infty C(t)\, dt$$

and so that

$$C_\lambda(t) \equiv \frac{1}{\lambda^2} C\left(\frac{t}{\lambda^2}\right). \tag{14.4}$$

Our interest will be in the limit of $U_t^{(\lambda)} = \mathcal{T} \exp\left\{-i \int_0^t ds\, \Upsilon_s^{(\lambda)}\right\}$ as $\lambda \to 0$, and we would like to realize the limit as a quantum stochastic process. For each $\lambda > 0$ fixed, we define the collective creator

$$A^+(f, \lambda) \triangleq \int_0^\infty f(t)\, a_t(\lambda)\, dt$$

for square-integrable functions f, and similarly $A^-(f, \lambda) = A^+(f, \lambda)^*$. Likewise, we define the collective Weyl displacement operator as

$$D(f, \lambda) \triangleq \exp\left\{A^+(f, \lambda) - A^-(f, \lambda)\right\}. \tag{14.5}$$

We aim to study the limit of matrix elements of $U_t^{(\lambda)}$ between collective exponential vectors of the type $D(f, \lambda)\Omega$ and show that they agree with those of a unitary quantum stochastic process. Let us note the following facts that show this is reasonable.

Lemma 14.1.1 *Let $f_1, f_2 \in L^2(\mathbb{R}_+, dt)$, then*

$$\lim_{\lambda \to 0}\left[A^-(f_1, \lambda), A^+(f_2, \lambda)\right] = \int_0^\infty f_1(t)^* f_2(t)\, dt.$$

Proof For finite $\lambda > 0$, we have

$$\left[A^-(f_1, \lambda), A^+(f_2, \lambda)\right] = \int_0^\infty dt_1 \int_0^\infty dt_2 f_1(t_1)^* C_\lambda(t_1 - t_2) f_2(t_2),$$

and the result follows from the limiting delta-function assumption for the two-point function. $\qquad\square$

Corollary 14.1.2 *Let* $f_1, \ldots, f_n \in L^2(\mathbb{R}_+, dt)$ *for some integer* $n > 0$, *then*

$$\lim_{\lambda \to 0} \langle \Omega | D(f_1, \lambda) \ldots D(f_n, \lambda) \Omega \rangle = \langle \Omega | D(f_1) \ldots D(f_n) \Omega \rangle,$$

where the right-hand side is taken on the Fock space $\Gamma_+ \left(L^2(\mathbb{R}_+, dt) \right)$ – *we denote the vacuum again by* Ω.

The corollary readily follows from the lemma by virtue of the Gaussian moment generating function formulas.

We therefore have a quantum probabilistic limit theorem that says that the $A^{\pm}(f, \lambda)$ are converging in law to limit fields $A^*(f)$ and $A(f)$. The hope then is that limits of the form

$$\lim_{\lambda \to 0} \langle \phi_1 \otimes D(f_1, \lambda)\Omega | \, U_t^{(\lambda)} \, \phi_2 \otimes D(f_2, \lambda)\Omega \rangle \tag{14.6}$$

will exist and take the form $\langle \phi_1 \otimes D(f_1)\Omega | \, U_t \, \phi_2 \otimes D(f_2)\Omega \rangle$ for arbitrary vectors $\phi_j \in \mathfrak{h}$ and $f_j \in L^2(\mathbb{R}_+, dt)$ and where U_t is a well-defined quantum stochastic process on $\mathfrak{h} \otimes \Gamma_+ \left(L^2(\mathbb{R}_+, dt) \right)$. We will also try to establish a similar result for the Heisenberg evolution $J_t^{(\lambda)}(X) = U_t^{(\lambda)\dagger}(X \otimes 1_R)U_t^{(\lambda)}$, for fixed bounded observables $X \in \mathcal{B}(\mathfrak{h}_S)$.

The formal Dyson series development of $U_t^{(\lambda)}$ involves the multiple time integrals[1]

$$U_t^{(\lambda)} = \sum_{n=0}^{\infty} (-i)^n \int_{\Delta_n(t)} ds_n \ldots ds_1 \, \Upsilon_{s_n}(\lambda) \ldots \Upsilon_{s_1}(\lambda). \tag{14.7}$$

14.1.1 The Dyson Series Expansion of $U_t^{(\lambda)}$

Substituting the Dyson series (14.7) into (14.6), we obtain a series expansion:

$$\langle \phi_1 \otimes D(f_1, \lambda)\Omega | \, U_t^{(\lambda)} \, \phi_2 \otimes D(f_2, \lambda)\Omega \rangle$$

$$= \sum_{n=0}^{\infty} (-i)^n \int_{\Delta_n(t)} ds_n \ldots ds_1$$

$$\times \langle \phi_1 \otimes D(f_1, \lambda)\Omega | \, \Upsilon_{s_n}(\lambda) \ldots \Upsilon_{s_1}(\lambda) \, \phi_2 \otimes D(f_2, \lambda)\Omega \rangle. \tag{14.8}$$

Lemma 14.1.3 *We have the identity*

$$\langle \phi_1 \otimes D(f_1, \lambda)\Omega | \, \Upsilon_{s_n}(\lambda) \ldots \Upsilon_{s_1}(\lambda) \, \phi_2 \otimes D(f_2, \lambda)\Omega \rangle$$

$$= \langle \phi_1 \otimes \Omega | \, \tilde{\Upsilon}_{s_n}(\lambda) \ldots \tilde{\Upsilon}_{s_1}(\lambda) \, \phi_2 \otimes \Omega \rangle \, \langle D(f_1, \lambda)\Omega | D(f_2, \lambda)\Omega \rangle, \tag{14.9}$$

---

[1] We recall our notation: for $\sigma \in \mathfrak{S}_n$, we have the simplex $\Delta_n^{\sigma}(t) \triangleq \{(s_n, \ldots, s_1) : t > s_{\sigma(n)} > \cdots > s_{\sigma(1)} > 0\}$, and $\Delta_n(t)$ in (5.1) is the simplex corresponding to the identity permutation.

where $\tilde{\Upsilon}_s(\lambda)$ is

$$\tilde{\Upsilon}_s(\lambda) = \tilde{E}_{\alpha\beta}(t,\lambda) \otimes [a_t^+(\lambda)]^\alpha [a_t^-(\lambda)]^\beta, \tag{14.10}$$

where

$$\tilde{E}_{00}(t,\lambda) = E_{00} + E_{10}f_1(t,\lambda) + E_{01}f_2^*(t,\lambda) + E_{11}f_1(t,\lambda)f_2^*(t,\lambda);$$
$$\tilde{E}_{10}(t,\lambda) = E_{10} + f_2^*(t,\lambda)E_{11};$$
$$\tilde{E}_{01}(t,\lambda) = E_{01} + f_1(t,\lambda)E_{11};$$
$$\tilde{E}_{11}(t,\lambda) = E_{11}; \tag{14.11}$$

with

$$f_j(t,\lambda) = \int_0^\infty f_j(u)C_\lambda(t-u)\,du. \tag{14.12}$$

Proof On the left-hand side, we have the exponential vector

$$D(f_2,\lambda)\Omega = e^{-\frac{1}{2}\int\int f_2^*(s)C_\lambda(s-u)f_2(u)ds\,du}e^{A^+(f_2,\lambda)}\,\Omega.$$

If we try to move the operator $e^{A^+(f_2,\lambda)}$ to the far left, commuting past the operators $\Upsilon_{s_n}(\lambda)\ldots\Upsilon_{s_1}(\lambda)$ we find that we effectively implement the replacement of the annihilators as

$$a_t^-(\lambda) \to a_t^-(\lambda) + f_2^*(t,\lambda), \tag{14.13}$$

with the creators being unchanged. Similarly, doing the same procedure with $D(f_1,\lambda)$ we implement the change $a_t^+(\lambda) \to a_t^+(\lambda) + f_1(t,\lambda)$, with the annihilators being unaffected. The overall result is to replace each Hamiltonian term $\Upsilon_s^{(\lambda)}$ with $\tilde{\Upsilon}_s(\lambda) = E_{\alpha\beta} \otimes [a_t^+(\lambda) + f_1(t,\lambda)]^\alpha [a_t^-(\lambda) + f_2^*(t,\lambda)]^\beta$, which is then rearranged to give (14.10). $\qquad\square$

Up to the factor $(-i)^n\langle D(f_1,\lambda)\Omega | D(f_2,\lambda)\Omega\rangle$, we see that the n-th term in the Dyson series expansion (14.8) of the matrix element is

$$\int_{\Delta_n(t)} ds_n\ldots ds_1\,\langle\phi_1|\,\tilde{E}_{\alpha_n\beta_n}(s_n,\lambda)\ldots\tilde{E}_{\alpha_1\beta_1}(s_1,\lambda)\,\phi_2\rangle$$
$$\times\langle\Omega|\,[a_{s_n}^+(\lambda)]^{\alpha_n}[a_{s_n}^-(\lambda)]^{\beta_n}\ldots[a_{s_1}^+(\lambda)]^{\alpha_1}[a_{s_1}^-(\lambda)]^{\beta_1}\,\Omega\rangle. \tag{14.14}$$

The vacuum expectation in (14.14) can be computed using (9.93). Let us recall the identity

$$\sum_{\alpha,\beta\in\{0,1\}^n}\langle\Omega|\,[A^+(f_n)]^{\alpha(n)}[A^-(g_n)]^{\beta(n)}\ldots[A^+(f_1)]^{\alpha(1)}[A^-(g_1)]^{\beta(1)}\Omega\rangle$$
$$= \sum_{\pi\in\mathrm{Part}(n)}\prod_{\{i(1),\ldots,i(k)\}\in\pi}\langle g_{i(k)}|f_{i(k-1)}\rangle\cdots\langle g_{i(3)}|f_{i(2)}\rangle\langle g_{i(2)}|f_{i(1)}\rangle \tag{14.15}$$

for $f_1, g_1, \ldots, f_n, g_n \in f.$, where we take the various sets (parts of the partition) $\{i(1), \ldots, i(k)\} \in \pi$ to be ordered so that $i(1) < i(2) < \cdots < i(k)$ and if the set is a singleton it is given the factor of unity.

We again resort to a diagrammatic convention in order to describe the Dyson series expansion into sums of integrals of products of two-point functions. There is a one-to-one correspondence between the diagrams appearing in the n-th term of the Dyson series and set of partitions of the n vertices. The diagram pictured in Figure 9.3 would contribute a weight of

$$(-i)^{17} \int_{\Delta_{17}(t)} \tilde{E}^{01}(t_{17}, \lambda) \tilde{E}^{00}(t_{16}, \lambda) \ldots \tilde{E}^{10}(t_1, \lambda) \, C_\lambda(t_{17} - t_{11}) \ldots C_\lambda(t_2 - t_1)$$

to the series.

14.1.2 Principal Terms (in the Dyson Series)

A standard technique in perturbative quantum field theory and quantum statistical mechanics is to develop a series expansion and argue on physical grounds that certain "principal terms" will exceed the other terms in order of magnitude (Abrikosov et al., 1963). Often it is possible to resum the principal terms to obtain a useful representation of the dominant behavior. Mathematically, the problem comes down to showing that the remaining terms are negligible in the limiting physical regime being considered.

Let us consider a typical diagram. We shall assume that within the diagram there are n_1 singleton vertices $[\ldots, \underline{\quad\bullet\quad}, \ldots]$, n_2 contraction pairs $[\ldots \underline{\quad\overparen{\bullet\quad\bullet}\quad} \ldots]$, n_3 contraction triples $[\ldots \underline{\quad\overparen{\bullet\quad\overparen{\bullet\quad\bullet}\quad}\quad} \ldots]$, and so on. This yields a set of occupation numbers $\mathbf{n} = (n_j)$ and a diagram has a total of $E(\mathbf{n}) = \sum_j j n_j$ vertices, which are partitioned into $N(\mathbf{n}) = \sum_j n_j$ connected subdiagrams. We see that the total number of diagrams contributing to the n-th level of the Dyson series will be given by the Bell number B_n.

The resulting terms can be split into two types: **type I** will survive the $\lambda \to 0$ limit; **type II** will not. They are distinguished as follows:

Type I: Terms involving contractions of time consecutive annihilator/creator pairs only. (That is, under the time-ordered integral in (14.14), an annihilator $a_{s_{j+1}}^-(\lambda)$ must be contracted with the creator $a_{s_j}^+(\lambda)$.)

Type II: All others cases.

The terminology used here is due to Accardi et al. (1990).

Let n be a positive integer and $m \in \{0, \ldots, n-1\}$. Let $\{(p_j, q_j)\}_{j=1}^m$ be contractions pairs over indices $\{1, \ldots, n\}$ such that if $P = \{p_1, \ldots, p_m\}$ and

$Q = \{q_1, \ldots, q_m\}$, then P and Q are both nondegenerate subsets of size m and we require that $p_j < q_j$ for each j and that Q be ordered so that $q_1 < \cdots < q_m$. We understand that $(p_j, q_j)_{j=1}^m$ is *type I* if $q_j = p_j + 1$ for each j and *type II* otherwise. The following result is an extension of lemma 4.2 in Accardi et al. (1990).

Lemma 14.1.4 *Let* $(p_j, q_j)_{j=1}^m$ *be a set of m pairs of contractions over the set of indices* $\{1, \ldots, n\}$, *then*

$$\left| \int_{\Delta_n(t)} ds_1 \ldots ds_n \prod_{j=1}^m C_\lambda \left(s(p_j) - s(q_j) \right) \right| \leq \frac{1}{2^m} \frac{t^{n-m}}{(n-m)!}. \quad (14.16)$$

Moreover, as $\lambda \to 0$,

$$\int_{\Delta_n(t)} ds_1 \ldots ds_n \prod_{j=1}^m C_\lambda \left(s(p_j) - s(q_j) \right) \to \begin{cases} \frac{1}{2} \frac{t^{n-m}}{(n-m)!}, & \text{type I;} \\ 0, & \text{type II.} \end{cases} \quad (14.17)$$

Proof Let $q = q_1$ and set $t(q) = [s(p) - s(q)]/\lambda^2$, then

$$\left| \int_{\Delta_n(t)} ds_1 \ldots ds_n \prod_{j=1}^m \langle a_{s(p_j)}^-(\lambda) a_{s(q_j)}^+(\lambda) \rangle \right| =$$

$$\left| \int_0^t ds(1) \cdots \int_0^{s(q-2)} ds(q-1) \int_{[s(p)-s(q-1)]/\lambda^2}^{s(p)/\lambda^2} dt(q) \right.$$

$$\left. \int_0^{s(p)-\lambda^2 t(q)} ds(q+1) \cdots \int_0^{s(n-1)} ds(n) \, C\left(t(q)\right) \prod_{j=2}^m C_\lambda \left(s(p_j) - s(q_j) \right) \right|.$$

However, we have that $s(p) - \lambda^2 t(p) < s(q-1)$, and so we obtain the bound

$$\left| \int_0^t ds(1) \cdots \int_0^{s(q-2)} ds(q-1) \int_{-\infty}^\infty dt(q) \int_0^{s(p)-\lambda^2 t(q)} ds(q+1) \right.$$

$$\left. \cdots \int_0^{s(n-1)} ds(n) \, C\left(t(q)\right) \prod_{j=2}^m C_\lambda \left(s(p_j) - s(q_j) \right) \right|.$$

And so, working inductively we obtain (14.16).

Suppose now that the pairs are of *type I*, then $p = q - 1$ and so the lower limit of the $t(q)$-integral is zero. Consequently, we encounter the sequence of integrals

$$\int_0^{s(q-2)} ds(q-1) \int_0^{s(q-1)/\lambda^2} dt(q) \int_0^{s(q-1)-\lambda^2 t(q)} ds(q+1) C\left(t(q)\right) \ldots.$$

This occurs for each q-variable, and so we recognize the limit as stated in (14.17) for *type I* terms.

For *type II* pairs, on the other hand, let $j = \min\{k\colon p_k < q_k - 1\}$; setting $q = q_k$, we encounter the sequence of integrals

$$\int_0^{s(q-2)} ds(q-1) \int_{[s(p)-s(q-1)]/\lambda^2}^{s(p)/\lambda^2} dt(q) \int_0^{s(q-1)} ds(q+1)\, C\left(t\left(q\right)\right)\dots.$$

But now, with respect to the variables $s(1),\dots,s(p),\dots,s(q-1)$, we have that, since $s(p) \neq s(q-1)$, the lower limit $[s(p) - s(q-1)]/\lambda^2$ of the $t(q)$-integral is almost always negative and so, as $t \mapsto C(t)$ is continuous, we have the dominated convergence of the whole term to zero.   □

Clearly *type II* terms do not contribute to the n-th term in the series expansion in the limit. However, we must establish a uniform bound for all these terms when the sum over all terms is considered. We do this in the next section.

Before proceeding, let us remark that expression (14.14) is bounded by

$$C_{\alpha_n\beta_n}\dots C_{\alpha_1\beta_1}\,\|\phi_1\|\,\|\phi_2\|$$

$$\times \int_{\Delta_n(t)} ds_n\dots ds_1 \langle\Omega|[a_{s_n}^+(\lambda)]^{\alpha_n}[a_{s_n}^-(\lambda)]^{\beta_n}\dots[a_{s_1}^+(\lambda)]^{\alpha_1}[a_{s_1}^-(\lambda)]^{\beta_1}\Omega\rangle,$$

$$(14.18)$$

where

$$C_{11} = \|E_{11}\|;$$

$$C_{10} = \|E_{10}\| + \|E_{11}\|c_2; \quad C_{01} = \|E_{01}\| + \|E_{11}\|c_1;$$

$$C_{00} = \|E_{00}\| + \|E_{10}\|c_1 + \|E_{01}\|c_2 + \|E_{11}\|c_1 c_2 \qquad (14.19)$$

and $c_j = \int_{-\infty}^{\infty} du\, |C(u) f_j(u)\|$.

We will make the assumption that

$$\frac{1}{2}C_{11} < 1 \quad \text{and that} \quad C = \max\{C_{11}, C_{10}, C_{01}, C_{00}\} < \infty.$$

We need to do some preliminary estimation. We employ the occupation numbers introduced in Section 1.6.1. The number of times that we will have $(\alpha, \beta) = (1, 1)$ in a particular term will be $\sum_{j>2}(j - 2)n_j$ (that is, singletons and pairs have none, triples have one, quadruples have two, etc.), and this equals $E(\mathbf{n}) - 2N(\mathbf{n}) + n_1$. Therefore, we shall have

$$C_{\alpha_n\beta_n}\dots C_{\alpha_1\beta_1} \le C_{11}^{E(\mathbf{n})-2N(\mathbf{n})+n_1} C^{2N(\mathbf{n})-n_1}. \qquad (14.20)$$

14.1.3 Generalized Pulé Inequalities

We shall denote by Part($\mathbf{n}$) the set of all partitions having the same occupation number sequence $\mathbf{n}$. Given a partition $\pi \in$ Part($\mathbf{n}$) we use the convention

$q(j, k, r)$ to label the r-th element of the k-th j-tuple. A simple example of a partition in Part($\mathbf{n}$) is given by selecting in order from $\{1, 2, \ldots, E(\mathbf{n})\}$, first of all n_1 singletons, then n_2 pairs, then n_3 triples, and so on. The labeling for this particular partition will be denoted as $\bar{q}(\cdot, \cdot, \cdot)$, and explicitly we have

$$\bar{q}(j, k, r) = \sum_{l<j} l\, n_l + (k-1)n_j + r. \tag{14.21}$$

We shall denote by $\mathfrak{S}_\mathbf{n}^0$ the collection of **Pulé permutations**, that is, $\rho \in \mathfrak{S}_n$, $E(\mathbf{n}) = n$, such that $q = \rho \circ \bar{q}$ again describes a partition in Part($\mathbf{n}$). Specifically, $\mathfrak{S}_\mathbf{n}^0$ consists of all the permutations ρ for which the following requirements are met:

- the order of the individual j-tuples is preserved for each j:

$$\rho(\bar{q}(j, k, 1)) < \rho(\bar{q}(j, k', 1)) \qquad \forall j, 1 \le k < k' \le n_j; \tag{14.22}$$

- creation always precedes annihilation in time for any contraction pair:

$$\rho(\bar{q}(j, k, 1)) < \rho(\bar{q}(j, k, 2)) < \cdots < \rho(\bar{q}(j, k, j)) \qquad \forall j, 1 \le k \le n_j. \tag{14.23}$$

In these notations, we may rewrite the identity (9.93) as

$$\sum_{\alpha, \beta \in \{0,1\}^n} \langle \Omega | \, [A^+(f_n)]^{\alpha(n)}[A^-(g_n)]^{\beta(n)} \ldots [A^+(f_1)]^{\alpha(1)}[A^-(g_1)]^{\beta(1)} \Omega \rangle$$
$$= \sum_{\substack{\mathbf{n} \\ E(\mathbf{n})=n}} \sum_{\rho \in \mathfrak{S}_\mathbf{n}^0} \prod_{j\ge 2} \prod_{k=1}^{n_j} \prod_{r=1}^{j-1} \langle g_{\rho(\bar{q}(j,k,r+1))} | f_{\rho(\bar{q}(j,k,r))} \rangle. \tag{14.24}$$

To better understand this, we return to our diagram conventions. Given an arbitrary diagram, we wish to construct the Pulè permutation putting it to the basic form. For instance, we might have an initial segment of a diagram looking like that shown in Figure 14.1.

There will exist a permutation σ of the n vertices that will reorder the vertices so that we have the singletons first, then the pair contractions, then the triples, and so on, so that we obtain a picture of the type shown in Figure 14.2.

Figure 14.1 Initial diagram.

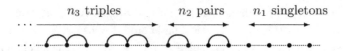

n_3 triples n_2 pairs n_1 singletons

Figure 14.2 Reordered version.

The permutation is again unique if we retain the induced ordering of the first emission times for each connected block.

Putting all this together, we get the bound

$$
C_{\alpha_n \beta_n} \ldots C_{\alpha_1 \beta_1} \int_{\Delta_n(t)} ds_n \ldots ds_1
$$

$$
\times \, \langle \Omega | \, [a_{s_n}^+(\lambda)]^{\alpha_n} [a_{s_n}^-(\lambda)]^{\beta_n} \ldots [a_{s_1}^+(\lambda)]^{\alpha_1} [a_{s_1}^-(\lambda)]^{\beta_1} \, \Omega \rangle
$$

$$
\leq \sum_{\mathbf{n}}^{E(\mathbf{n})=n} \sum_{\rho \in \mathfrak{S}_{\mathbf{n}}^0} C_{11}^{E(\mathbf{n})-2N(\mathbf{n})+n_1} C^{2N(\mathbf{n})-n_1}
$$

$$
\times \int_{\Delta_n(t)} ds_n \ldots ds_1 \prod_{j\geq 2} \prod_{k=1}^{n_j} \prod_{r=1}^{j-1} C_\lambda(s_{\rho(\bar{q}(j,k,r+1))} - s_{\rho(\bar{q}(j,k,r))}), \quad (14.25)
$$

where we use the estimate (14.20) and we obtain the sum over all relevant terms by summing over all admissible permutations of the basic $\bar{q}$ term. To estimate the simplicial integral, we generalize an argument due to Pulé (Pulé, 1974: lemma 3). Let $\tilde{\rho}$ be the induced mapping on $\mathbb{R}^n$ obtained by permuting the Cartesian coordinates according to $\rho \in \mathfrak{S}_{\mathbf{n}}^0$. Then the bound in (14.25) can be written as

$$
\sum_{\mathbf{n}}^{E(\mathbf{n})=n} C_{11}^{E(\mathbf{n})-2N(\mathbf{n})+n_1} C^{2N(\mathbf{n})-n_1}
$$

$$
\times \int_R ds_n \ldots ds_1 \prod_{j\geq 2} \prod_{k=1}^{n_j} \prod_{r=1}^{j-1} C_\lambda(s_{\bar{q}(j,k,r+1)} - s_{\bar{q}(j,k,r)}), \quad (14.26)
$$

where $R = \cup\{\tilde{\rho}\Delta_n(t) : \rho \in \mathfrak{S}_{\mathbf{n}}^0\}$. This is down to the fact that the image sets $\tilde{\rho}\Delta_n(t)$ will be distinct for different $\rho \in \mathfrak{S}_{\mathbf{n}}^0$. Now the region, R, of integration is a subset of $[0, t]^n$ for which the variables $s_{\bar{q}(j,k,1)}$ are ordered primarily by the index j and secondarily by the index k. Moreover, each of the variables

$$
u_{\bar{q}(j,k,r)} \overset{\triangle}{=} s_{\bar{q}(j,k,r+1)} - s_{\bar{q}(j,k,r)} \quad (14.27)
$$

are positive, $(\forall j; k = 1, \ldots n_j; r = 1, \ldots, j-1)$. (These properties of R are implicit from the choice of the ordering $\bar{q}$ and of the nature of the permutations $\rho \in \mathfrak{S}_{\mathbf{n}}^0$.) Consider the change of variables

$$(s_1, \ldots, s_n) \mapsto (s_{\bar{q}(j,k,1)}; u_{\bar{q}(j,k,r)}), \qquad (14.28)$$

where the ordering is first by the j, second by the k, and for the u's finally by the $r = 1, \ldots, j-1$. This defines a volume-preserving map that will take R into $\Delta_{n_1}(t) \times \Delta_{n_2}(t) \times \cdots \times [0, \infty)^{n_2} \times [0, \infty)^{2n_3} \times \cdots$. From this, we are able to find the upper estimate on (14.26) of the form

$$\sum_{\mathbf{n}}^{E(\mathbf{n})=n} C_{11}^{E(\mathbf{n})-2N(\mathbf{n})+n_1}$$

$$\times \ C^{2N(\mathbf{n})-n_1} \frac{(t \vee 1)^{n_1}}{n_1!} \frac{(t \vee 1)^{n_2}}{n_2!} \cdots \left[\int_0^\infty |C_\lambda(s)| ds \right]^{n_2 + 2n_2 + \cdots}$$

$$= \sum_{\mathbf{n}}^{E(\mathbf{n})=n} C_{11}^{E(\mathbf{n})-2N(\mathbf{n})+n_1} C^{2N(\mathbf{n})-n_1} \frac{(t \vee 1)^{N(\mathbf{n})}}{n_1! n_2! \ldots} \left(\frac{1}{2} \right)^{E(\mathbf{n})-N(\mathbf{n})}$$

$$\leq \sum_{\mathbf{n}}^{E(\mathbf{n})=n} \frac{e^{AE(\mathbf{n})+BN(\mathbf{n})}}{n_1! n_2! \ldots}, \qquad (14.29)$$

where $A = \ln(\frac{1}{2} C_{11})$ and $B = \ln(t \vee 1) + \ln(C^2 \vee 1) + \ln(C_{11}^{-2} \vee 1)$.

The restriction to those sequences $\mathbf{n}$ with $E(\mathbf{n}) = n$ can be lifted and the following estimate for the entire series obtained:

$$\Omega(A, B) = \sum_{\mathbf{n}} \frac{e^{AE(\mathbf{n})+BN(\mathbf{n})}}{n_1! n_2! \ldots} = \prod_{k=1}^\infty \sum_{n_k=0}^\infty \frac{e^{(kA+B)n_k}}{n_k!} = \exp \left\{ \frac{e^{A+B}}{1-e^A} \right\}. \qquad (14.30)$$

Here we use the $\sum \prod \longleftrightarrow \prod \sum$ lemma, (Lemma 1.2.1). The requirement for convergence is that $e^A < 1$, or equivalently, that $\frac{1}{2} C_{11} < 1$.

14.1.4 Limit Transition Amplitudes

We are now ready to resum the Dyson series. First of all, observe that the functions $f_j(t, \lambda)$ defined in (14.12) will have the limits

$$f_j(t) \equiv \lim_{\lambda \to 0} f_j(t, \lambda). \qquad (14.31)$$

Likewise, we obtain $\tilde{E}_{\alpha\beta}(t) = \lim_{\lambda \to 0} \tilde{E}_{\alpha\beta}(t, \lambda)$, which will be just the expressions in (14.11) with the $f_j(t, \lambda)$ replaced by their limits. Explicitly, we have

$$\tilde{E}_{11}(t) = E_{11}, \quad \tilde{E}_{01}(t) = E_{\alpha 1}[f_1(t)]^\alpha,$$
$$\tilde{E}_{10}(t) = E_{1\beta}[f_2^*(t)]^\beta, \quad \tilde{E}_{00}(t) = [f_1(t)]^\alpha E_{\alpha\beta}[f_2^*(t)]^\beta. \qquad (14.32)$$

Secondly, only *type I* terms will survive the limit. This means that, for the n-th term in the Dyson series, the only sequences $\alpha_1, \beta_1, \alpha_2, \beta_2, \ldots, \alpha_n, \beta_n$ appearing will be those for which $0 = \alpha_n = \beta_1$ and $\beta_l = \alpha_{l+1}$ for $l = 1, \ldots, n-1$.

Thirdly, we encounter the following limit of the two-point function. Let f and g be Schwartz functions, then we will have the limit

$$\int_0^T dt_2 \int_0^{t_2} dt_1 \, C_\lambda(t_2 - t_1) f(t_2) g(t_1) \to \frac{1}{2} \int_0^T ds f(s) g(s).$$

Therefore, employing Lemma 14.1.4, we find

$$\lim_{\lambda \to 0} \langle \phi_1 \otimes D(f_1, \lambda)\Omega | \, U(t/\lambda^2, \lambda) \, \phi_2 \otimes D(f_2, \lambda)\Omega \rangle = \langle D(f_1)\Omega | D(f_2)\Omega \rangle$$

$$\times \sum_n (-i)^n \int_{\Delta_n(t)} ds_n \ldots ds_1 \prod_{l=1}^{n-1} \left[\frac{1}{2} \partial_+ + (s_{l+1} - s_l) \right]^{\beta_l}$$

$$\times \sum_{\beta \in \{0,1\}^{n-1}} \langle \phi_1 | \tilde{E}_{0\beta_{n-1}}(s_n) \ldots \tilde{E}_{\beta_2\beta_1}(s_2) \tilde{E}_{\beta_1 0}(s_1) \, \phi_2 \rangle, \qquad (14.33)$$

where we introduce the symbol ∂_+ for a one-sided delta-function, that is, $\int \partial_+(t-s) f(s) ds = f(t^+)$.

We now develop this series. Suppose that we have $\beta_{k+1} = 0 = \beta_k$, that is, there are no contractions to the k-th term, then we encounter the factor $\tilde{E}_{00}(s) = [f_1(s)]^\alpha E_{\alpha\beta} [f_2^*(s)]^\beta$, where $s = s_k$.

Otherwise, if we have contractions on the terms associated to *consecutive* variables $s_{k+r}, \ldots, s_{k+1}, s_k$ and we assume that s_{k+r} is not paired to s_{k+r+1}, nor s_k to s_{k-1}, in such cases we then encounter the factor $\tilde{E}_{01}(s_{k+r})\tilde{E}_{11}(s_{k+r-1}) \ldots$ $\tilde{E}_{11}(s_{k+1}) \, \tilde{E}_{10}(s_k)$ with the variables $s_{k+r}, \ldots, s_{k+1}, s_k$ all forced equal to a common value such as s. This factor will then be $[f_1(s_k)]^\alpha E_{\alpha 1}(E_{11})^{r-2}$ $E_{1\beta}[f_2^*(s_k)]^\beta$.

Now (14.33) involves a sum over all consecutive pairings: the corresponding partition will have all parts consisting of consecutive labels. We can list these parts in increasing order, say from 1 to m if there are m of them, and let r_j be the size of the j-th part. The number of contractions will be $\sum \beta_l$, and this will be $n - m = \sum_{j=1}^m (r_j - 1)$. With these observations, we see that (14.33) becomes

$$\langle D(f_1 \otimes 1_{[S_1,T_1]})\Omega | D(f_2 \otimes 1_{[S_2,T_2]})\Omega \rangle$$

$$\times \sum_n \sum_m \sum_{\substack{r_1 + \cdots + r_m = n \\ r_m, \ldots r_1 \geq 1}} \int_{\Delta_m(t)} ds_m \ldots ds_1 \, (-i)^{\sum_{j=1}^m r_j} \left(\frac{1}{2} \right)^{\sum_{j=1}^m (r_j - 1)}$$

$$\langle \phi_1 | E_{\alpha_m, \beta_m}^{(r_m)} \ldots E_{\alpha_1, \beta_1}^{(r_1)} \, \phi_2 \rangle \, [f_1(s_m)]^{\alpha_m} [f_2^*(s_m)]^{\beta_m} \ldots [f_1(s_1)]^{\alpha_1} [f_2^*(s_1)]^{\beta_1},$$

$$(14.34)$$

where we set

$$E_{\alpha,\beta}^{(r)} \triangleq \begin{cases} E_{\alpha\beta}, & r = 1; \\ E_{\alpha 1}(E_{11})^{r-2}E_{1\beta}, & r \geq 2. \end{cases} \qquad (14.35)$$

In the following, we shall encounter the coefficients

$$G_{\alpha\beta} \triangleq -i\sum_{r=1}^{\infty}\left(-\frac{i}{2}\right)^{r-1} E_{\alpha,\beta}^{(r)} = -iE_{\alpha\beta} - \frac{1}{2}E_{\alpha 1}\frac{1}{1+\frac{i}{2}E_{11}}E_{1\beta}. \qquad (14.36)$$

To see this diagrammatically, let us retain only type I terms in the series:

We see the first appearance of scattering in the last term in the third term of the series: such terms however eventually outproliferate diagrams with no scattering. The terms have been grouped by vertex number, however, so it also possible to group them by effective vertex number (equal to the number of parts, or equivalently the original simplex degree minus the number of contractions) to give

where now each box is the following sum over all effective one-vertex contributions:

which is analogous to the expression of the self-energy in quantum field theory: as a sum over irreducible terms. (As we have seen, one-vertex contributions terminate at second order when there is no scattering: as this is a form of cumulant expansion, the emission/absorption problem is Gaussian, while allowing scattering means that we must have cumulants to all orders!)

If the limit effective one-vertex label is t, then its weight is

$$-i\tilde{E}_{00}(t) + \frac{(-i)^2}{2}\tilde{E}_{01}(t)\tilde{E}_{10}(t) + \left(\frac{(-i)^3}{2}\right)^2 \tilde{E}_{01}(t)\tilde{E}_{11}(t)\tilde{E}_{10}(t) + \cdots$$

$$= -i\tilde{E}_{00}(t) - \frac{1}{2}\tilde{E}_{01}(t)\frac{1}{1+\frac{i}{2}E_{11}}\tilde{E}_{10}(t)$$

$$\equiv [f_1^*(t)]^{\alpha} G_{\alpha\beta}[f_2(t)]^{\beta},$$

where the geometric series can be summed since $\|\frac{1}{2}E_{11}\| < 1$.

Theorem 14.1.5 *Suppose the system operators $E_{\alpha\beta}$ are bounded and that $\frac{1}{2}\|E_{11}\| < 1$. Let $\phi_1, \phi_2 \in \mathfrak{h}_S$ and $f_1, f_2 \in L^2(\mathbb{R}_+, dt)$. Then*

$$\lim_{\lambda \to 0} \langle \phi_1 \otimes D(f_1, \lambda)\Omega | U_t^{(\lambda)} \phi_2 \otimes D(f_2, \lambda)\Omega \rangle$$
$$= \langle \phi_1 \otimes D(f_1)\Omega | U_t \phi_2 \otimes D(f_2)\Omega \rangle,$$

where $(U_t : t \geq 0)$ is a unitary adapted quantum stochastic process on $\mathfrak{h}_S \otimes \Gamma(L^2(\mathbb{R}^+))$ given by the Stratonovich time-ordered exponential

$$U_t \equiv \vec{T}_S \exp\left\{ -i \int_0^t E_{\alpha\beta} \otimes dA_t^{\alpha\beta} \right\}. \tag{14.37}$$

Proof Suppose we have the QSDE $dU_t = G_{\alpha\beta} U_t \otimes dA_t^{\alpha\beta}$, with initial condition $U_0 = 1$, then we have the expansion

$$U_t = \sum_{m \geq 0} \int_{\Delta_m(t)} G_{\alpha(m)\beta(m)} \cdots G_{\alpha(1)\beta(1)} \otimes dA_{s(m)}^{\alpha(m)\beta(m)} \cdots dA_{s(1)}^{\alpha(1)\beta(1)} \tag{14.38}$$

and so $\langle \phi_1 \otimes D(f_1)\Omega | U_t \phi_2 \otimes D(f_2)\Omega \rangle$ can be expressed as

$$\langle D(f_1)\Omega | D(f_2)\Omega \rangle \sum_{m \geq 0} \langle \phi_1 | G_{\alpha(m)\beta(m)} \cdots G_{\alpha(2)\beta(2)} G_{\alpha(1)\beta(1)} \phi_2 \rangle$$

$$\times \int_{\Delta_m(t)} ds_m \cdots ds_1 \, ([f_1(s_m)]^{\alpha(m)} [f_2^*(s_m)]^{\beta(m)}) \cdots ([f_1(s_1)]^{\alpha(1)} [f_2^*(s_1)]^{\beta(1)}).$$

By inspection, this evidently agrees with (14.34) with the coefficients $G_{\alpha\beta}$ given by (14.36). However, comparison with Proposition 13.8.3 shows that this is the Stratonovich time-ordered exponential (14.37). $\qquad\square$

14.1.5 Dynamical Evolutions

Let X be a bounded operator on the system state space $\mathfrak{h}_S$. We define its Heisenberg evolute to be

$$J_t^{(\lambda)}(X) \triangleq U_t^{(\lambda)\dagger} [X \otimes 1_R] U_t^{(\lambda)}. \tag{14.39}$$

In addition, what we term the coevolute is defined to be

$$K_t^{(\lambda)}(X) \triangleq U_t^{(\lambda)} [X \otimes 1_R] U_t^{(\lambda)\dagger}. \tag{14.40}$$

We wish to study the limits of $J_t^{(\lambda)}$ and $K_t^{(\lambda)}$ as quantum processes taken relative to the Fock vacuum state $\Omega \in \mathfrak{h}_R$ for the Bose reservoir. To this end, we note the developments

$$K_t^{(\lambda)}(X) = \sum_n (-1)^n \int_{\Delta_n(t)} ds_n \dots ds_1 \, \mathcal{X}_{\Upsilon_{s_n}^{(\lambda)}} \circ \dots \circ \mathcal{X}_{\Upsilon_{s_1}^{(\lambda)}}(X \otimes 1_R), \quad (14.41)$$

$$J_t^{(\lambda)}(X) = \sum_{n,\hat{n}} (-i)^{n+\hat{n}} \int_{\Delta_n(t)} ds_n \dots ds_1 \int_{\Delta_{\hat{n}}(t)} dt_{\hat{n}} \dots dt_1$$

$$\times \Upsilon_{s_1}^{(\lambda)} \dots \Upsilon_{s_n}^{(\lambda)} [X \otimes 1_R] \, \Upsilon_{t_{\hat{n}}}^{(\lambda)} \dots \Upsilon_{t_1}^{(\lambda)}, \quad (14.42)$$

where $\mathcal{X}_H(.) \triangleq \frac{1}{i}[., H]$.

We note that the coevolution has the simpler form when iterated. The evolution itself requires a separate expansion of the unitaries. (This disparity is related to the proof of unitarity for quantum stochastic processes in Hudson and Parthasarathy (1984), where the isometric property requires some work while the coisometric property is established immediately.) In fact, the same inequalities as used to establish the convergence of $U_t^{(\lambda)}$ suffice for the coevolution: in both cases, we have a Picard-iterated series. We likewise have the expansion

$$\langle \phi_1 \otimes D(f_1, \lambda)\Omega | J_t^{(\lambda)}(X) \, \phi_2 \otimes D(f_2, \lambda)\Omega \rangle$$

$$= \sum_{n,\hat{n}} (-i)^{n-\hat{n}} \int_{\Delta_n(t)} ds_n \dots ds_1 \int_{\Delta_{\hat{n}}(t)} dt_{\hat{n}} \dots dt_1$$

$$\langle \phi_1 | \tilde{E}_{\alpha_1 \beta_1}(s_1, \lambda) \dots \tilde{E}_{\alpha_n \beta_n}(s_n, \lambda) \, X \, \tilde{E}_{\mu_{\hat{n}} \nu_{\hat{n}}}(s_{\hat{n}}, \lambda) \dots \tilde{E}_{\mu_1 \nu_1}(s_1, \lambda) \, \phi_2 \rangle$$

$$\times \langle \Omega | [a_{s_1}^+(\lambda)]^{\alpha_1} [a_{s_1}^-(\lambda)]^{\beta_1} \dots [a_{s_n}^+(\lambda)]^{\alpha_n} [a_{s_n}^-(\lambda)]^{\beta_n}$$

$$\dots [a_{t_{\hat{n}}}^+(\lambda)]^{\mu_{\hat{n}}} [a_{t_{\hat{n}}}^-(\lambda)]^{\nu_{\hat{n}}} \dots [a_{t_1}^+(\lambda)]^{\mu_1} [a_{t_1}^-(\lambda)]^{\nu_1} \Omega \rangle. \quad (14.43)$$

The vacuum average of the reservoir operators can be expressed as a sum of products of two-point functions with each summand representable as a partition of $n + \hat{n}$ vertices. Our strategy is similar to before. We shall use diagrams to describe the individual contributions and attempt to obtain a uniform estimate. The Heisenberg diagrams are more involved than last time due to the scattering; however, the general idea goes through again.

Let us consider an arbitrary Heisenberg diagram. If we considered only the $t - t$ contractions and ignored everything else, then we would have a partition of the n t–variables, let's say with occupation numbers $\mathbf{n} = (n_j)$. Likewise, if we looked at only the $s - s$ contractions, then we have a partition of the n' s–variables, say with occupation numbers $\mathbf{n}' = (n'_j)$. At this stage, we can then take the $s - t$ contractions into account. The following diagram shows a quartet of s variables joined to a triplet of t variables:

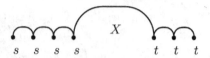

Let l_{jk} be the number of $s - t$ contractions joining a part of j s's to a part of k t's: here we use an obvious abuse of terminology, as technically they are all in the same part! We also introduce the occupation numbers $\mathbf{l} = (l_j)$, $\mathbf{l}' = (l'_j)$, where $l_k = \sum_j l_{jk}$ and $l'_j = \sum_k l_{jk}$. (When no scattering was present, we only had the possibility that l_{11}, previously denoted as l, could be nonzero.) It is convenient to introduce the occupation numbers $\mathbf{m} = (m_j)$ and $\mathbf{m}' = (m'_j)$, where $m_j = n_j - l_j$ and $m'_j = n'_j - l'_j$. Here m_j counts the number of parts of t-variables of size j having no elements contracted with an s-variable.

The procedure adopted in the last chapter is now repeated. We consider equivalence classes of Heisenberg diagrams leading to the same set of sequences $\mathbf{n}, \mathbf{n}', \mathbf{l}, \mathbf{l}'$, or equivalently $\mathbf{m}, \mathbf{m}', \mathbf{l}, \mathbf{l}'$ as in Chapter 13. We can choose a basic Heisenberg diagram as the representative of each class, and there will be permutations $\rho \in \mathfrak{S}^0_{\mathbf{n}}$ and $\rho' \in \mathfrak{S}^0_{\mathbf{n}'}$ of the t and s variables respectively, which will allow us to reorganize the basic Heisenberg diagram into any other element of the the class. (We omit the explicit choice of a basic Heisenberg diagram and leave its specification to the reader as an exercise.)

Now for each diagram in a given class there will be chronologically ordered blocks of sizes $m_1, m_2, \ldots, m'_1, m'_2, \ldots, l_1, l_2, \ldots$ and by the type of argument encountered before we arrive at the following upper bound for the sum of absolute values for all the diagrams:

$$\sum_{\mathbf{m}, \mathbf{m}', \mathbf{l}} C_{11}^{E(\mathbf{m}+\mathbf{m}'+\mathbf{l}+\mathbf{l}')-2N(\mathbf{m}+\mathbf{m}'+\mathbf{l}+\mathbf{l}')+m_1+m'_1+l_1+l_2}$$

$$\times C^{2N(\mathbf{m}+\mathbf{m}'+\mathbf{l}+\mathbf{l}')-(m_1+m'_1+l_1+l_2)} \left(\frac{1}{2}\right)^{E(\mathbf{m}+\mathbf{m}'+\mathbf{l}+\mathbf{l}')-N(\mathbf{m}+\mathbf{m}'+\mathbf{l}+\mathbf{l}')}$$

$$\times \frac{(t \vee 1)^{N(\mathbf{m}+\mathbf{m}'+\mathbf{l})}}{(m_1! m_2! \ldots)(m'_1! m'_2! \ldots)(l_1! l_2! \ldots)}. \tag{14.44}$$

Here we add sequences of occupation numbers componentwise, i.e. $\mathbf{m} + \mathbf{m}'$ is $(m_j + m'_j)$, and so on, and we note that $N(\mathbf{l}) = N(\mathbf{l}')$. Recalling the constants A and B from before, and introducing $B' = \frac{1}{2} \ln(t \vee 1) + \ln(C^2 \vee 1) + \ln(C_{11}^{-2} \vee 1)$, we sum the series to get the upper bound

$$\exp\left\{2\frac{e^{A+B}}{1-e^A} + \frac{e^{2A+2B'}}{1-e^{2A}}\right\}, \tag{14.45}$$

which is again convergent as $e^A < 1$.

We now wish to determine the limit $\lambda \to 0$. Once again, only diagrams having time-consecutive $s - s$ and $t - t$ contractions, as well as non-crossing $s - t$ contractions, are going to contribute to the limit. The presence of scattering now means that we have more diagrams; however, we can reduce this using the effective vertex method and, once again we can arrive at a simple recursive formula. This time, we have

Here we meet new effective vertices in the final diagram. On the right, we have

which for vertex time t corresponds to the operator weight

$$
- i\tilde{E}_{10}(t) + \frac{(-i)^2}{2}\tilde{E}_{11}(t)\tilde{E}_{10}(t) + \frac{(-i)^3}{2^2}\tilde{E}_{11}(t)\tilde{E}_{11}(t)\tilde{E}_{10}(t) + \cdots
$$
$$
= -i\frac{1}{1 + \frac{i}{2}E_{11}}\tilde{E}_{10}(t)
$$
$$
\equiv G_{1\beta}[f_2(t)]^\beta.
$$

While on the left we have

which has the weight

$$
i\tilde{E}_{01}(t) + i^2\frac{1}{2}\tilde{E}_{01}(t)\tilde{E}_{11}(t) + i^3\left(\frac{1}{2}\right)^2\tilde{E}_{01}(t)\tilde{E}_{11}(t)\tilde{E}_{11}(t) + \cdots
$$
$$
= i\tilde{E}_{01}(t)\frac{1}{1 - \frac{i}{2}E_{11}}
$$
$$
\equiv [f_1(t)^*]^\alpha G_{1\alpha}^\dagger.
$$

The recursion relation here is

$$\langle \phi_1 \otimes \exp(f_1) | U_t^\dagger [X \otimes 1] U_t \phi_2 \otimes \exp(f_2) \rangle$$
$$= \langle \phi_1 \otimes \exp(f_1) | [X \otimes 1] \phi_2 \otimes \exp(f_2) \rangle$$
$$+ \int_0^t ds [f_1(s)^*]^\alpha \langle \phi_1 \otimes \exp(f_1) |$$
$$U_s^\dagger [X G_{\alpha\beta} + G_{\beta\alpha}^\dagger X + G_{1\alpha}^\dagger X G_{1\beta}] \otimes 1 U_s \phi_2 \otimes \exp(f_2) \rangle [f_2(s)]^\beta$$
$$= \langle \phi_1 \otimes \exp(f_1) | \Big\{ I$$
$$+ \int_0^t U_s^\dagger [(X G_{\alpha\beta} + G_{\beta\alpha}^\dagger X + G_{1\alpha}^\dagger X G_{1\beta}) \otimes dA_s^{\alpha\beta}] U_s \phi_2 \otimes \exp(f_2) \rangle$$
$$\equiv \langle \phi_1 \otimes \exp(f_1) | \Big\{ I + \int_0^t U_s^\dagger \mathcal{L}_{\alpha\beta}(X) U_s dA_s^{\alpha\beta} \Big\} \phi_2 \otimes \exp(f_2) \rangle,$$

and this is the form we want!

To summarize, the prelimit flow $J_t^{(\lambda)} : \mathcal{B}(\mathfrak{h}_S) \mapsto \mathcal{B}(\mathfrak{h}_S \otimes \mathfrak{h}_R)$ given by $J_t^{(\lambda)}(X) \triangleq U_t^{(\lambda)\dagger} (X \otimes 1_R) U_t^{(\lambda)}$ converges in the sense of weak matrix elements, for fixed $X \in \mathcal{B}(\mathfrak{h}_S)$, to the limit process $J_t(X) = J_t^\dagger (X \otimes 1) J_t$. We find that $(J_t)_{t \geq 0}$ determines a quantum stochastic flow on $\mathfrak{h}_S \otimes \Gamma(L^2(\mathbb{R}^+, \mathfrak{k}_\omega))$ and from the quantum stochastic calculus we obtain the quantum Langevin, or stochastic Heisenberg, equation:

$$dJ_t(X) = J_t(\mathcal{L}_{\alpha\beta}(X)) \otimes dA_t^{\alpha\beta}.$$

The superoperators $\mathcal{L}_{\alpha\beta}(X) = X G_{\alpha\beta} + G_{\beta\alpha}^\dagger X + G_{1\alpha}^\dagger X G_{1\beta}$ are known as the **Evans–Hudson maps** – see Evans and Hudson (1988) – and these can be written in the standard form

$$\mathcal{L}_{11}(X) = (S^\dagger X S - X);$$
$$\mathcal{L}_{10}(X) = S^\dagger [X, L]; \quad \mathcal{L}_{01}(X) = -[X, L^\dagger] S;$$
$$\mathcal{L}_{00}(X) = \frac{1}{2} [L^\dagger, X] L + \frac{\gamma}{2} L^\dagger [X, L] - i[X, H]. \tag{14.46}$$

In particular, $\mathcal{L}_{00}$ is a generator of the GKS-Lindblad type (Lindblad, 1976); (Gorini et al., 1976). We shall give a more detailed treatment of the convergence in the next section.

14.1.6 The Convergence of the Heisenberg Evolution

We now wish to determine the limit $\lambda \to 0$ of (14.43). We have an integration over a double simplex region, and the main features emerge from examining the vacuum expectation of the product of creation and annihilation operators.

Evidently, the vacuum expectation can be decomposed as a sum over products of two-point functions, and it is here that Lemma 14.1.4 becomes important. What must happen for a term to survive the limit? If we have any contractions between vertices labeled by the t's, then the term will vanish if the times are not consecutive. The same is true for contractions between vertices labeled by the s's. From our estimate in the previous section, we can ignore the terms that do not comply with this.

As a result, contractions between the s's, say, will come in time-consecutive blocks: for instance, we will typically have m blocks of sizes $r_1, r_2, \ldots, r_m$ (these are integers 1, 2, 3, ..., and $\sum_{j=1}^{m} r_j = n$). With a similar situation for the t's, we obtain the expansion

$$\langle \phi_1 \otimes D(f_1, \lambda)\Omega | \, J_t^{(\lambda)}(X) \, \phi_2 \otimes D(f_2, \lambda)\Omega \rangle$$

$$= \sum_{n, \hat{n}} (-i)^{n-\hat{n}} \sum_{m, \hat{m}} \sum_{r_1, \ldots, r_m}^{\sum r = n} \sum_{l_1, \ldots, l_{\hat{m}}}^{\sum l = \hat{n}} \int_{\Delta_n(t)} ds_n \ldots ds_1 \int_{\Delta_{\hat{n}}(t)} dt_{\hat{n}} \ldots dt_1$$

$$\times \langle \phi_1 | \, \tilde{E}_{\alpha_1 \beta_1}^{(r_1)}(s_1^{(1)}, \ldots s_{r_1}^{(1)}; \lambda) \ldots \tilde{E}_{\alpha_m \beta_m}^{(r_m)}(s_1^{(m)}, \ldots s_{r_m}^{(m)}; \lambda)$$

$$\times X \tilde{E}_{\mu_{\hat{m}} \nu_{\hat{m}}}^{(l_{\hat{m}})}(t_1^{(\hat{m})}, \ldots, t_{l_{\hat{m}}}^{(\hat{m})}; \lambda) \ldots \tilde{E}_{\mu_1 \nu_1}^{(l_1)}(t_1^{(1)}, \ldots, t_{l_1}^{(1)}; \lambda) . \phi_2 \rangle$$

$$\times \prod_{j=1}^{m} \prod_{k=1}^{r_j} C_\lambda (s_{k+1}^{(j)} - s_k^{(j)}) \times \prod_{\hat{j}=1}^{\hat{m}} \prod_{\hat{k}=1}^{l_j} C_\lambda (t_{\hat{k}+1}^{(\hat{j})} - t_{\hat{k}}^{(\hat{j})})$$

$$\times \langle \Omega | \, [a_{s_1^{(1)}}^+(\lambda)]^{\alpha_1} [a_{s_{r_1}^{(1)}}^-(\lambda)]^{\beta_1} \ldots [a_{s_1^{(m)}}^+(\lambda)]^{\alpha_m} [a_{s_{r_m}^{(m)}}^-(\lambda)]^{\beta_m}$$

$$[a_{t_{l_{\hat{m}}}^{(\hat{m})}}^+(\lambda)]^{\mu_{\hat{m}}} [a_{t_1^{(\hat{m})}}^-(\lambda)]^{\nu_{\hat{m}}} \ldots [a_{t_{l_1}^{(1)}}^+(\lambda)]^{\mu_1} [a_{t_1^{(1)}}^-(\lambda)]^{\nu_1} . \Omega \rangle + \text{negligible terms},$$

$$(14.47)$$

where we relabel the times as

$$s_k^{(j)} \triangleq s_{r_1 + \cdots + r_{j-1} + k}, \qquad 1 \le k \le r_j;$$

$$t_k^{(j)} \triangleq t_{l_1 + \cdots + l_{j-1} + k}, \qquad 1 \le k \le l_j;$$

and introduce the block product of system operators

$$\tilde{E}_{\alpha\beta}^{(r_j)}(s_1^{(j)}, \ldots s_{r_j}^{(j)}; \lambda) \triangleq \tilde{E}_{\alpha 1}(s_1^{(j)}; \lambda) \tilde{E}_{11}(s_2^{(j)}; \lambda) \ldots \tilde{E}_{11}(s_{r_j-1}^{(j)}; \lambda) \tilde{E}_{1\beta}(s_{r_j}^{(j)}; \lambda).$$

We now examine the limit of (14.47). The estimate on the series expansion of the Heisenberg evolute given in the previous section shows that we can ignore the so-called negligible terms in (14.47). The limit is rather difficult to see at this stage. However, what we can do is to recast the expression that we claim will be the limit,

$$\langle \phi_1 \otimes D(f_1 \otimes 1_{[S_1,T_1]})\Omega| J_t(X) \phi_2 \otimes D(f_2 \otimes 1_{[S_2,T_2]})\Omega\rangle, \qquad (14.48)$$

with $J_t(X) = U_t^\dagger(X \otimes 1)U_t$, in a more explicit form.

Recall the chaotic expansion of the process U_t given in (14.38); expression (14.48) then becomes

$$\sum_{m,\hat{m}} \int_{\Delta_m(t)} \int_{\Delta_{\hat{m}}(t)} \sum_{r_1,\dots,r_m} \sum_{l_1,\dots,l_{\hat{m}}} (i)^{\sum r - \sum l} \left(\frac{1}{2_-}\right)^{\sum r - m} \left(\frac{1}{2_+}\right)^{\sum l - \hat{m}}$$
$$\times \langle \phi_1| \tilde{E}^{(r_1)}_{\alpha_1\beta_1} \dots \tilde{E}^{(r_m)}_{\alpha_m\beta_m} X \tilde{E}^{(l_{\hat{m}})}_{\mu_{\hat{m}}\nu_{\hat{m}}} \dots \tilde{E}^{(l_1)}_{\mu_1\nu_1} .\phi_2\rangle$$
$$\times \langle D(f_1 \otimes 1_{[S_1,T_1]})\Omega| dA^{\alpha_m\beta_m}_{s_m} \dots dA^{\alpha_1\beta_1}_{s_1}$$
$$\times dA^{\mu_{\hat{m}}\nu_{\hat{m}}}_{t_{\hat{m}}} \dots dA^{\mu_1\nu_1}_{t_1} D(f_2 \otimes 1_{[S_2,T_2]})\Omega\rangle.$$

Now the expectation between the states $D(f_j \otimes 1_{[S_j,T_j]})\Omega$ can be converted into an expectation between the Fock vacuum state Ω if we make the following replacements:

$$dA^{11} \rightarrow dA^{11} + f_2^* dA^{10} + f_1 dA^{01} + f_1 f_2^* dA^{10},$$
$$dA^{10} \rightarrow dA^{10} + f_1 dA^{00},$$
$$dA^{01} \rightarrow dA^{01} + f_2^* dA^{00},$$
$$dA^{00} \rightarrow dA^{00}, \qquad (14.49)$$

where $f_j(t) = 1_{[S_j,T_j]}(f_j|g)$ as in (8.1). This leads to the development

$$\sum_{m,\hat{m}} \int_{\Delta_m(t)} \int_{\Delta_{\hat{m}}(t)} \sum_{r_1,\dots,r_m} \sum_{l_1,\dots,l_{\hat{m}}} (i)^{\sum r - \sum l} \left(\frac{1^*}{2}\right)^{\sum r - m} \left(\frac{1}{2}\right)^{\sum l - \hat{m}}$$
$$\times \langle \phi_1| \tilde{E}^{(r_1)}_{\alpha_1\beta_1}(s_1) \dots \tilde{E}^{(r_m)}_{\alpha_m\beta_m}(s_m) X \tilde{E}^{(l_{\hat{m}})}_{\mu_{\hat{m}}\nu_{\hat{m}}}(t_{\hat{m}}) \dots \tilde{E}^{(l_1)}_{\mu_1\nu_1}(t_1) .\phi_2\rangle$$
$$\times \langle \Omega| dA^{\alpha_m\beta_m}_{s_m} \dots dA^{\alpha_1\beta_1}_{s_1} dA^{\mu_{\hat{m}}\nu_{\hat{m}}}_{t_{\hat{m}}} \dots dA^{\mu_1\nu_1}_{t_1} \Omega\rangle, \qquad (14.50)$$

where the operators $\tilde{E}^{(r)}_{\alpha b}(t)$ are given by

$$\tilde{E}^{(r)}_{\alpha\beta}(t) = \begin{cases} \tilde{E}_{\alpha\beta}(t), & r = 1; \\ \tilde{E}_{\alpha 1}(t)(\tilde{E}_{11}(t))^{r-2}\tilde{E}_{1\beta}(t), & r \geq 2. \end{cases}$$

Again we note that the operators $\tilde{E}_{\alpha\beta}(t)$ have been introduced in (14.32).

It remains to be shown that the limit of (14.47) will be (14.50). We observe that

$$\lim_{\lambda \to 0} \langle \phi_1 \otimes D(f_1, \lambda)\Omega | \, J_t^{(\lambda)}(X) \, \phi_2 \otimes D(f_2, \lambda)\Omega \rangle$$

$$= \sum_{n,\hat{n}} (-i)^{n-\hat{n}} \sum_{m,\hat{m}} \sum_{r_1,\dots,r_m}^{\sum r = n} \sum_{l_1,\dots,l_{\hat{m}}}^{\sum l = \hat{n}} \int_{\Delta_m(t)} ds_m \dots ds_1 \int_{\Delta_{\hat{m}}(t)} dt_{\hat{m}} \dots dt_1$$

$$\times \langle \phi_1 | \, \tilde{E}_{\alpha_1 \beta_1}^{(r_1)}(s_1) \dots \tilde{E}_{\alpha_m \beta_m}^{(r_m)}(s_m) \, X \, \tilde{E}_{\mu_{\hat{m}} \nu_{\hat{m}}}^{(l_{\hat{m}})}(t_{\hat{m}}) \dots \tilde{E}_{\mu_1 \nu_1}^{(l_1)}(t_1) \, . \phi_2 \rangle$$

$$\times \left(\frac{1}{2} \right)^{\sum r - m} \times \left(\frac{1}{2} \right)^{\sum l - \hat{m}}$$

$$\times \lim_{\lambda \to 0} \langle \Omega | \, [a_{s_1}^+(\lambda)]^{\alpha_1} [a_{s_1}^-(\lambda)]^{\beta_1} \dots [a_{s_m}^+(\lambda)]^{\alpha_m} [a_{s_m}^-(\lambda)]^{\beta_m}$$

$$[a_{t_{\hat{m}}}^+(\lambda)]^{\mu_{\hat{m}}} [a_{t_{\hat{m}}}^-(\lambda)]^{\nu_{\hat{m}}} \dots [a_{t_1}^+(\lambda)]^{\mu_1} [a_{t_1}^-(\lambda)]^{\nu_1} \, . \Omega \rangle. \tag{14.51}$$

We now require the fact that

$$\lim_{\lambda \to 0} \int_R ds_m \dots ds_1 dt_{\hat{m}} \dots dt_1 \, \langle \Omega | \, [a_{s_1}^+(\lambda)]^{\alpha_1} [a_{s_1}^-(\lambda)]^{\beta_1} \dots$$

$$\times [a_{s_m}^+(\lambda)]^{\alpha_m} [a_{s_m}^-(\lambda)]^{\beta_m} [a_{t_{\hat{m}}}^+(\lambda)]^{\mu_{\hat{m}}} [a_{t_{\hat{m}}}^-(\lambda)]^{\nu_{\hat{m}}} \dots$$

$$\times [a_{t_1}^+(\lambda)]^{\mu_1} [a_{t_1}^-(\lambda)]^{\nu_1} \, . \Omega \rangle f(s_m, \dots, s_1, t_{\hat{m}}, \dots, t_1)$$

$$= \int_R \langle \Omega | \, dA_{s_m}^{\alpha_m \beta_m} \dots dA_{s_1}^{\alpha_1 \beta_1} dA_{t_{\hat{m}}}^{\mu_{\hat{m}} \nu_{\hat{m}}} \dots dA_{t_1}^{\mu_1 \nu_1} \, \Omega \rangle f(s_m, \dots, s_1, t_{\hat{m}}, \dots, t_1)$$

for f continuous and R a bounded region in $m + \hat{m}$ dimensions, which is the union of simplices. This is readily seen, of course, by expanding the Ω-expectation as a sum of products of two-point functions and reassembling the limit in terms of the Ω-expectations of the processes $A_t^{\alpha\beta}$. This is evident from Lemma 14.1.4 and Theorem 14.1.5 quoted earlier and from the quantum Itō calculus Hudson and Parthasarathy (1984).

We therefore see that the limit form as given in (14.51) agrees with the stated limit.

Theorem 14.1.6 *Suppose that $E_{\alpha\beta}$ are bounded with $K\|E_{11}\| < 1$, as before. Let $\phi_1, \phi_2 \in \mathfrak{h}_S$ and $f_1, f_2 \in L^2(\mathbb{R}_+, dt)$. Then, for $X \in B(\mathfrak{h}_S)$,*

$$\lim_{\lambda \to 0} \langle \phi_1 \otimes D(f_1, \lambda)\Omega | \, J_t^{(\lambda)}(X) \, \phi_2 \otimes D(f_2, \lambda)\Omega \rangle$$

$$= \langle \phi_1 \otimes D(f_1)\Omega | \, J_t(X) \, \phi_2 \otimes D(f_2)\Omega \rangle.$$

To summarize, the prelimit flow $J_t^{(\lambda)}: \mathcal{B}(\mathfrak{h}_S) \mapsto \mathcal{B}(\mathfrak{h}_S \otimes \mathfrak{h}_R)$ given by $J_t^{(\lambda)}(X) \triangleq U_t^{(\lambda)\dagger}(X \otimes 1_R) U_t^{(\lambda)}$ converges in the sense of weak matrix elements, for fixed $X \in \mathcal{B}(\mathfrak{h}_S)$, to the limit process $J_t(X) = U_t^\dagger(X \otimes 1) U_t$. We find that $(J_t)_{t \geq 0}$ determines a quantum stochastic flow on $\mathfrak{h}_S \otimes \Gamma(L^2(\mathbb{R}^+, dt))$, and from the quantum stochastic calculus (Hudson and Parthasarathy, 1984), we obtain the quantum Langevin, or stochastic Heisenberg, equation

$$dJ_t(X) = J_t(\mathcal{L}_{\alpha\beta}(X)) \otimes dA_t^{\alpha\beta}, \tag{14.52}$$

where the $\mathcal{L}_{\alpha\beta}$ are the Evans–Hudson maps encountered in (14.46).

The analogous result will hold for the coevolution. Though, as mentioned before, there is a more immediate proof using the original estimates.

14.2 A Microscopic Model

We shall consider a quantum mechanical system S with Hilbert space $\mathfrak{h}_S$ coupled to a Bose quantum field reservoir R over a one-particle space $\mathfrak{h}_R^1$, so that the appropriate Hilbert space is $\mathfrak{H}_R = \Gamma(\mathfrak{h}_R^1)$. We shall take the reservoir to be in the Fock vacuum state Ω. The interaction between the system and the reservoir will be given by the formal Hamiltonian

$$H^{(\lambda)} = H_S \otimes 1_R + 1_S \otimes d\Gamma(H_R^1) + H_{\text{Int}}^{(\lambda)}, \tag{14.53}$$

where the operators H_S and H_R^1 are self-adjoint and bounded below on $\mathfrak{h}_S$ and $\mathfrak{h}_R^1$, respectively. The interaction is taken to be

$$
\begin{aligned}
H_{\text{Int}}^{(\lambda)} = {}& E_{11} \otimes A^+(g)A^-(g) + \lambda E_{10} \otimes A^+(g) \\
& + \lambda E_{01} \otimes A^-(g) + \lambda^2 E_{00} \otimes 1_R,
\end{aligned}
\tag{14.54}
$$

where $E_{\alpha\beta}$ are bounded operators on $\mathfrak{h}_S$ with E_{11} and E_{00} self-adjoint and $E_{10} = E_{01}^\dagger$. The operators $A^+(g)$ and $A^-(g)$ are the creation and annihilation operators with test function $g \in \mathfrak{h}_R^1$. (The parameter λ is real and will later emerge as a rescaling parameter in which we hope to obtain a Markovian limit.) We shall also assume the following harmonic relations:

$$
\begin{aligned}
e^{+i\tau H_S} E_{\alpha\beta}\, e^{-i\tau H_S} &= e^{i\omega\tau(\beta-\alpha)}\, E_{\alpha\beta}; \\
e^{+i\tau H_R} A_R^\pm(g)\, e^{-i\tau H_R} &= A_R^\pm(\theta_\tau g),
\end{aligned}
\tag{14.55}
$$

where $(\theta_\tau : \tau \in \mathbb{R})$ will be the one-parameter group of unitaries on $\mathfrak{h}_R^1$ with Stone generator H_R^1.

We transfer to the interaction picture with the help of the unitary

$$U(\tau, \lambda) = e^{+i\tau(H_S \otimes 1_R + 1_S \otimes H_R)}\, e^{-i\tau H^{(\lambda)}}. \tag{14.56}$$

Rather than the S-matrix limit $t \to \infty$, we shall be interested in the more subtle van Hove (1955), or weak coupling, limit where we rescale time as t/λ^2. Here the interaction is weak, but over suitable time scales one accumulates an appreciable effect.

In the weak coupling regime, we are interested in the behavior at long time scales $\tau = t/\lambda^2$, and from our earlier specifications we see that $U_t^{(\lambda)} = U(t/\lambda^2, \lambda)$ satisfies the interaction picture Schrödinger equation

$$\frac{\partial}{\partial t} U_t^{(\lambda)} = -i \, \Upsilon_t(\lambda) \, U_t^{(\lambda)} \tag{14.57}$$

with $\Upsilon_t(\lambda)$ given by

$$\begin{aligned}
\Upsilon_t^{(\lambda)} &= E_{11} \otimes a_t^+(\lambda) a_t^-(\lambda) + E_{10} \otimes a_t^+(\lambda) + E_{01} \otimes a_t^-(\lambda) + E_{00} \otimes 1 \\
&= E_{\alpha\beta} \otimes [a_t^+(\lambda)]^\alpha [a_t^-(\lambda)]^\beta.
\end{aligned} \tag{14.58}$$

Here we meet the time-dependent rescaled reservoir fields

$$a_t^\pm(\lambda) \triangleq \frac{1}{\lambda} e^{\mp i\omega t/\lambda^2} A^\pm(\theta_{t/\lambda^2} g). \tag{14.59}$$

The $a_t^\pm(\lambda)$ are creation/annihilation fields having vacuum correlation function

$$\langle a_t^+(\lambda) a_s^-(\lambda) \rangle = \frac{1}{\lambda^2} C\left(\frac{t-s}{\lambda^2}\right) \tag{14.60}$$

where $C(\tau) = \langle g, \theta_\tau g \rangle e^{i\omega\tau}$ is assumed to be integrable in τ. In the sense of Schwartz distributions, we have

$$\lim_{\lambda \to 0} \langle a_t^-(\lambda) a_s^+(\lambda) \rangle = \gamma \, \delta(t-s),$$

where

$$\gamma = \int_{-\infty}^{+\infty} dt \, C(t) = \int_{-\infty}^{+\infty} d\tau \, \langle g, e^{i\tau(H_R^1 - \omega)} g \rangle = 2\pi \, \langle g, \delta(H_R^1 - \omega) g \rangle. \tag{14.61}$$

We shall assume that $\gamma \equiv 1$. This normalization may always be achieved by absorbing factors into the $E_{\alpha\beta}$ if necessary. We will also be interested in the constants

$$\kappa_+ \triangleq \int_0^\infty dt \, C(t), \quad \kappa_- \triangleq \int_{-\infty}^0 dt \, C(t), \quad K \triangleq \int_0^\infty dt \, |C(t)|. \tag{14.62}$$

Specifically, we have

$$\kappa_+ = \left\langle g, \frac{1}{i(H_R^1 - \omega - i0^+)} g \right\rangle = \frac{1}{2}\gamma - i \, \mathrm{PV} \left\langle g, \frac{1}{(H_R^1 - \omega)} g \right\rangle,$$

where PV denotes the principle value part.

We shall assume that $C(-t) = C(t)^*$ so that $\kappa_\pm \equiv \frac{1}{2} \pm i\sigma$.

The limit $\lambda \to 0$ for the preceding, the two-point function becomes delta-correlated. However, it is vital to have a mathematical framework in which to interpret the limit states and observables.

For convenience, we set

$$\theta_\tau^\omega \triangleq \exp\{i\tau(H_R^1 - \omega)\}. \tag{14.63}$$

We assume the existence of a nonzero subspace, $\mathfrak{k}_\omega$, of $\mathfrak{h}_R^1$ called the "mass-shell" Hilbert space and where

$$\int_{-\infty}^{\infty} |\langle f_j, \theta_u^\omega f_k \rangle| du < \infty$$

whenever $f_j, f_k \in \mathfrak{k}_\omega$. The mass shell Hilbert space is constructed as follows: we first introduce a sesquilinear form defined by

$$(f_j | f_k)_\omega \triangleq \int_{-\infty}^{\infty} \langle f_j, \theta_v^\omega f_k \rangle \, du \equiv 2\pi \, \langle f_j, \delta(H_R^1 - \omega) f_k \rangle, \tag{14.64}$$

and we then quotient out the null elements for this form; the completed Hilbert space will then be $\mathfrak{k}_\omega$ and $(.|.)_\omega$ will be its inner product.

The test vector g appearing in the interaction must belong to $\mathfrak{k}_\omega$ so that the constant $\gamma \equiv (g|g)_\omega$ is finite. Recall that for convenience, we have assumed that $\gamma \equiv 1$. As a concrete example, take the one particle space to be $\mathfrak{h}_R^1 = L^2(\hat{\mathbb{R}}^3, d^3k)$, $A_R^+ = \int g(\mathbf{k}) a(\mathbf{k}) \, d^3k$ and $H_R^1 = \int \omega(\mathbf{k}) a^*(\mathbf{k}) a(\mathbf{k}) \, d^3k$, where we have the CCR $[a(\mathbf{k}), a^*(\mathbf{k}')] = \delta(\mathbf{k} - \mathbf{k}')$. In this case, we have

$$\gamma \equiv 2\pi \int \delta(\omega(\mathbf{k}) - \omega) |g(\mathbf{k})|^2 d^3k, \quad \sigma \equiv -i\mathrm{PV} \int \frac{1}{(\omega(\mathbf{k}) - \omega)} |g(\mathbf{k})|^2 d^3k.$$

The rescaled fields $a_t(\lambda)$ will converge, in a sense to be spelled out in the following, to quantum white noises a_t: more correctly, integrated versions of these fields converge to the fundamental quantum stochastic processes (Hudson and Parthasarathy, 1984). It was originally suggested by Spohn (1980) that the weak coupling limit should be properly considered as a Markovian limit underscored by a functional central limit. The rigorous determination of irreversible semigroup evolutions has been given for specific models (Pulé, 1974).

The program now is to start from the microscopic model for a system-reservoir interaction and then obtain by some Markovian limit procedure, such as the weak coupling limit, a quantum stochastic evolution. It was first noted by Waldenfels (1986) that stochastic models successfully describe the weak coupling limit regime for the Wigner–Weisskopf atom. Later, Accardi et al. (1990) showed how to do this for an interaction of emission/absorption type $\Upsilon_t^{(\lambda)} = E_{10} \otimes a_t^+(\lambda) + E_{01} \otimes a_t^-(\lambda)$, where E_{10} and E_{01} are bounded, mutually adjoint operators on the system space $\mathfrak{h}_S$.

The limit takes the form

$$dU_t = (S - 1)U_t \otimes dA_t^{11} + LU_t \otimes dA_t^{10}$$
$$- L^\dagger S U_t \otimes dA_t^{01} - \left(\frac{1}{2}L^\dagger L + iH\right) U_t \otimes dA_t^{00},$$

where

$$S = \frac{1 - i\kappa_- E_{11}}{1 + i\kappa_+ E_{11}} \quad \text{(unitary)},$$

$$L = -i(1 + i\kappa_+ E_{11})^{-1} E_{10} \quad \text{(bounded)},$$

$$H = E_{00} + \text{Im}\left\{\kappa_+ E_{01} \frac{1}{1 + i\kappa_+ E_{11}} E_{10}\right\} \quad \text{(self-adjoint)}. \qquad (14.65)$$

From the quantum stochastic calculus (Hudson and Parthasarathy, 1984), we have that the process U_t defined as the solution of (14.65) exists and is an adapted, unitary process.

References

Abrikosov, A. A., Gorkov, L. P., and Dzyaloshinski, I. E. 1963. *Methods of Quantum Field Theory in Statistical Physics*. New York: Dover Publications.

Accardi, L., Frigerio, A., and Lu, Y. G. 1989. On the Weak Coupling Limit Problem. *Quantum Probability and Applications, IV (Rome, 1987), Lecture Notes in Math.*, **1396**, 20–58.

Accardi, L., Frigerio, A., and Lu, Y. G. 1990. Weak Coupling Limit as a Quantum Functional Central Limit Theorem. *Commun. Math. Phys.*, **131**, 537–570.

Accardi, L., Lu, Y. G., and Obata, N. 1996. Towards a Non-Linear Extension of Stochastic Calculus. *Sūrikaisekikenkyūsho Kōkyūroku*, **957**, 1–15.

Accardi, L., Lu, Y. G., and Volovich, I. 2000. A White-Noise Approach to Stochastic Calculus. *Acta Applicandae Mathematicae*, **63**, 3–25.

Accardi, L., Lu, Y. G., and Volovich, I. 2002. *Quantum Theory and Its Stochastic Limit*. Berlin: Springer-Verlag.

Araki, H., and Woods, E. J. 1963. Representations of the Canonical Commutation Relations Describing a Non-relativistic Infinite Bose Gas. *J. Math. Phys.*, **4**, 637–642.

Attal, S., and Pautrat, Y. 2006. From Repeated to Continuous Quantum Interactions. *Annales H. Poincaré*, **7**(1), 59–104.

Baez, J. C., Segal, I. E., and Zhou, Z. 1992. *Introduction to Algebraic and Constructive Quantum Field Theory*. Princeton: Princeton University Press.

Bargmann, V. 1961. On a Hilbert Space of Analytic Functions and an Associated Integral Transform, Part I. *Comm. Pure Appl. Math.*, **14**, 187–214.

Belavkin, V. P. 1988. A New Form and $\star$-algebraic Structure of Quantum Stochastic Integrals in Fock Space. *Rend. Sem. Mat. Fis. Milano*, **58**, 177–193.

Bogoliubov, N. N., and Shirkov, D. V. 1959. *Introduction to the Theory of Quantized Fields*. Interscience Monographs and Texts in Physics and Astronomy, Vol. 3. New York: Wiley-Interscience.

Bogoliubov, N. N., and Shirkov, D. V. 1983. *Quantum Fields*. Advanced book program. Reading, MA: Benjamin/Cummings.

Bogoliubov, N. N., Logunov, A. A., Oksak, A. I., and Todorov, I. T. 1990. *General Principles of Quantum Field Theory*. Dortrecht: Kluwer.

Bourbaki, N. 1974. *Elements of Mathematics: Algebra I, chap. 1-3*. Paris: Hermann.

Bozejko, M., Kümmerer, B., and Speicher, R. 1997. *q*-Gaussian Processes: Noncommutative and Classical Aspects. *Commun. Math. Phys.*, **185**, 129–154.

Bozejko, M., and Yoshida, H. 2006. Generalized *q*-deformed Gaussian Random Variables. *Quantum Probability, Banach Center Publications*, **73**, 127–140.

Bremermann, H. J., and L. Durand, III. 1961. On Analytic Continuation, Multiplication, and Fourier Transformations of Schwartz Distributions. *J. Math. Phys.*, **2**, 240–258.

Caianiello, E. R. 1973. *Combinatorics and Renormalization in Quantum Field Theory.* Frontiers in Physics. Reading, Mass.: W. A. Benjamin.

Callen, H. B. 1985. *Thermodynamics and an Introduction to Thermostatics.* New York: John Wiley and Sons.

Chebotarev, A. M. 1997. Symmetric Form of the Hudson–Parthasarathy Equation. *Math. Notes*, **61**, 2510–2518.

Cook, J. M. 1961. Asymptotic Properties of a Boson Field with Given Source. *J. Math. Phys.*, **2**, 33–45.

Dembo, A., and Zeitouni, O. 1998. *Large Deviations Techniques and Applications.* Applications of Mathematics, Vol. 38. New York: Springer-Verlag.

Dieudonné, J. 1972. *Treatise on Analysis*, Vol. III. New York: Academic Press. Appendix, pp. 347–377.

Eckmann, J. P., and Epstein, H. 1979. Time-ordered Products and Schwinger Functions. *Commun. Math. Phys.*, **64**, 95–130.

Emch, G. G. 1972. *Algebraic Methods in Statistical Mechanics and Quantum Field Theory.* New York: Wiley-Interscience.

Emery, M. 1989. *Stochastic Calculus in Manifolds, with an Appendix by P.A. Mayer.* Universitext. Berlin: Springer-Verlag.

Evans, M., and Hudson, R. L. 1988. Multidimensional Quantum Diffusions. *Quantum Probability III*, Lecture Notes in Mathematics, **1303**, 69–88.

Eyink, G. L. 1996. Action Principle in Nonequilibrium Statistical Dynamics. *Phys. Rev. E*, **54**(4), 3419–3435.

Faris, W. G. 2009/11. Combinatorial Species and Feynman Diagrams. *Sém. Lothar. Combin.*, **61A**, 37.

Farré, M., Jolis, M., and Utzet, F. 2008. Multiple Stratonovich Integral and Hu-Meyer Formula for Lévy Processes. *The Annals of Probability*, **38**, 2136–2169.

Feldman, J. S., and Osterwalder, K. 1976. The Wightman Axioms and the Mass Gap for Weakly Coupled $(\varphi^4)_2$ Quantum Field Theories. *Ann. Phys.*, **97**, 80–135.

Feynman, R. P. 1949. The Theory of the Positron. *Phys. Rev.*, **76**, 749–759.

Freidlin, M. I., and Wentzell, A. D. 1998. *Random Perturbations of Dynamical Systems.* Fundamental Principles of Mathematical Sciences, Vol. 260. 2nd edn. New York: Springer-Verlag.

Gallavotti, G. 1985. Renormalization Theory and Ultraviolet Stability for Scalar Fields via Renormalization Group Methods. *Rev. Mod. Phys.*, **57**, 471–562.

Gelfand, I. M., and Shilov, G. E. 1964. *Generalized Functions. Vol. 1: Properties and Operations.* New York: Academic Press.

Gell-Mann, M., and Low, F. 1951. Bound States in Quantum Field Theory. *Phys. Rev.*, **84**, 350–354.

Glimm, J., and Jaffe, A. 1987. *Quantum Physics: A Functional Integral Point of View.* 2nd edn. New York: Springer.

Gorini, V., Kossakowski, A., and Sudarshan, E. C. G. 1976. Completely Positive Dynamical Semigroups on N-level Systems. *Journ. Math. Phys.*, **17**, 821–825.

Gough, J. E. 1997. Non-Commutative Ito and Stratonovich Noise and Stochastic Evolutions. *Theor. Math. Phys.*, **113**, 276–284.

Gough, J. E. 2004. Holevo-Ordering and the Continuous-Time Limit for Open Floquet Dynamics. *Lett. Math. Phys.*, **67**, 207–221.

Gough, J. E. 2005. Quantum Flows as Markovian Limit of Emission, Absorption and Scattering Interactions. *Commun. Math. Phys.*, **254**, 489–512.

Graham, R. L., Knuth, D. E., and Patashnik, O. 1988. *Concrete Mathematics: A Foundation for Computer Science*. Boston: Addison-Wesley.

Greub, W. 1978. *Multilinear Algebra*. 2nd edn. New York: Springer.

Guichardet, G. 1970. *Symmetric Hilbert Spaces and Related Topics*. Lect. Notes Math., Vol. 261. Berlin and Heidelberg: Springer-Verlag.

Haag, R. 1958. Quantum Field Theories with Composite Particles and Asymptotic Conditions. *Phys. Rev.*, **112**, 669–673.

Hegerfeldt, G. C. 1974. Remark on Causality and Particle Localization. *Phys. Rev. D*, **10**, 3320–3321.

Hegerfeldt, G. C., and Ruijsenaars, S. N. M. 1980. Remarks on Causality, Localization, and Spreading of Wave Packets. *Phys. Rev. D*, **22**, 377–384.

Heisenberg, W., and Pauli, W. 1929. Zur Quantendynamik der Wellenfelder. *Z. Physik*, **56**, 1–61.

Heisenberg, W., and Pauli, W. 1930. Zur Quantendynamik der Wellenfelder II. *Z. Physik*, **59**, 168–190.

Hepp, K. 1965. On the Connection between the LSZ and Wightman Quantum Field Theory. *Commun. Math. Phys.*, **1**, 95–111.

Hida, T. 2008. *Lectures on White Noise Functionals*. Singapore: World Scientific Publishing Company.

Hille, E., and Phillips, R. S. 1957. *Functional Analysis and Semigroups*. Providence: American Math Society.

Holevo, A. S. 1989. Stochastic Representation of Quantum Dynamical Semigroups. *Trudy Mat. Inst. Steklov*, ed. L. Accardi. **191**, 130–139.

Holevo, A. S. 1992. Time-ordered Exponentials in Quantum Stochastic Calculus. *Quantum Probability and Related Topics VII*, ed. L. Accardi. 175–202.

Holevo, A. S. 1996. Exponential Formula in Quantum Stochastic Calculus. *Proc. Roy. Soc. Edinburgh, section A Math.*, **126**, 375–389.

Hudson, R. L., and Streater, R. F. 1981. Ito's Formula Is the Chain Rule with Wick Ordering. *Phys. Lett. A*, **86**, 277–279.

Hudson, R. L., and Parthasarathy, K. R. 1984. Quantum Itô's Formula and Stochastic Evolutions. *Commun. Math. Phys.*, **93**, 301–323.

Hudson, R. L. 2012. The Early Years of Quantum Stochastic Calculus. *Commun. Stoch. Anal.*, **6**, 111–123.

Itô, K. 1951. Multiple Wiener Integral. *J. Math. Soc. Japan*, **3**, 157–169.

Itzykson, C., and Zuber, J.-B. 1980. *Quantum Field Theory*. New York: McGraw-Hill.

Jona-Lasinio, G. 1983. Large Fluctuations of Random Fields and Renormalization Group: Some Perspectives. *Scaling and Self-Similarity in Physics*, ed. J. Frölich, **7**, 11–28.

Joyal, A. 1981. Une théorie combinatoire des séries formelles. *Advances in Mathematics*, **42**, 1–82.

Kastler, D. 1961. *Introduction à l'Électrodynamique Quantique*. Travaux et Recherches Mathématiques. Paris: Dunod.

Kupsch, J., and Smolyanov, O. G. 1998. Functional Representations for Fock Superalgebras. *Infin. Dimens. Anal. Quantum Probab. Relat. Top.*, **1**(2), 285–324.

Kupsch, J., and Smolyanov, O. G. 2000. Hilbert Norms for Graded Algebras. *Proc. Amer. Math. Soc.*, **128**, 1647–1653.

Landsman, N. P., and Weert, Ch. G. van. 1987. Real- and Imaginary-time Field Theory at Finite Temperature and Density. *Phys. Rep.*, **145**, 141–249.

Lehmann, H., Symanzik, K., and Zimmermann, W. 1955. Zur Formulierung quantisierter Felder. *Nuovo Cimento*, **1**(1), 205–225.

Lehmann, H., Symanzik, K., and Zimmermann, W. 1957. On the Formulation of Quantized Field Theories II. *Nuovo Cimento*, **6**(2), 319–333.

Lindblad, G. 1976. On the Generators of Completely Positive Semi-groups. *Commun. Math. Phys.*, **48**, 119–130.

Lindsay, J. M., and Maassen, H. 1988. An Integral Kernel Approach to Noise. *Quantum Probability III*, Eds. L. Accardi and W. von Waldenfels, Lecture Notes in Mathematics, **1303**, 192–208.

Maassen, H. 1987. Quantum Markov processes on Fock space described by integral kernels. *Quantum Probability II, eds. L. Accardi and W. von Waldenfels, Lecture Notes in Mathematics*, **1136**, 361–374.

Magnus, W., Oberhettinger, F., and Soni, R. P. 1966. *Formulas and Theorems for the Special Functions of Mathematical Physics*. 3rd edn. New York: Springer.

Maurin, K. 1968. *General Eigenfunction Expansions and Unitary Representations of Topological Groups*. Warszawa: PWN – Polish Scientific Publishers.

Methée, P. D. 1954. Sur les distributions invariantes dans la groupe des rotations de Lorentz. *Comment. Math. Helv.*, **28**, 225.

Meyer, Paul-André. 1993. *Quantum Probability for Probabilists*. Lecture Notes in Mathematics, 1538. Berlin: Springer-Verlag.

Nakano, T. 1959. Quantum Field Theory in Terms of Euclidean Parameters. *Prog. Theor. Phys.*, **21**, 241–259.

Nelson, E. 1973. The Free Markoff Fields. *J. Funct. Anal.*, **12**, 211.

Nica, A., and Speicher, R. 2006. *Lectures on the Combinatorics of Free Probability*. London Mathematical Society Lecture Note Series, Vol. 335. Cambridge: Cambridge University Press.

Obata, N. 1994. *White Noise Calculus and Fock Space*. Lecture Notes in Math., Vol. 1577. Berlin: Springer.

Øksendal, B. 1992. *Stochastic Differential Equations*. Universitext. Berlin: Springer-Verlag.

Osterwalder, K., and Schrader, R. 1973. Axioms for Euclidean Green's Functions. *Commun. Math. Phys.*, **31**, 83–112.

Osterwalder, K., and Schrader, R. 1975. Axioms for Euclidean Green's Functions II. *Commun. Math. Phys.*, **42**, 281–305.

Parthasarathy, K. R. 1992. *An Introduction to Quantum Stochastic Calculus*. Basel: Birkhäuser.

Peccati, G., and Taqqu, M. S. 2011. *Wiener Chaos: Moments, Cumulants and Diagrams.* Universitext. Milano: Bocconi and Springer.

Pulé, J. V. 1974. The Bloch Equations. *Commun. Math. Phys.*, **38**, 241–256.

Reed, M., and Simon, B. 1972. *Methods of Modern Mathematical Physics I: Functional Analysis.* New York: Academic Press.

Reed, M., and Simon, B. 1975. *Methods of Modern Mathematical Physics II: Fourier Analysis, Self-Adjointness.* New York: Academic Press.

Rivers, R. J. 1987. *Path Integral Methods in Quantum Field Theory.* Cambridge: Cambridge University Press.

Rota, G.-C., and Wallstrom, T. C. 1997. Stochastic Integrals: A Combinatorial Approach. *Annals of Probability*, **25**, 1257–1283.

Saitoh, N., and Yoshida, H. 2000a. A *q*-deformed Poisson Distribution Based on Orthogonal Polynomials. *Journal of Physics A: Mathematical and General*, **33**(7), 1435–1444.

Saitoh, N., and Yoshida, H. 2000b. *q*-deformed Poisson Random Variables on q-Fock Space. *Journal of Mathematical Physics*, **41**(8), 5767–5772.

Schweber, S. S. 1961. *An Introduction to Relativistic Quantum Field Theory.* New York: Harper & Row.

Schwinger, J. 1958. On the Euclidean Structure of Relativistic Field Theories. *Proc. N. A. S.*, **44**, 956–965.

Segal, I. E. 1956. Tensor Algebras over Hilbert Spaces. I. *Trans. Amer. Math. Soc.*, **81**, 106–134.

Segal, I. E. 1962. Mathematical Characterization of the Physical Vacuum for a Linear Bose–Einstein Field. *Illinois J. Math.*, **6**, 500–523.

Spohn, H. 1980. Kinetic Equations from Hamiltonian Dynamics: Markovian Limits. *Rev. Mod. Phys.*, **53**, 569–615.

Steinmann, O. 1968. A Rigorous Formulation of LSZ Field Theory. *Commun. Math. Phys.*, **10**, 245–268.

Streater, R. F., and Wightman, A. S. 1964. *PCT, Spin & Statistics, and All That.* New York: W. A. Benjamin.

Stueckelberg, E. C. G., and Rivier, D. 1949. Causalité et structure de la matrice S. *Helv. Phys. Acta*, **23**, 215–222.

Summers, S. J. 2012. *A Perspective on Constructive Quantum Field Theory.* arXiv:1203.3991.

Symanzik, K. 1966. Euclidean Quantum Field Theory. I. Equations for a Scalar Model. *J. Math. Phys.*, **7**, 510–525.

Touchette, H. 2009. The Large Deviation Approach to Statistical Mechanics. *Phys. Rep.*, **478**, 1–69.

van Hove, L. 1952. Les difficultés de divergences pour un modèle particulier de champ quantifié. *Physica*, **18**, 145–159.

van Hove, L. 1955. Quantum Mechanical Perturbations Giving Rise to a Statistical Transport Equation. *Physica*, **21**, 617–640.

Voiculescu, D., Dykema, K. J., and Nica, A. 1975. *Free Random Variables.* Vol. 1. Providence: American Math. Soc.

Waldenfels, W. Von. 1986. Itō Solution of the Linear Quantum Stochastic Differential Equation Describing Light Emission and Absorption. *Lecture Notes in Mathematics*, **1055**, 384–411.

Weinberg, S. 1964. Feynman Rules for Any Spin. *Phys. Rev.*, **133**, B1318–B1332.

Weinberg, S. 1995. *The Quantum Theory of Fields I. Foundations*. Cambridge: Cambridge University Press.

Wick, G. C. 1950. The Evaluation of the Collision Matrix. *Phys. Rev.*, **80**(2), 268–272.

Wick, G. C. 1954. Properties of Bethe–Salpeter Wave Functions. *Phys. Rev.*, **96**, 1124–1134.

Wiener, N. 1930. Generalized Harmonic Analysis. *Acta Math.*, **55**, 117–258.

Wightman, A. S. 1956. Quantum Field Theory in Terms of Vacuum Expectation Values. *Phys. Rev.*, **101**, 860–866.

Wightman, A. S., and Garding, L. 1964. Fields as Operator-valued Distributions in Relativistic Quantum Theory. *Arkiv för Fysik*, **28**, 129–184.

Yang, C. N., and Feldman, D. 1950. The S-matrix in the Heisenberg Representation. *Phys. Rev.*, **79**, 972–978.

Zinoviev, Y. M. 1995. Equivalence of Euclidean and Wightman Field Theories. *Commun. Math. Phys.*, **174**, 1–27.

Index